D0861465

The

C-SPAN

Revolution

The

C-SPAN

Revolution

By

Stephen Frantzich and John Sullivan

University of Oklahoma Press : Norman and London

Library of Congress Cataloging-in-Publication Data

Frantzich, Stephen E.
    The C-span revolution / Stephen Frantzich and John Sullivan.
      p.   cm.
    Includes bibliographical references and index.
    ISBN 0–8061–2870–4 (alk. paper)
    1. C-SPAN (Television network). 2. Television programs, Public
service—United States. I. Sullivan, John. II. Title.
HE8700.79.U6F73   1996
384.55′532—dc20   96-18189
                                                    CIP

Text design by Debora Hackworth.

The paper in this book meets the guidelines for permanence and durabil-
ity of the Committee on Production Guidelines for Book Longevity of the
Council on Library Resources, Inc. ⊗

1 2 3 4 5 6 7 8 9 10

To the over four hundred participants
in the C-SPAN Professors' Seminars
whom we were supposed to teach
but from whom we learned a great deal.

Contents

Illustrations

Graphs

Tables

Preface

For almost two decades C-SPAN has been bringing the American public an in-depth and inside look at public affairs. This book is the story of how C-SPAN grew from a vague idea in the 1970s to a recognizable factor in the political process by the 1990s. It traces the interests both in Congress and in the cable industry that converged and allowed the idea to become a reality. The book follows the network's struggle to grow, articulates its mission, and describes both the audience C-SPAN attracts and the impact the network has had on American political and journalistic life. Most of the previous accounts of C-SPAN have been relatively brief, concentrating on the persona of its founder, Brian Lamb, rather than on the network itself. Relatively little has been written to provide a comprehensive picture of the history, operations, and influence of C-SPAN. This book is an attempt to rectify that situation. It was written to be read both inside and outside the Washington beltway, for the students who examine the political process as well as for the citizens who participate in this process. It also is intended for those who simply want to know more about C-SPAN.

At the outset the authors recognized that personal perspectives shape and limit what ones sees, what one views as important, and how one judges. Perspectives give value. That is why two of us from different academic disciplines came together to work on this project. As a political scientist, Stephen Frantzich has long been interested in political institutions and the impact that new technology has on the distribution of power and the political process. As someone trained in American studies and rhetoric, John Sullivan has dedicated his work to understanding the nature and state of public discourse in a society that is wary both of the spoken word and of its members' abilities to make sound judgments based on that word. We have written a book that neither of us could have, or would have, written alone. At times our differences in telling the C-SPAN story have been fundamental. We have attempted to allow those differences to show. The book may be longer and less coherent as a result but is very likely better for it.

The authors first met at the initial "C-SPAN in the Classroom Seminar for Professors." Frantzich originated the idea and served as a presenter; Sullivan arrived as a skeptic. Over the last eight years we have come to work closely with the content and formatting of this ongoing series of seminars designed to help professors understand C-SPAN and use it more effectively in teaching public affairs. In the process we came to know the network and those who work there. It is telling for this book, and for C-SPAN, that the content and the format of each seminar have been left to the academics who staffed it. Our loyalty from the beginning has been to the free exchange of ideas, not to C-SPAN.

It is telling too that a network whose programming format is "based on an overriding respect of the ability of viewers to reach their own conclusions about what they see" cooperated completely with the project—not even considering any right of review. Those at C-SPAN will see these pages for the first time when readers do. For the most part, we have had access to all the information for which we had the savvy to ask. Naturally, personnel files that might involve privacy issues were closed. We were allowed, in fact invited, to watch C-SPAN at work, sitting in on formal meetings and informal gatherings. What we did not glean from files and meetings we picked up in the hallways or on shoots with crews. Interview subjects were forthcoming; to those questions for which they had no answer, they generally replied, "I'll find out." We owe a special debt of gratitude to James Lardner, who made available his notes from interviews he conducted in preparation for an article on C-SPAN in the *New Yorker*. His research both confirmed and supplemented what we had already found. Traditional documentary research was augmented with extensive computerized searches of LEXIS-NEXIS, a full-text database of over three thousand newspapers and magazines. As would be appropriate for research on television, some of the material came from videotape transcriptions of C-SPAN programming.

Because C-SPAN deemphasizes personalities, so have we in this text, but the staffers at the network who have contributed to this effort deserve our personal thanks. Without them the book could not have been written. We have attempted to tell the C-SPAN story as we saw it and heard about it from those staffers. The errors in fact and interpretation are ours alone, and we take full responsibility for them.

Our task in preparing this book has been made easier by a number of people. The staff of C-SPAN and other interview subjects were generous with their time and expertise. Brian Lamb set the tone for his staff, showing continuing interest in the project and regularly asking, "Are you getting everything you need?" Although personality traits and experience made some people less forthcoming, alternative sources of information were usually readily available.

Stephen Frantzich's efforts were augmented by research assistants, such as Micah Demby, Scott Cantor, and Aime Scott, from the Anne Arundel County mentorship program. Ensign Robert Rivera checked footnotes and bibliography entries. Frantzich's secretary, Ann Cribbs, helped balance the demands of his "day job" to allow completion of the project. Barbara Breeden and Ruth Hennessy, of the Nimitz Library, were the tip of the iceberg of the library staff who contributed to the project. His wife, Jane Frantzich, served as the editor of his rough drafts, gently—and sometimes not so gently—lobbying for revisions. He received research funding from the Dirksen Center, the Joan Shorenstein Barone Center at Harvard University's Kennedy School of Government, and the U.S. Naval Academy.

Dean Raymond Nelson, of the Faculty of Arts and Sciences at the University of Virginia, provided research leave to John Sullivan at a critical point in the development of this project. Professor Richard Cornell and his wife, Bo, gave of their time to transcribe several of the interviews and, more important, had faith in this project from its inception. Carolyn Anderson was there at the beginning, and Patty DeCourcy and Judy Birckhead provided valued help when push came to shove. Throughout, Janet Sullivan suffered through sometimes illegible scrawl to make the transcripts readable. Nancy Sullivan provided a sail when the ship seemed dead in the water. She and her sister, Maury, may not have made their father computer literate, but both were there in their own way when needed.

Kim Wiar, of the University of Oklahoma Press, showed considerable interest in the project and provided early guidance. Alice Stanton managed the development of the project, while Teddy Diggs provided expert editing. Finally we would both like to thank our students and friends who patiently listened to more about C-SPAN than they wanted to hear throughout this project.

Just as C-SPAN attempts to be a "fly on the wall" at public events, we hope that the readers will get the feeling they are on the inside looking out. C-SPAN is a moving target, changing and evolving month to month. We have attempted to focus on those abiding aspects that have long-term significance. The C-SPAN "revolution" is not one of violence or conflict. It is rather a series of subtle changes in the coverage of public affairs by the media and in the use of public affairs coverage by political activists. It is the merging of new technologies and political self-interest to create an environment of broader access to public affairs for those who want to take an interest or to use its opportunities to promote causes and political careers. C-SPAN is part of a larger revolution in communications and information access premised on the general principles that more information is better than less and that decentralized information sources serve democracy better than centralized ones. At times C-SPAN has led the change, and at times it has simply become part of the larger flow of events. Brian Lamb's simple goal of opening up the political process spawned a number of unintentional consequences just as important to the revolution as the initial goals. For many members of the public and for some political activists, the revolution has passed them by. But for those capitalizing on this revolution, it is hard to imagine going back to a world of closed meetings, limited access, and full reliance on the traditional mass media sources. Even though we as authors have some reservations about the revolution in public affairs coverage, as well as about the consequences of that revolution, we see its basic contours as inevitable and its general thrust as positive. We hope both to chronicle C-SPAN's part in the development of this revolution and to provide insights that will grant all participants an understanding of C-SPAN's importance, enabling them to make educated judgments about how to capitalize on C-SPAN for the benefit of democracy. Let the revolution proceed.

The

C-SPAN

Revolution

C-SPAN's MISSION STATEMENT

To provide C-SPAN's audience access to the live gavel-to-gavel proceedings of the U.S. House of Representatives and the U.S. Senate and to other forums where public policy is discussed, debated and decided—all without editing, commentary or analysis and with a balanced presentation of points of view;

To provide elected and appointed officials and others who would influence public policy a direct conduit to the audience without filtering or otherwise distorting their points of view;

To provide the audience, through the call-in programs, direct access to elected officials, other decision makers and journalists on a frequent and open basis;

To employ production values that accurately convey the business of government rather than distract from it;

To conduct all other aspects of C-SPAN operations consistent with these principles.

Prologue

Setting the Stage for C-SPAN

Laws and institutions must go hand in hand with the progress of the human mind. . . . With the change of circumstances, institutions must advance also to keep pace with the times. We might just as well require a man to wear still the coat which fitted him as a boy as civilized society to remain ever under the regimen of their barbarous ancestors.

Thomas Jefferson, *Writings*

Those who say they give the public what it wants begin by underestimating taste, and end by debauching it.

Pilkington Report, 1960

A continuing struggle in all societies and their primary institutions concerns finding a balance between capitalizing on positive forces for change and holding out against those negative aspects of change that may well undermine the institutions and the society of which these forces are a part. The pressure for change comes from many quarters. New technologies carry with them

3

an almost inherent logic that they *must* be used but provide less direction as to *how* they should be used. Television emerged as largely an entertainment medium driven by consumer tastes. Almost four decades passed before changing political conditions and further waves of technological change allowed television to be more fully harnessed to support a better understanding of public affairs. The ability to watch sessions of Congress, tune into congressional committees, participate in a nationally televised call-in program, see foreign legislatures operate, and view any public affairs events gavel-to-gavel would have been seen as highly unlikely in the predominant media environment up to the late 1970s. Public affairs programming has come a long way in two decades, and C-SPAN (Cable-Satellite Public Affairs Network) has played a major role in that change.

Technological changes based on the integration of computers and television continue to provide challenges and opportunities for an even more informed citizenry. The precise nature of the next round of changes is still an open question. What is clearer is that millions of Americans already possess public affairs viewing options unheard of just a few years ago. This book is an attempt to chronicle the history, operations, and impact of C-SPAN's crucial role in this revolutionary change in information access. The story of C-SPAN is one premised on changes in technology, driven by a mix of idealism and pragmatism, brought to fruition by dedicated practitioners, made significant by clever politicians, and guaranteed a future by forward-looking leaders. It is also the story of self-serving initiatives, missed opportunities, personal failures, and some inherent dangers of information and misinformation. The saga of C-SPAN is more than a story of one network with a relatively small niche market. It highlights the process of technological innovation, the principles of organizational management, the changing nature of the public, and the role of technology in the political process. The story recognizes the principle that "power and influence in society depend on the control and strategic use of information."[1]

Assessing the factors that allowed the growth and development of C-SPAN is difficult. A variety of technological and societal changes flowed together like tributaries of a river, affecting each other in a constantly changing process and creating a new environment—itself only transitory. At a minimum, we should be informed by an understanding of the media environment, the political environment, the technological environment, and the changing nature of the public affairs audience. Our analysis will transport us back to the 1970s, when the idea of C-SPAN was germinating.

The Media Environment

The dominant role of television is a relatively recent phenomenon. Until the 1960s, Americans considered newspapers more important than television as a source of news.[2] Along with an increased dependence on television as a news source, Americans changed their evaluations of the veracity of various sources. In the early 1960s, people receiving conflicting reports on television and in newspapers were equally likely to believe either source. By the 1970s, however, over 50 percent of Americans believed television news when it conflicted with the information provided in the newspaper. Only 20 percent believed newspapers over television in such a situation—a difference that remains today.[3]

Television in the 1970s was considerably different from television today. The three national networks called the shots. In an attempt to guarantee their commercial success, the government undergirded NBC, ABC, and CBS, making it impossible for other national outlets to emerge.[4] In C-SPAN founder Brian Lamb's view, "The dominance of the three commercial networks during the past half century was an aberration that disrupted a normal flow of public affairs information, overcentralized communication patterns, and deprived the American people of free choice."[5]

In the 1970s, national television news was in its infancy. The first regular fifteen-minute national newscasts had not begun until the late 1950s, when each of the three networks offered an eleven-minute national newscast wrapped around four minutes of commercials. The television medium was and is strategically conservative. The tendency of most businesses, if they are making money, is not to change, and television was no different. Many media executives felt that it was the fifteen-minute local news programs that drew the audience. When CBS executives decided to challenge NBC's team of Chet Huntley and David Brinkley in 1963 in an effort to gain the number-one position, the idea of expanding to a half hour of national news was seen as risky. Many asked whether the public would have the patience for such a long program, and the local affiliates objected to the infringement on their territory and revenue. But the gambit worked for CBS, and the other networks quickly followed. "Within a year, television had become the dominant source of news in America . . . [and] a sea change in the culture had begun."[6] As Brian Lamb saw it: "From 1963 to about 1983, the commercial television news organizations had an overwhelming amount of power. That was probably not the way the forefathers would have liked it. We were force-feeding the masses with three newscasts because they wanted to watch TV, not because they wanted to watch news."[7]

Even with the expansion to a half hour of news, the trend toward sound-bite coverage of politics was well on its way. Commentators were given more airtime than were public officials, and both were crowded into stories that seldom lasted more than two minutes. A geographical and topical spread of stories on the evening news substituted for any attempt at depth. Network officials operated on the almost religious belief that the American public lacked the tolerance for extended political discourse. Network spokespeople often admitted their shortcomings but argued that the expectations were unrealistic, since television news was not designed to serve as a viewer's sole basis for political understanding. But even though other sources remained, the appeal of television took its toll. With newspaper and newsmagazine readership in decline, the public became increasingly dependent on network television news. In 1960, 44 percent of Americans reported regularly following the election in a newspaper. In each succeeding presidential election year, that figure dropped, until it was 28 percent by 1976.[8]

Television news viewers had little choice other than the networks for national news and were seldom challenged to go beyond the summaries. "Commercial television, so brilliant at giving people what they want, has always been cautious, not to say craven, about giving people more than they know they want or stimulating them to want more."[9] Perhaps the networks' limited commitment to expansive coverage of public affairs is best represented by their willingness to drop gavel-to-gavel coverage of the national political conventions. From the 1950s to 1980, the networks had cleared their schedules, upsetting their regular pattern of commercial advertisements, to cover both parties' national conventions in their entirety. But as soon as other outlets (CNN and C-SPAN) began to take on that task in 1984, the networks moved to truncated and edited summary programs, accepting the assumption that full coverage was passé and would simply increase the "snooze-factor" among viewers.[10]

Coverage of national events such as party conventions and even presidential speeches was now subject to increasingly selective, strategic decision making based on audience factors as opposed to newsworthiness.[11] Such programming decisions had the potential for leaving the American public less informed. The temptation to follow ratings rather than substance proved difficult to overcome.

The 1970s generated concern about the networks' commitment not only to the quantity of public affairs programming but also to the quality of that coverage. Challenging the often ill-defined "establishment" became a popular cause during the 1960s and 1970s. Depending on one's perspective, the media were either part of that establishment or, at the least, willing conspirators. In 1971, Robert Cirino took the media to task for eroding democracy by limiting the number of viewpoints that were broadcast. He argued that government officials and the networks

> prevented real public participation by not allowing all ideas to compete fairly for public acceptance. They have allowed free speech, but rendered it useless by not allowing anti-establishment voices to have *equal* access to the technology of persuasion. . . . A person speaking to eighty million people has quite an advantage over someone with a conflicting view talking to a few thousand people in an auditorium or ten people on a street corner. The idea that gets amplification and extension through the media—not necessarily the most reasonable idea—is the one that wins the endorsement of the people.[12]

Both liberals and conservatives found some ammunition in analyses such as Cirino's. Liberals would define the media as part of "the establishment," which underreported the antiwar, civil rights, and environmental movements. Conservatives in the White House of Richard Nixon would read the criticism as evidence that the media bypassed the silent majority in favor of eastern liberals. Thus both liberals and conservatives saw themselves on the outside trying to fight for the attention of the media.

Somewhat less conspiratorial criticisms of network coverage of public affairs recognized the structural characteristics of television. As a visual medium, it was expected that television would utilize its visual components. Producers and technicians were taught to take full advantage of lighting, camera angles, and action. Television journalists began to accept the dictum that "there is no story without pictures." The denigration of "talking heads" and the emphasis on action eliminated much of what politics and public affairs are about. The great—and often lengthy—speeches of the masters of public affairs, speakers such as Thucydides, James Madison, and Thomas Jefferson, may well have been left in the editing room if the mainstream television media had covered their events.

Certain stories were never told because they were not good television. "Good television" often became defined by producers and directors as an action-filled story with personable characters. To reduce the time taken to introduce a story—particularly within the time confines of the evening news—producers centered their stories on well-known figures or caricatures. Public affairs often became a set of melodramas pitting the "good guys" against the "bad guys." Alternatively, the news hook for a story became the well-known—albeit not the most important—person who repeatedly entered the scene and through whom the story could at least partially be told. It was assumed that the viewers could not keep track of too many stories or too many characters, so each was limited. Such an approach was not new to television, but the power of the medium to engage the public heightened its influence.

Since television executives suspected that excitement and action generated more audience interest than explanation and analysis, certain portions of stories were emphasized. Political campaigns were portrayed on television like horse races, with journalists focusing on the leaders of the pack and asking: Who is ahead? Who is falling behind? What campaign strategies account for this? Discussions of broad issues and policy prescriptions took a backseat in most reporting.[13] To a large degree the print media followed what was happening on television, reporting on the internal machinations of the campaign and including long, personal profiles of the candidates. Theodore White, covering each presidential campaign from 1960 to 1972, created a new genre of campaign book: "a wide-screen thriller with full-blooded heroes and white-knuckled suspense on every page."[14]

Policy battles in Congress, the executive branch, and the courts were—and continue to be—difficult for the television media to cover. The steps in the policy process have no clear beginning and no clear end. Drama is often missing; in fact, the policy process has the nerve to be downright dull at times. As former Representative Bill Frenzel (R-MN) pointed out, "Congress is a process, not an event."[15] Much the same can be said of the other branches. "The media are good at 'institutional' scorekeeping: they keep us posted about what happened to major policy proposals and bills. But they do less to explain the process that produced the outcome or to keep 'individual' scores on the members' performance."[16] The continuing problem of covering the policy process is exemplified by the coverage of two key events of the early 1990s. Although lax regulation and the default of many savings-and-loan companies will have long-term financial effects for all Americans, this complex story of public policy was greatly overshadowed by stories with considerably more limited impact. The "House Bank Scandal," which emerged at about the same time as the savings-and-loan story, generated a great deal more media attention, since overdrawing one's bank account was understandable to most viewers and since it was possible to clearly identify the key offenders. Lost somewhere in the reporting were the facts that no public money was at risk and that all the unsecured advances were repaid.

Part of the problem for television during the 1970s was the tyranny of the visual. Television cameras were seldom allowed to record official congressional proceedings either on or off the floor (see chapter 2). The president of the United States, as the head of the executive branch, was more amenable to television coverage but served as only the tip of the policy-making iceberg. The courts rejected most television coverage as unseemly and disruptive.

It was not so much that network television executives were smug and self-confident about their performance in broadcasting public affairs but rather that they lacked the motivation to change. They had initiated the movement of the nation, transforming it from relatively isolated communities and nonoverlapping reservoirs of information into a national community that could share information and perspectives at a moment's notice. Their increasing dominance as an information source seemed beyond challenge. In their view, the system was working, and few industry leaders were willing to undermine their position. The locus of change would have to come from elsewhere.

The Political Environment: Growing Dissatisfaction

Throughout history, politicians have recognized the importance of the media as crucial bulwarks of democracy. Jefferson argued: "The people are the only censors of their governors. . . . The way to present these interpositions of the people is to give them full information of their affairs through the channel of public papers, and to contrive that those papers should penetrate the whole mass of the people."[17] A century later, William Jennings Bryan reinforced these ideas by saying: "The government being the people's business, it necessarily follows that its operation should be at all times open to public view. Publicity is therefore as essential to honest administration as freedom of speech is to representative government."[18]

American political culture and operations have long shown a commitment to open government. Although the Founders met in secret, they established, in the Constitution and in early government, a number of procedures that would guarantee public access and public records. Some decisions—such as open public galleries, established for the House from the outset (1789) and adopted later in the Senate (1795)—allowed for direct, real-time observation by the general public and the press. The publication of committee and chamber proceedings provided a delayed record of legislative activity.

Throughout our history, few Americans have experienced the national policy process directly. The size and complexity of the policy-making process makes it difficult to grasp. The demands of everyday life lead people to make practical choices of how to spend leisure time, and public affairs seldom make it to the top of the priority list. Even people with an extraordinary interest in politics have found it difficult to penetrate the decision-making process.

During policy-making meetings in both Congress and the executive branch in the 1970s, the doors were largely closed to the public. In Congress, less than 60 percent of committee meetings were open to the public until the reforms of 1973.[19] The few hundred seats in the House and Senate galleries did little to make the system more open. New technology that could have opened the legislative process to the public was prevented from having a significant impact by a set of arcane congressional rules. Members fought against installing microphones for debate and refused to allow radio or television into the chambers. Even common courtesies to the press, such as allowing cameras in and around the Capitol, faced stringent rules and often outright prohibition. As NBC correspondent Linda Ellerbee described the challenge for electronic journalists, "Congress was able to enforce a handful of stupid rules it would have never tried with print journalism . . . not only could I not take my camera crew—my equivalent of a pen and notebook—into the chamber of the House or Senate, I could not take my camera crew some places in and around the Capitol that were, at the same time, open to tourists with their cameras."[20]

The official written records of Congress did not solve the problem. They were—and remain—difficult to interpret and decipher. The *Congressional Record*, with its policy of giving congressional members liberal rights to "revise and extend" their remarks, created an official record based on reporting what members *wished* they had said rather than what they had *actually* said. Speeches were regularly edited for grammar, facts, and content. Comments never spoken on the floor showed up as literate and persuasive arguments, while misstatements or embarrassing phraseology never saw the printed page. Since the courts have ruled that the written *Record* is the basis for determining legislative intent, serious observers recognize its importance. The decision to accept the substantive limits of the *Record* did not solve the observers' problems. Its lack of user-friendly aids, such as good indexing and an understandable organization, made the proceedings difficult for even the most diligent observer to follow. Committee reports were often difficult to track down and often did not become available until months after policy decisions had been made. The dial-up bill-status office told where a bill was in the process but was forbidden from telling how individual members had voted.

In many ways, both the House of Representatives—the "people's chamber"—and the intentionally more remote Senate developed a "guild" attitude. Specialized language, complex procedures, and denial of access helped maintain the mystery of the institutions and their members. Unaware of the "secret handshake" and denied relevant information, much of the public was kept politically impotent. While publicly supporting citizen access, both the House and the Senate acted as if they felt that what the public did not know would not hurt them—and the "them" was clearly the members. Criticism could be met with the simple response, "You just don't understand how Congress works."

Lacking the time and ability to get direct access, the public was forced to rely on the media. By the early 1970s, politicians in general and members of Congress in particular began expressing considerable frustration with the way the media portrayed their efforts. Although some distinction was made between electronic journalists and print journalists, both came under fire for their shortcomings. Television was often viewed as a heightened caricature of print failures.

The media were criticized for both the quantity and the quality of their coverage. Some critics simply objected to the media's unwillingness to tell the story of Congress the way members wanted it told. More thoughtful criticism reflected on the conflicting needs of the two institutions. As one member of Congress put it: "The man-bites-dog syndrome is a real affliction. What makes news is the exception and not the rule. And, clearly, what makes news is what feeds public suspicions . . . if ten congressmen are misbehaving, the public thinks the condition is epidemic and forgets that there are 425 others who are living with their wives, keeping their hands out of the cookie jar, and probably doing a good job of trying to represent their constituencies."[21] A colleague in the House made a similar point: "The number one problem is superficiality. [The media] emphasize the conflict rather than the true mood of the House on many issues. . . . They [play] to the extremes and [ignore] much of the debate because the main issue at hand was not that visually exciting."[22]

Although spoken by members of Congress, the above criticisms reflect the outlook of most elected officials. President Harry Truman, knowing that the media seldom got stories "right," expressed frustration that he would have to rely on the media once he left office. Perhaps expressing a tougher skin than members of Congress, Truman outlined his cynical rule of thumb: "Whenever the press quits accusing me, I know I'm in the wrong pew."[23] President John F. Kennedy participated in numerous jocular interchanges with the media, but his humor had a bite when he was asked to give his evaluation of media coverage of his administration. He said, "I am reading more and enjoying it less. . . . I have not complained nor do I plan to make any general complaints. . . . I think that they are doing their task, as a critical branch. . . . And I am attempting to do mine. And we are going to live together for a period, and then go our separate ways."[24] Some politicians recognized the potential of the emerging electronic media. After one of his first chances to circumvent the media by speaking directly to the public on television, President Kennedy called *Washington Post* editor Ben Bradlee and gleefully proclaimed, "When we don't have to go through you bastards, we can really get the story to the American people."[25]

Particularly grating to Congress was the imbalance in coverage of the different branches. Until Franklin Roosevelt's presidency, Congress had dominated the print coverage of national public affairs.[26] But the slow and complex decision-making process of Congress, involving a large cast of actors, was difficult for the electronic media to interpret for the American public. The expectation of action-filled video and recognizable personalities was lacking. As one Capitol Hill television producer put it: "You have to sell a story to the editor, and the editor wants action, not process. . . . And that isn't what Congress is about. Congress is about process. . . . So [we] are constantly beating our heads against the wall, caught in battles both with sources in Congress and the producers in New York."[27] The evidence is clear. Congress has increasingly lost out to the more focused, action-oriented, and photogenic presidency when it comes to coverage on network news programs (see graph 1).

For individual members of Congress the diminished visibility was even more of a problem. Over half the House members were never mentioned on the national evening news during a typical year, and over one-third of the senators received one mention or less.[28] The local media covered their own members more extensively and favorably in their own market areas but

Graph 1

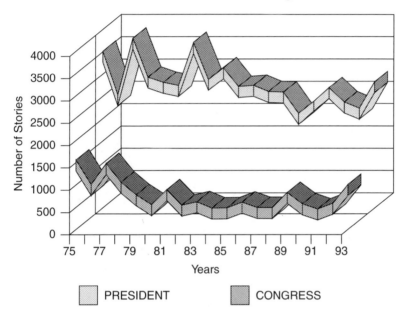

Presidental Domination of the Evening News

Despite variations in historial events, the gap in media coverage between Congress and the president remains wide. *Source:* Calculated
by the authors from the *Vanderbilt Television News Index.*

lacked the status and potential impact of the national media. The fact that
local audiences were shifting from local to national news sources increased
politicians' insecurity. In a political world in which visibility is interpreted
as importance, members of Congress and other politicians were interested in
discovering new ways of becoming known and of getting their message out.

The 1970s will go down in history as a turning point in American cynicism toward politics and those associated with it. With the Vietnam conflict,
the Watergate affair, and a series of congressional scandals, the major institutions of government began to fall into disrepute. Since 1970, seldom
more than 20 percent of the public have expressed a "great deal of confidence" in Congress. The mainstream media developed a heightened pattern

of cynicism, which seemed to permeate their coverage and colored public perceptions. This gave politicians of all political stripes the motivation to accept alternative options for publicizing their activities.

The media, vainly attempting to assert that they simply reported the events, found it difficult to disassociate themselves from the growing cynicism. Far from being seen as neutral reporters of the emerging problems, the media began to be seen as part of the problem—no doubt urged on by the criticisms of those receiving negative publicity. The media's self-image as "white knights of truth" more often became, in the public mind, a strategy of condemning the messenger for the message (see graph 2). Although trust in television news increased at the same time trust in newspapers declined, even trust in television news began to level off. Clearly, the public was ready for new avenues to public affairs information.

The Technological Environment: Arrival of Cable

Considering the pervasiveness of television in contemporary society, it is difficult to imagine that television barely existed fifty years ago. Both of the authors of this book vividly remember when the first families in their neighborhoods purchased television sets. The number of television sets in the United States increased from five thousand in 1946 to almost ten million by 1950.[29] But owning a television set gained a person little more than a prestigious plant stand unless there were signals to receive. Commercial television stations developed most rapidly in large urban areas, which had significant potential viewing audiences for their programming—and especially for their commercials. Small towns and areas unable to receive adequate signals were bypassed in the first wave of the television revolution.

Necessity is the mother of invention, and inventiveness knows no geographical boundaries. These two dictums tell a great deal about the cable revolution of the 1970s. Cable originated in a most unlikely place. Mahanoy City, Pennsylvania, was a rather bleak coal town in 1948. Part-time appliance store owner John Walson had a problem. His customers wanted television but could not get Philadelphia stations directly because of the mountains. Topography began as his enemy and eventually became his friend. His first solution—simply driving customers to the top of the mountain to see television brought in by the antenna he had installed—was not practical, so he decided to string wires and bring the signal into town; along the way he wired eight homes. By 1949 he was charging customers two dollars per month to

Graph 2

Public Trust in the Media

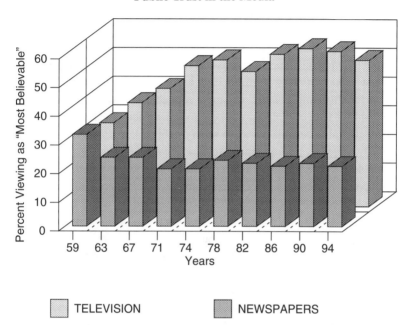

TELEVISION   NEWSPAPERS

While public trust in newspapers has declined over the last thirty-five years, trust in television has increased. *Source:* Norman J. Ornstein, Thomas E. Mann, and Michael J. Malbin, *Vital Statistics on Congress: 1995–1996* (Washington, D.C.: Congressional Quarterly Press, 1996), p. 68.

bring in the three Philadelphia stations, and his sales figures for television sets went up. At the same time in Astoria, Oregon, Ed Parsons—also motivated by the desire to overcome the barriers of the mountains—built an antenna and began distributing signals. In Lansford, Pennsylvania, Robert Tarleton expanded the reach of cable by experimenting with new cable and amplification procedures. In 1950, Panther Valley TV Company became the first commercially viable cable system by using a utility-like approach to selling its services.[30] The story was the same all over the country. Driven by the desire to overcome physical barriers to the reception of good television signals, thousands of mom-and-pop cable operations emerged. Even the terminology was significant. Calling the new endeavor CATV (community-antenna television) rather than "cable television" reflected the view that cable was simply the result of American ingenuity and cooperative effort; it was the logical extension of the personal "rabbit ear" antennas used by most viewers to directly access commercial stations.

The nascent cable industry received help from an unexpected quarter. The Federal Communications Commission (FCC) put a freeze on the creation of new broadcast channels from 1948 to 1952 while it worked out procedures for allocating channels in a more equitable manner. Since the pent-up demand for television could not be satisfied by additional local broadcast channels, cable became a way to satisfy public interest in those areas not already served by a commercial broadcast channel.

Cable television began as a "simple neighborhood convenience for people with poor TV reception."[31] The technology was relatively rudimentary, but virtually everyone was pleased. Customers were willing to pay for television if this was the only way to receive it. The commercial stations appreciated the ability to extend the range of their signals, and the FCC was willing to leave CATV companies unregulated. Most of the cable companies were mom-and-pop operations serving a few thousand households. The early cable operators were pioneers "working out of dingy offices in the back of shopping malls or industrial parks."[32] They worked hard, improvised equipment, and seldom had much capital to spare. The early system owners "sold cable door-to-door" and knew "how to climb a pole to patch up a blown amplifier."[33]

During the 1950s and 1960s, CATV systems began to expand their availability into areas served by broadcast television. To be competitive, they developed their own programming, which potential customers could not get for free. The commercial broadcast stations began to see cable operators as competitors. Again the technology was rather simple. Videotapes were sent from one system to the next by mail in a process called "bicycling." As an early pioneer in the business remembered: "It was fine when we extended the viewing audience of a broadcast station out into another area. . . . In that case they loved us. But the minute we started to come into their markets, it was quite a different story."[34]

Later cable companies went another step and began to transmit programming from distant channels via microwave relay. In 1958, one CATV operator was sued for "pirating" a commercial signal. Although the operator won the right to carry the commercial signal, the decades of battles in court and in Congress had just begun. The FCC increasingly began to assert its jurisdiction over cable, and cable began to acquire a vast array of worthy opponents in the political arena. Both commercial broadcasters and the movie industry saw cable companies as competition. State and local governments saw them as a source of revenue. Viewers working through elected officials asserted their rights to "free television." Some industries might have folded given the opposition, but the rugged individualists of the cable fraternity had a different view. As cable pioneer Bill Daniels put it: "If all these big gunners are trying to stop us, we must have something. If we didn't have anything, they could care less about us."[35]

By 1966, the FCC—under pressure from broadcasters and the movie industry—had become more aggressive. Cable companies were required to carry all local broadcasting and were prohibited from importing programming from distant points. Later regulations limited cable to showing movies over ten years old and outlawed cable competition in broadcasting sports events. Despite the limitations, the number of cable companies grew from 1,325 in 1965 to nearly 2,500 in 1970 and to almost 8,000 in 1983.[36]

The arrival of commercial satellite transmission of television signals in the mid-1970s came at a fortuitous time for cable and changed the complexion of the industry. Now cable had an efficient method of distributing its own unique programming. In 1975, Home Box Office (HBO) became the first satellite network. It was followed by Pat Robertson's Channel 27, Showtime, ESPN, and Ted Turner's Channel 17. Transponder positions on satellites and earth stations for sending signals cost as much as $2 million. Local cable companies needed to install dishes that initially cost about $100,000.[37]

Some of the initial programming was both limited and questionable. Turner's Channel 17 spoofed newscasts during late-night hours. The comedian-anchor sometimes used his German shepherd as cohost. "The dog, wearing a shirt and tie chewed a mouthful of peanut butter and seemed to be reading news copy."[38] It became clear that cable needed to create high-quality programming of its own as well as retransmitting commercial broadcast channels with quality as high as what customers were used to receiving free.

Only one major roadblock stood in the way of making cable offerings complete. FCC rules had denied cable stations the right to import distant signals. Taking a cue from Nixon's dislike of the broadcast media and his "less government is best" deregulation philosophy, the White House Office of Telecommunications Policy (OTP) recommended the virtual deregulation of cable. A young press officer, Brian Lamb, served in the press office of the OTP, spreading the doctrine and in the process learning the ins and outs of telecommunications policy. By 1972, the FCC reversed itself, permitting cable to import distant signals once gain.

The desire for unique programming grew during the 1970s as cable operators went head to head in local communities while attempting to get exclusive franchises. In some areas, a virtual bidding war erupted over each franchise, with cable operators offering more and better programming—as well as a number of "deal sweeteners" (such as contributions to local government projects) that had nothing to do with the provision of a cable system.[39]

Running the cable did not guarantee an audience. In an environment in which viewers were used to "paying" for television by watching commercials and buying advertised products, it was not clear that the bulk of the American public would pay for television with real money—and still be subject to the commercials. The percentage of people who had to get cable to overcome problems with traditional reception was not large enough to sustain a vibrant cable industry. Viewers had to be convinced that they could get better and more diverse programming with higher technical quality than the programming that the commercial networks could provide.

The cost of running cable and providing converter boxes forced cable companies to focus their initial efforts on upscale neighborhoods that could afford their services. Although cabling took time and effort, the figures are dramatic. By 1995, less than twenty years after the active creation and promotion of the new technology, almost 70 percent of U.S. homes were connected to cable (see graph 3).

In the 1970s the dramatic growth pattern for cable was neither anticipated nor guaranteed. Cable operators took an active role in creating a demand for their product. Despite some bad blood between cable operators over franchise bidding excesses, most of them recognized that their future lay in expanding programming. They came to realize "that programming, not a transmission medium, was their principal product and that if they intended to be long-term cable operators they had better establish some control over their product."[40] By the late 1970s, cable had a transmission vehicle, almost twenty millions homes wired,[41] and a series of unique programs for which people would pay. Expanded satellite transponders were available, and the technology would support more programs. Transmission equipment and the televisions in subscribers' homes could handle more channels. By the late 1970s, cable operators saw expanded programming as the method of continuing their growth. They were ready for viable programming suggestions.

Graph 3

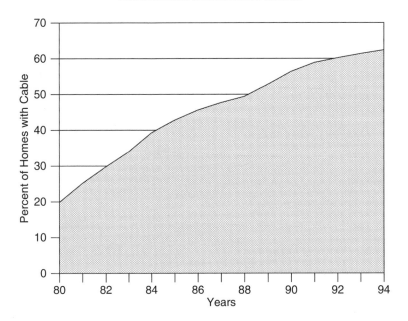

Cable Access in American Homes

The percentage of homes with cable access has increased from less than 20 percent to over 60 percent from 1980 to 1994. *Source:* Harold Stanley and Richard Niemi, *Vital Statistics on American Politics*, 4th ed. (Washington, D.C.: Congressional Quarterly Press, 1994), p. 47.

The Public Affairs Audience: A Moving Target

It is one thing to have access to cable and another to use it. The clearest evidence of subscriber usage is indirect. Once cable access approached 50 percent, the network share of the prime-time audience began to plummet dramatically.[42] Since viewing time increased, the audience was going somewhere, and that somewhere was cable.[43] After surveys and rating services began to monitor cable viewership, it was clear that some of that audience was moving to public affairs programming such as the Cable News Network (CNN), the first cable service to be rated. But we are getting ahead of our story, since CNN was not launched until *after* C-SPAN. Even the most ardent supporters of public affairs programming, or C-SPAN, would probably not have predicted that by the 1990s, less than half of the television viewers tuned in to the evening news would be watching the networks (see graph 4). In the 1970s, betting on cable as a distribution vehicle was still a matter of faith.

The evidence during the 1970s and early 1980s concerning the existence of a viable audience for public affairs was even less clear. Politics and government are, at best, of tangential interest to most Americans. Although presidential elections are the Olympics of politics, only about half of the eligible U.S. population bothers to vote. As the scene was being set for C-SPAN, the evidence was somewhat disheartening. The 1976 National Election Study reported that only 37 percent of the adult population indicated they were "very interested" in the presidential campaign. When it came to more direct involvement, only 27 percent claimed to have watched "a good many" television programs on the campaign, and only 11 percent had attended a political meeting, had worked for a candidate, or had given money to a political campaign.[44]

Audience research clearly indicated that the network television audience was switching to cable, but evidence concerning political interest and involvement did not clearly indicate that public affairs programming would be a primary destination. In fact the data seemed to indicate a very different direction. There was a core of self-identified highly interested citizens, but the question remained open as to whether they would transfer their viewing loyalty to C-SPAN.

Conclusion

By the mid-1970s the key pieces seemed to be in place. Not only observers of the media but also, increasingly, members of the media themselves were aware of the limitations and shortcomings of public affairs coverage. The established commercial media showed little interest in making wholesale changes in their coverage of public affairs. Uncomfortable with negative portrayals, public officials—the subjects of public affairs coverage—were responsive to new options. The public began to distrust the established media almost as much as they distrusted policymakers. The arrival of cable television created a new distribution mechanism, changing viewing patterns and creating a need for expanded programming options. Cable operators were not above public criticism, however. Paying customers—especially in an environment of free commercial channels—could be quite demanding. The public showed intense displeasure over disruptions in services and over

Graph 4

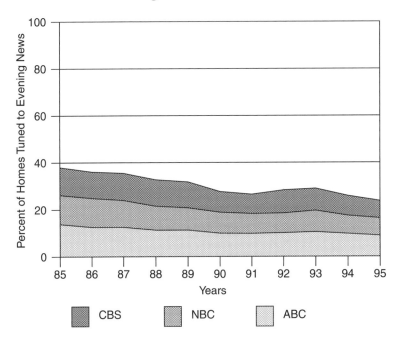

The Declining Network News Audience Share

The percentage of households watching network evening news programs has declined regularly during the last decade. *Source:* Nielson ratings reported in the LEXIS-NEXIS database.

price increases. Each new dissatisfaction led to calls for more government regulation.

From a practical perspective, cable operators were looking for expanded programming for which customers would be willing to pay. If that programming was seen as a public service, public relations payoffs would be even greater. Idealistically, a number of individuals within the cable industry recognized the value of more diverse public affairs programming and saw the role that cable could play in its development. The technological, political, and organizational stages were set for a grand experiment.

CHAPTER ONE

C-SPAN

An Idea Whose Time Had Come

"Opening day" has connotations of balloons, banners, and hoopla. The very phrase "launching a new service" implies excitement and activity. The first day of coverage of the House of Representatives on March 19, 1979, was like any other business day in the C-SPAN offices. Whereas over three million households could now see the House proceedings live, the signal did not reach across the Potomac River to C-SPAN's cramped two-room office in Crystal City. Like most other Americans, the people in the C-SPAN office did not hear Representative Al Gore (D-TN) make the maiden speech extolling "the marriage of the medium and our open debate . . . to revitalize representative democracy."[1] Newt Gingrich (R-GA), who had been elected to the House a few months earlier, had not yet discovered the medium that would catapult him into power sixteen years later.

In reality, there was hardly room in the C-SPAN offices for a television set. They were so small that the only technical staff person worked out of a closet. Jana Fay, the first paid employee, remembered typing on a thirty-year-old typewriter that would rip the page if she tried to underline. Her typing table was so rickety that she had to wedge herself in and hold the table with one knee in order for it to be stable enough for typing. As one early staff member put it, "We made the term 'fly by night operation' look good."[2] There was not much feeling of being part of a historical event or furthering the accomplishment of an idealistic goal. Everyone was so busy getting the signal up and going and making sure the bills were paid that there was little time for reflection. Everything was touch and go. The initial signal almost did not get broadcast because bad weather had delayed the building of the earth station necessary for sending the signal.[3] With significant effort, the building was completed literally hours before the first broadcast. Even if reflection had been desired, the physical separation between the C-SPAN office and its Hill staff made joint contemplation difficult. The two-person staff on the Hill seldom had anything but phone contact with the business office. It would be almost a year before the C-SPAN office was moved to Arlington, Virginia, and the staff could see the signal they were sending. Fay remembered, "It was not real in my mind until a few months after the first broadcast when we had a big press conference on the Hill, complete with a slide show."[4] Such modest beginnings not only describe how many good ideas can grow from nothing but also reflect the approach of Brian Lamb, C-SPAN's founder.

Brian Lamb: Merging Idealism with Opportunity

It is impossible to think about C-SPAN without understanding Brian Lamb. Although the network is self-consciously antipersonality and Lamb is genuinely self-effacing, there would be no C-SPAN without his vision and efforts. A modest midwesterner from Indiana, Lamb eschews personal promotion and the star system of modern media. He cooperates with the media when they attempt to understand what he is doing, but he always tries to redirect the interest to C-SPAN rather than profiling himself as a person. Having never spoken his own name on the air, Lamb would probably omit this section of the book if he were the author. But without an idea of what makes Lamb tick, there is no understanding of what makes C-SPAN tick.

Brian Lamb loves to learn. In some ways C-SPAN allows him to live his fantasy of constantly learning in the "University of the Real World." He learns from the best professors—people directly involved in public affairs on a daily basis. Lamb is quick to admit, "My reason to get into C-SPAN was fairly selfish: I wanted an alternative on my television set."[5]

Most revolutions begin with a simple idea and an individual who believes in it. To some degree, the simpler the idea, the more revolutionary it is. Brian Lamb's idea was very simple: show the American people what their government is doing, and let them see the action undiluted. Just when we begin to believe that modern and complex American society leaves little room for an individual entrepreneur with a revolutionary idea, stories like that of C-SPAN emerge. As journalist Colman McCarthy put it: "How many people can say that if I wasn't doing it, it wouldn't be happening[?] . . . Brian Lamb can say that."[6] One of Lamb's early C-SPAN colleagues pointed out: "It was like 'Mr. Smith goes to Washington.' Brian is really a populist and a democrat with a small 'd.' He really believes in the system. . . . Brian is one of the least cynical people you'll ever meet. . . . He was careful not to let them [his cable colleagues] know how strongly he felt about it—how idealistic it was. . . . My advice to him was . . . [t]urn it down a little, because you are going to scare these guys off."[7]

An early backer from the cable industry agreed: "Brian had a burning intensity and a vision that was more in his gut than in his head. . . . There are few times when one person makes a difference and Brian Lamb did that."[8] Lamb's former boss, Bob Titsch, the editor of *Cablevision* magazine, remembered: "His theory was that if the people were informed, they would then act to change the politicians. I thought the concept was wonderful. Brian is one of the last of the Boy Scouts."[9]

The story of C-SPAN is really one of an idealist with a vision, a responsive environment, and an opportunity to make that vision a reality. Although all of the components were necessary, nothing would have happened without the initial vision.

THE ORIGIN OF A VISION

We are all products of our abilities, upbringing, and life experiences. In some ways, Brian Lamb is an unlikely person to have delivered on his vision. A modest, personally conservative product of Lafayette, Indiana, and Purdue University, Lamb distinguished himself more as a hard worker than as a sparkling personality or a brilliant scholar. As a youngster he had become enamored with radio, and he later hosted a college television program called *Dance Date.*

Lamb's service in the U.S. Navy during the early 1960s left an indelible mark. As a curious twenty-four-year-old public affairs officer stationed at the Pentagon during the tumultuous time of the Vietnam conflict, Lamb observed the media at close range. His experiences changed his view of the media, the military, and the war protesters. In his role as a public affairs officer, he saw the facts distorted by political considerations. He privately complained, "I was only able to give the answers I was given." Facts and figures about the number of planes lost or soldiers missing were "delayed for days and days on purpose. . . . And then they were maneuvered around at the highest echelon, so they would play in the news in a certain way."[10] Lamb's naïvéte about what the public is allowed to learn about government was beginning to show significant cracks.

At one point Lamb was assigned to monitor an antiwar protest at the Pentagon, and what he saw disturbed him. A group of demonstrators were lounging around, showing no activity until the cameras appeared. "The minute the camera started rolling, the kids got up, the placards came out, and they yelled and screamed, 'Stop the war!' " One reporter was having difficulty getting his lines out and had to try again and again. As if "by spoken agreement between the demonstrators and the news crew—the lying down-and-jumping-up routine repeated itself with each take."[11] Lamb realized that what he was going to see on the evening news was much different from what had actually happened. He remembered being really angry about the power of the three networks. "When initial licenses were given out for television, the government was so worried that someone would fail that they established limited broadcast rights to three networks and created regulations to protect them. The result was similar coverage and a lack of diversity."[12] Although the intentions of the initial rules to undergird competition and provide expanded access to information were laudatory, the practical implications were more questionable. Lamb developed a long-term feeling that the public was "being treated unfairly by the television news."[13] He felt that the most grating limitation was the lack of variety: "I'd been taught all my life that this was a democracy, and that many voices were better than few voices. . . . I kept learning by being part of the system that there were very few voices."[14]

Some might write off Lamb's dissatisfaction with the media as simply an extension of the Nixon administration's diatribes against the "nattering nabobs of negativism" and the "eastern liberal press," but Lamb's frustration seemed to truly cross ideological and partisan boundaries. Pasted over his clear midwestern conservatism is a broader concern for hearing the full story. Early on, Washington became his political playground, offering him the opportunity to experience a wide variety of events firsthand.

Volunteering to serve as a social aide at the White House of Lyndon Johnson provided Lamb with the opportunity to vicariously experience power in Washington and whet his appetite for national politics. Lamb's curiosity led him beyond the official corridors of power in an attempt to understand the broader currents that were affecting American politics. A defining moment for Lamb occurred after a visit to a black Baptist church to hear the fiery civil rights leader Stokely Carmichael. Lamb remembered a thoughtful and intelligent thirty-minute speech, only a couple of minutes of which were incendiary. The event looked quite different when he watched David Brinkley on the NBC news that night. "What made it on [the air] was the fire and brimstone." The longer he viewed Washington, the more he saw such misrepresentation as the norm. Lamb explained, "I've always had a strong need to see things for myself, and I figured other people felt the same need."[15] Lamb began to reframe some of his basic questions. He moved from the naive question of "Are the media biased?" to the more sophisticated question of "*How* are the media biased, and what is the consequent effect on our interests and values?"[16]

After his service in the Navy, Lamb returned to Indiana to work as the assistant manager of a television station and to start a stint in the real world of politics. Despite his growing commitment to linking politicians to the people, Lamb's initial attempt to create such a linkage turned out to be a sham. During the 1968 presidential campaign, he took an assignment with the campaign of Richard Nixon and Spiro Agnew to tape-record people at shopping centers and town meetings in the Midwest, providing them the opportunity to "Speak to Nixon-Agnew." He was told that the interviews were boiled down and excerpts sent to the campaign plane. As he later found out, "It was all hooey, a gimmick to attract the attention of the evening news and plant firmly in the minds of the public that Nixon-Agnew wanted to listen to the people."[17] Rather than becoming a disappointed cynic and giving up, Lamb filed away the experience and looked for a way that such interaction between the public and the politicians could be done more legitimately.

Though a product of the Midwest, Lamb's heart was in Washington, D.C. The common malady of "Potomac fever" had claimed another victim. Returning to the capital, he worked for UPI Audio as a Washington reporter. Later he worked as press secretary for Senator Peter Dominick (R-CO) and "saw firsthand how difficult it is for any public official to be heard fairly over the electronic media."[18] The next step was a stint as a press spokesman for the White House Office of Telecommunications Policy (OTP), a White House think tank designed to develop presidential policy initiatives. Lamb became a spokesman for an office that promoted the deregulation of the media and the advantages of competition and private funding. The experience refined his preference for the private enterprise basis of the media while heightening his frustration with the approach taken by existing media outlets. While at the OTP he also made some good friends with other staff members. A young lawyer by the name of Antonin Scalia asked Lamb to be the godfather for one of his children, leading to an intriguing potential conflict of interest later, when Scalia was appointed to the Supreme Court. Lamb became a key supporter of televising oral arguments before the Court, an idea Scalia opposed. Although the friendship remains, the two have "agreed to disagree on this issue."[19] Another young OTP lawyer, Henry Goldberg, would provide Lamb with invaluable service when Lamb decided to challenge the existing media mix.

When asked about his White House experience, Lamb is quick to point out that he did not meet Nixon until many years later. The link with the Nixon White House caused some in Washington to initially question his objectivity. In May 1974, Lamb went back to print journalism, as the Washington bureau chief for *Cablevision* magazine. It was in this job that Lamb got to know the cable industry and its leaders.

Seeing the media from a number of different perspectives changed Lamb. He concluded, "My experiences with Washington coverage at the Pentagon and on Capitol Hill led me to believe that there had to be a better way to cover public affairs."[20] He recognized that network television was not showing the nation what was really happening in national politics. He stated his "gut instinct" at the time: "If I'm interested enough to want to know more of what's going on behind the scenes, there's got to be some more who feel the same way."[21]

Lamb is uncomfortable being called a visionary. He is much more modest, explaining that his interest "came more out of frustration than any great vision."[22]

A RESPONSIVE ENVIRONMENT

Lamb's vision of improved public affairs programming solidified at about the time cable television was experiencing dramatic growth. He noticed that cable operators were hungry for programming, and he came up with a first step to satisfy their needs and his vision. Lamb's initial ideas were on a much smaller scale than the creation of a public affairs network. He initially thought about doing interviews with political leaders and simply shipping out the tapes to cable networks and for broadcast.[23] In 1977, Lamb agreed to a reduced workload and salary cut from $42,000 per year to $21,000 at his *Cablevision* job to pursue a new venture called "Cable Video." With Bob Titsch's help, he sold the concept to fifteen cable system operators and began to conduct fifteen-minute interviews with members of Congress. Lamb pressed his fellow *Cablevision* reporter colleague Pat Gushman into serving as cameraman. The distribution was pretty basic. After running the tape, a cable system would send it on to the next station, in a process called "bicycling."[24]

But the process served neither Lamb's desire to inform the public nor cable's broader needs. The cable industry needed more than programming. It needed the legitimacy that public service broadcasting might bring. In the mid-1970s, cable was seen as stealing broadcast signals, offering morally questionable programming, and providing little public service.[25] If Lamb could promote his ideas of public affairs programming as a demonstrable public service with good public relations payoffs, he might be able to get cable owners to listen.

Thoughts of launching a new network benefited from the permissive legal environment promoted by the OTP, for which Lamb had worked. In 1972 the OTP had engineered, among commercial broadcasters and cable companies, a compromise ensuring a free-market approach to cable television. Lamb used his position as a writer for the major cable industry publication to plant the idea that cable could gain great benefits by offering public affairs programming. Now he needed an opportunity to make the idea a reality.

CAPITALIZING ON AN OPPORTUNITY

It is hard to pinpoint the exact point at which C-SPAN moved from an abstract concept to an operational goal. The change consisted of a series of moments, one of which occurred in July 1977, at a luncheon meeting of Brian Lamb and John Evans, president of Arlington Cable Partners. Evans, always looking for new programming, mentioned he had heard about black-and-white cameras in the House of Representatives and commented, "Gee, if there were some way I could tap into the House camera and carry it on our cable system."[26] Lamb was excited. He had been musing about expanded public affairs programming. Initially the vision was limited to providing the House signal in Arlington, Virginia, where a large number of the members of the House lived. Soon the vision expanded. From that point, Evans and Lamb began to explore the mechanics of how to raise funds to get the operation up and going. Evans did more than contemplate and plan: he put his money where his mouth was, offering to provide the wiring and the "head end" needed for the cable connection.[27]

The ball was now in the congressional court. Lamb could prepare for broadcasting, but he needed Congress to make some decisions about what would be covered and how the coverage would be broadcast. As the House deliberated, Lamb attempted to get his plans in order so that he would be prepared for whatever the House decided.

Cameras in Congress

Over a decade after television had become the primary source of political news for most Americans, the House of Representatives was still debating when and how the cameras should be admitted.[28] For a number of years, Congress had played a cat-and-mouse game with the electronic medium, wary of its impact but afraid to ignore it. The result was a series of intricate political dances, with Congress going one step forward and then two steps backward. The first electronic coverage of Congress occurred by subterfuge. After Speaker of the House John Nance Garner (D-TX) banned radio coverage of the 1932 debate over repealing the Eighteenth Amendment (Prohibition), two enterprising journalists set up microphones in the doorway to the library. "The electricians turned up what they term 'the gain' and the radio ears were made so sensitive that the sound of the gavel, the ayes and nays, and other words from the floor were broadcast to the entire nation."[29] Although journalists were denied the official right to broadcast congressional sessions on the radio, a press gallery for radio reporters was established in 1939.[30]

Television cameras were allowed to cover the opening of the Eightieth Congress in 1947. A number of congressional committees were televised during the late 1940s and the early 1950s, including the famous House Un-American Activities Committee investigations of Alger Hiss and Whittaker Chambers, the Kefauver hearings on organized crime, and the Army-McCarthy hearings. The last hearings, chaired by Senator Joseph McCarthy from Wisconsin, were intended to ferret out Communists in the military but unintentionally revealed the power of television. In the long run, McCarthy's fanaticism did not play well to the television audience. By the time the hearings ended, "the power of television [had] extinguished McCarthyism as surely as the power of newspapers had helped create it."[31] McCarthy was eventually censured by the Senate for bringing the Senate into "dishonor and disrepute."

During Republican control of the House in 1953–54, television cameras were allowed to cover numerous committee hearings. Returning to the speakership in 1955 after two years of Republican control, Speaker Sam Rayburn (D-TX) once again banned cameras completely, both from the floor and from House committees. The cameras were later readmitted, to cover the opening of the Eighty-sixth Congress after Rayburn's departure. The Senate continued to allow coverage of some committees, but it was not until 1970 that committees were allowed to establish their own rules regarding televised coverage of their sessions. Televised hearings hit their stride with the extensive coverage of the Watergate hearings in 1973 (see box 1).

Behind the rhetoric about expanding the public gallery and opening the "people's House" to the people lay a more practical reason for televising House proceedings. During a 1974 hearing of the Joint Committee on Congressional Operations, Senator Lee Metcalf (D-MT) summarized many members' concerns: "A Congress unable to project its voice much beyond the banks of the Potomac . . . can be neither representative nor responsive. A Congress able only to whisper, no matter how intelligently, cannot check and balance the power of the Executive or safeguard the liberty of individual citizens."[32]

Many House members saw coverage of their chamber as gaining their body visibility relative to the Senate and circumventing the mainline television news programming that covered Congress via short—and often very negative—sound bites. A number of participants regarded televising Congress as emanating from Speaker Tip O'Neill (D-MA), who wanted to bring "the people closer to the House of Representatives."[33] O'Neill was influenced by younger members such as Al Gore, who urged him to push the idea. O'Neill's support was crucial but hesitant. Despite public statements, O'Neill "was convinced from the onset that the system would hurt him and the House leadership."[34] The Speaker was a realist, seeing how the votes were lining up.

A number of older members were nervous about the discomfort of bright lights and the potential for making the House and its members look fool-

Box 1.
A Chronology of House Television

1922   Representative Vincent Brennan's (R-MI) measure to allow radio coverage of the House and Senate fails in committee.

1924   Senator Robert Howell's (R-NE) proposal for radio coverage passes the Senate but is not acted on because of the cost.

1932   Despite the refusal of House Speaker John Nance Garner (D-TX) to allow radio broadcasts of the debate to repeal the Eighteenth Amendment (Prohibition), secretly placed microphones in a doorway pick up the debates and broadcast them to a wide audience.

1944   Senator Claude Pepper (D-FL) proposes radio coverage of Congress.

1947   Television cameras record the opening of the Eightieth Congress.

1948   The first congressional committees are covered by television: Senate Armed Services Committee (March); House Un-American Activities Committee (summer hearing on the State Department spy controversy involving Alger Hiss and Whittaker Chambers).

1951   The coverage of the Special Senate Committee to Investigate Organized Crime (Kefauver Commission) wins an Emmy Award.

1954   Senator Joseph McCarthy's Senate Permanent Investigating Subcommittee holds thirty-six days of televised hearing on suspected Communists (Army-McCarthy hearings).

1961   Television cameras record the opening of the Eighty-seventh Congress.

1970   The Legislative Reorganization Act allows House committees to broadcast their public hearings.

1973   The broadcast networks devote 319 hours to covering the Watergate hearings.

> ### Box 1 *(continued)*.
>
> 1974   Cameras are allowed in the Senate on a one-time basis to cover the swearing in of Vice-President Nelson Rockefeller.
>
> 1975   Representative John Anderson (R-IL) introduces a bill to conduct a test of televising House proceedings.
>
> 1977   Speaker Thomas (Tip) O'Neill (D-MA) unexpectedly announces, on March 2, a ninety-day, live, in-House test of televised coverage of floor proceedings.
>
> 1978   The House decides to control its own television system and begins purchasing equipment.
>
> 1979   The House allows public telecast of its sessions, and C-SPAN goes on the air on March 19.
>
> Adapted from Ronald Garay, *Congressional Television: A Legislative History* (Westport, Conn.: Greenwood Press, 1984), and Casey Peters, "The Cable Satellite Public Affairs Network: C-SPAN in the First Decade of Congressional Television," master's thesis, UCLA, 1992.

ish. Three basic arguments undergirded the opposition: (1) the work of Congress does not lend itself to television coverage and would produce only dull, uninformative, or even misleading television; (2) television coverage would disrupt congressional proceedings, turn members into performers, and encourage demagoguery; (3) broadcasters could not be trusted to present Congress fairly or accurately.[35]

There was relatively little Congress could do about the first concern aside from a few cosmetic changes in congressional proceedings. Debate in the House focused on the other two bases for hesitation. There was a real fear that roving cameras controlled by the media would seek out embarrassing scenes of someone sleeping or reading in the chamber. As then staff member and later Representative Tony Coehlo (D-CA) remembered, "Grandstanding was the public reason for opposition, nose picking was the real worry."[36] Proponents argued that their compatriots would either ignore the cameras or be on their best behavior.

The major sticking point in the House was not so much television or no television but rather who would control the cameras and how the signal would be distributed. Members felt that if they could not change the nature of House proceedings and there was no definitive evidence on how their colleagues would react to the cameras, they could at least design a distribution mechanism that would control any possible damage. The networks insisted on their journalistic right to control the cameras and determine the coverage. Some members had a vague conception of the House creating its own network. Others hoped that the established Public Broadcasting Service (PBS) would take responsibility for programming and distribution.

Speaker Tip O'Neill was wary of the networks and what they would do with television coverage, and he was able to convince his colleagues that the House would have to control its own cameras. This House decision offended journalists of all stripes. Even PBS—the preferred outlet for congressional coverage by a number of the key players—felt that the House was dictating coverage. Tony Coehlo, then Representative Bernard Sisk's (D-CA) chief of staff, remembered: "I begged PBS to take the House signal, because they were the logical folks to do it. It was there to have."[37] PBS's unwillingness to broadcast House proceedings stemmed from a fear that since PBS depended on funding by Congress, the network might be subject to manipulation in the way it broadcast the congressional signal.[38]

House members knew that the networks would not cover large blocks of House debate, and they did not want to get into the broadcasting business themselves. The solution was to produce a House "feed" that anyone could use. But House control of the cameras and the availability of a gavel-to-gavel feed still did not solve the problem of getting the signal out to the public.

As the debate proceeded, Brian Lamb indirectly helped break the logjam. Lamb's position as bureau chief for *Cablevision* magazine provided him with an entrée to key decision makers. On October 27, 1977, after a formal interview with Representative Lionel Van Deerlin (D-CA), a former television news commentator and chairman of the powerful House Subcommittee on Communications, Lamb broached his idea of cable distribution. Van Deerlin loved the idea and asked Lamb to write a speech for the congressman to deliver. Lamb returned to his office excited but feeling little time pressure. Later that afternoon, he received a call from Van Deerlin, who was in the House cloakroom. Van Deerlin told him: "You are not going to believe this, but this thing's on the floor right now. If you want this stuff on the record, give it to me over the phone, and I'll write it down and go out on the floor and make a speech."[39]

Lamb threw out ideas and made commitments about cable industry support, even though he had not received any formal approval from the cable operators. Van Deerlin broached the idea on the floor by saying: "Gavel to gavel coverage of the House and Senate proceedings, although they may create no new star competition among the performers, will be available at times and to an extent that no commercial station, certainly no network could or would provide. But they might easily be included within the new channel capability of a cable operator."[40] Van Deerlin's "impromptu but persuasive speech helped pass the resolution."[41] After the speech, one of Speaker O'Neill's top aides called Van Deerlin and said, "The Speaker wants to know what the hell you were talking about yesterday."[42] Van Deerlin put O'Neill's staff in touch with Lamb, and a series of meetings took place.

In January 1978, Lamb met with O'Neill about cable distribution. C-SPAN's commitment to gavel-to-gavel coverage provided comfort to those who wanted broader distribution, and the "chance to spit in the eye of network news people" appealed to the Speaker.[43] According to Gary Hymel, a key O'Neill aide, "C-SPAN's financial autonomy made it more acceptable as well. . . . It was important that any network remain separate from Congress."[44] Lamb left O'Neill's office with a handshake deal. Now he had to turn a concept into reality.

THE BALL ENTERS LAMB'S COURT

Lamb had made a number of commitments without having the formal backing of the cable industry. He now had to scramble to make cable distribution happen. The first problem was money. Initial reaction was cool. Lamb's fundraising difficulties with the cable industry were not surprising. Ken Gunter, one of the early cable operators, argued: "The cable industry is composed of a bi-polar group of people. It is made up of the most highly individualistic and aggressive businessmen you'll see anywhere. On the other hand, it has some of the stodgiest [and] most backward."[45]

The hard-nosed businessmen could not see the financial benefit of creating a channel that would cost money rather than make it. Unlike other channels, C-SPAN would provide no advertising revenue. In fact, the cable operator had to pay for the privilege of broadcasting the channel. Many cable operators with little foresight could not see the benefit of the new venture. However, a number of cable operators were convinced by the argument

The Meeting of Minds

Brian Lamb and Speaker Tip O'Neill meet during the early stages of deciding how to televise congressional sessions. (Photo courtesy of C-SPAN.)

from their chief lobbyist, Tom Wheeler, who said that offering such a public service would lend the industry some legitimacy and would help with local franchising and congressional regulation.[46]

Bob Rosencrans, owner of UA-Columbia, came to Lamb's rescue, writing a check for twenty-five thousand dollars and providing a demonstrable endorsement for Lamb to use as he tried to get up-front money from other operators. Rosencrans explained: "I was tired of knocking on congressmen's doors to explain what cable television was. . . . So if nothing else, I thought it would put cable on the map in Washington."[47] Rosencrans told Lamb, "I'll give you twenty-five thousand dollars and my name—see what you can do."[48] C-SPAN was sold to the cable industry not only as the right thing to do but also as a means of boosting its public image. As the first chairman of the C-SPAN board, Bob Rosencrans vigorously pursued his fellow cable operators for funding.

Perhaps symbolically, Rosencrans and his fellow initial contributor, Ken Gunter, are diametrical opposites when it comes to political beliefs. Bob

The First Dish

Bob Rosencrans and Brian Lamb stand at the site of C-SPAN's first satellite dish. (Photo courtesy of C-SPAN.)

Rosencrans, an eastern liberal from Connecticut, and Ken Gunter, a southern conservative from Texas, may not agree on many issues, but both were willing to make a tangible commitment. In Rosencrans's words: "Where we disagree on politics, we don't disagree on our system of government. We believe that everyone should have his day in court."[49] Each felt that by getting the word out and showing government in action, C-SPAN could persuade voters to switch to the other side of the political fence.[50] Both are still convinced that they were right.

Within a few weeks Lamb had gathered over four hundred thousand dollars from twenty-two operators, and his plans were ready to be set in motion once the House had its system up and going.[51] Lamb began hiring his small staff and lining up the necessary equipment.

Whereas direct financial support was important, a number of the early supporters found indirect ways to promote the fledgling network. Bob Titsch, publisher of *Cablevision*, kept Lamb on his staff for eighteen months while the network struggled to get on the air. The magazine's Washington bureau chief, Pat Gushman, shared an office with Lamb in Crystal City and worked to publicize the operation.[52] Although John Evans, an old friend of Lamb's from the navy days, was still trying to put together a cable system of his own in Arlington, Virginia, he provided twenty-five thousand dollars in labor.[53] Comm/Scope Marketing of Hickory, North Carolina, contributed the cable linkup between the Capitol and the C-SPAN earth station. Bob Rosencrans became the first chairman of the C-SPAN board and, through his company, helped C-SPAN secure a transponder spot that it would share with the Madison Square Garden Channel on RCA's Satcom I satellite for twenty-five thousand dollars.[54]

Most of the early supporters became members of the board of directors, helping set policy for the National Cable Satellite Corporation, C-SPAN's official name. For Lamb, the early supporters were like the nation's "founders," each making a different but significant contribution. Lamb calls Rosencrans his "George Washington."[55]

As Lamb prepared to get C-SPAN on the air, he turned to Henry Goldberg, his old colleague at the OTP, to serve as general counsel. C-SPAN benefited from the early free market approach to cable television championed by the OTP. Recently it has been argued that the current "regulatory scheme ensures that if Brian Lamb were trying to launch C-SPAN today, he'd be out of luck."[56] In 1978, things were different. Goldberg emphasized C-SPAN's public service component and received unusually fast approval from RCA and the Federal Communications Commission.[57]

The countdown to the initial broadcast became quite tense as the realities of getting the equipment up and running began to set in. Battles over right-of-way access to the satellite transmitter in Fairfax, Virginia, caused consternation. Poor weather slowed construction, and the facility was ready only hours before the first broadcast.

After some initial testing of equipment, the House went on the air on March 19, 1979, and C-SPAN was there. On the initial day of House coverage, the major networks used clips on the evening news. PBS bumped programs such as *Sesame Street* to cover the first two-and-one-half-hour session but did not plan to return on a regular basis.[58] C-SPAN was there for the duration. The first day was relatively routine. After Representative Al Gore's speech welcoming the coverage, the House proceeded with routine debate on the creation of a select committee to study committee jurisdictions. Few had any idea of what the next decade would bring and how a series of chance encounters, creative ideas, and hard work would change both the institution and the political process.

GROWING PAINS

House Speaker Newt Gingrich credits Brian Lamb not only with having a good idea but also with having the energy and discipline to maintain it. He noted, "An amazing number of creative innovators can't become sustainers, and Brian has understood that if he simply sustains C-SPAN that is probably as big a contribution as anybody needs to make in one lifetime."[59]

Getting C-SPAN on the air was only part of the challenge. Lamb and his staff still had to deal with two masters, the cable industry and Congress. Lamb spent much of his time trying to line up cable systems to carry C-SPAN and pay the monthly one cent per subscriber fee. Cable operators remember Lamb selling C-SPAN to reluctant owners. "[He was] like the oil-filter salesman who says if you don't get it for your car now, you'll have to get it later. . . . He [was] always on their backs."[60] Since many cable operations were small, lining up C-SPAN affiliates was a labor-intensive process of retail sales. Financial insecurity plagued C-SPAN's early years and brought out Lamb's personal conservatism. He hesitated to spend any money unless absolutely necessary. It was a period of "making do" with used equipment and basic necessities.[61]

A symbol of the early days was Brian Lamb's old yellow Toyota Corolla hatchback, which doubled as the C-SPAN equipment van and "limousine." Lamb used to save a few bucks in taxi fare by meeting board members at the airport in his hatchback. It would be years before the board could persuade him to purchase something more adequate.

After the initial launch of House coverage, the continuing story of C-SPAN's development is mostly one of small steps. Attempts to expand audience potential, develop new programming initiatives, and capitalize on new opportunities became regular parts of the C-SPAN approach. As we will see throughout this book, C-SPAN seldom simply rests on its laurels (see box 2).

From the outset, C-SPAN's relations with Congress were tenuous at best. Within a year Speaker O'Neill had lost his enthusiasm, calling House television "a disaster."[62] His major complaint was the extensive use of Special Orders speeches at the end of each legislative day, speeches that dramatically

> ### *Box 2.*
> #### C-SPAN Milestones: The Early Years
>
> | | |
> |---|---|
> | March 1977 | Initial idea of a public affairs network is proposed and seed money donated. |
> | March 1979 | C-SPAN begins televising the House to 3.5 million households on 350 cable systems. |
> | December 1979 | First regularly scheduled non-House programming is added: Close-Up Foundation student seminars. |
> | January 1980 | National Press Club speeches are added to the schedule; the first televised speaker is Paul Volcker, chairman of the Federal Reserve Board. |
> | October 1980 | Mutual Radio's *Larry King Show* is the subject of the first *Day in the Life* series. |
> | October 1980 | The first live viewer call-in is aired, with guests discussing communications issues. |
> | October 1980 | The first federal agency hearing is aired (Federal Communications Commission). |
> | January 1981 | The first congressional hearing is broadcast. |

expanded with television coverage. The number of speeches bothered him less than the Republicans' use of them to castigate the Democrats. O'Neill toyed with the idea of banning such speeches or even reversing the decision on television coverage, but he found little support among his colleagues.

C-SPAN's right to take the House signal and use it without interference was based on a verbal agreement with Speaker O'Neill. O'Neill turned over day-to-day oversight of House television to Charlie Rose (D-NC), who became chairman of the Speaker's Advisory Committee on Broadcasting.

If one's importance is known by the enemies one makes, C-SPAN can credit itself with at least one powerful enemy in Congress, Charlie Rose. As C-SPAN developed, Rose and his then wife and staff member Joan Teague Rose sought both access to viewer letters and editorial control over what C-SPAN covered. Brian Lamb rebuffed the attempts and made a long-term enemy.[63] A key sticking point revolved around C-SPAN's insistence on controlling which committee hearings to cover. Rose argued, "The power to pick what committee hearings members of Congress watch is the power to set your own legislative agenda."[64] Joan Teague Rose complained: "We gave free reign to them and they would not be cooperative. . . . They wouldn't even be here if it wasn't for us." She referred to Lamb as "stabbing Congress in the back after [it] had bent over backwards to cooperate."[65]

Another early disagreement concerned C-SPAN's control over the House signal itself. Although anyone could directly tap into the signal from the House, C-SPAN had developed a nationwide distribution system that made it easier for any television station to rebroadcast it. When Rose attempted to have C-SPAN provide its own produced programming free of charge, Lamb countered by asserting that the cable industry had a right to protect its investment. Lamb argued: "Anyone can pick up the same signal at no cost directly from the House. Once we get it and mix it with our own programming, it belongs to us."[66] In one of the first salvos, Rose threatened to copyright the House signal, thereby giving the House more control over its use. It soon became clear that there would be significant legal problems in reversing the long-term policy of not copyrighting any material developed by the government.

Whereas Rose could not force C-SPAN to cooperate—denying access to the House feed would face considerable opposition—he could hamper its operations *within* Congress. Rose denied the internal congressional television system access to the C-SPAN feed. House offices were limited to the direct feed from the floor and could not receive programming produced by C-SPAN, such as committee hearings. Rose's public criticisms varied. He asserted that C-SPAN had "taken a signal that should be free to everyone." He added, "Lamb is ripping everyone off." A Rose aide, arguing that C-SPAN was biased, noted that C-SPAN was a "Republican" network. "I don't think we should let a right-wing Republican like Brian Lamb decide what's on the House system."[67] Rose's characterization of Lamb and the network's bias holds little credence, however. Clearly, personal animosity colored reasonable judgment.

The logjam in congressional access to the C-SPAN signal was eventually broken by the intervention of Senator Barry Goldwater (R-AZ). He wrote to Senate Rules Committee Chairman Charles Mathias (R-MD) and asserted the Senate's need for access to C-SPAN. Mathias agreed and unilaterally made C-SPAN available to all Senate offices.[68] Thus, whereas the House was limited to the direct feed from the floor and was denied access to all other C-SPAN programming, the Senate was able to see all C-SPAN programming. Rose eventually backed down and allowed C-SPAN on the House cable system.

Early Operations

C-SPAN faced early challenges of technical necessities, audience needs, political challenges, and access issues. There was no assurance that the network would survive (see box 3).

TECHNICAL SUPPORT

A visitor to C-SPAN's Capitol Hill office, with its modern equipment and spectacular view of the Capitol, cannot imagine the shoestring operation that spawned the current prosperity. C-SPAN was initially run from a one-room technical studio in the basement of the Rayburn House Office Building on Capitol Hill and from a small business office in Crystal City, Virginia. The move of the business operations to a three-room suite in Arlington, Virginia, a half-hour from the Capitol, was a big step up.

The Arlington headquarters were far from luxurious. The call-in set served as the engineer's workbench when programs were not being conducted. The C-SPAN station logo letters were tacked to a beige-carpeted wall, giving the set a jerry-rigged look at best. Lamb recalled, "During the on-the-air interviews, I was always looking behind the guest because sometimes the 'C' would fall off our sign."[69]

The young staff was eager and flexible. "We called what we were doing 'guerilla television': C-SPAN was so new and different, we were making up the rules as we went along."[70] During the early years, C-SPAN staff saw themselves as being on a crusade. "We felt we were David fighting the commercial media which looked like Goliath."[71]

Equipment was expensive, and C-SPAN lacked the capital to buy the cameras it needed to do its own programming. Perhaps borrowing from entertainment television (*Let's Make a Deal*) or from Congress itself, Lamb decided to go beyond Congress for programming and beyond the cable operators for support. In 1979, he approached the Close-Up Foundation, which brings high school students to Washington to see government at work. "They wanted a television outlet; we needed equipment—we made a deal. They bought us two cameras, switching units, tape recorders, all that stuff. We did programs with these high school students talking to Members of Congress;

> **Box 3.**
> **C-SPAN Milestones: The 1980s**
>
> | | |
> |---|---|
> | October 1981 | C-SPAN studio and staff move to Capitol Hill. |
> | December 1981 | C-SPAN becomes available to some D.C. businesses via the Capitol Connection. |
> | April 1982 | "Friends of C-SPAN" is launched to encourage cable companies to carry C-SPAN, after a satellite transponder switch causes some operators to drop carriage. |
> | October 1982 | C-SPAN begins round-the-clock programming, seven days a week. |
> | October 1983 | President Ronald Reagan makes an exclusive appearance on C-SPAN's *Close-Up* broadcast and later telephones the follow-up call-in program. |
> | Summer 1984 | C-SPAN covers its first national party conventions. |
> | September 1984 | "Grassroots 84" launches a fourteen-city tour of election coverage. |
> | January 1986 | C-SPAN is added to the Senate cable system. |
> | June 1986 | The U.S. Senate goes on the air on C-SPANII. |
> | January 1987 | C-SPANII expands to twenty-four hours a day. |
> | November 1988 | C-SPAN begins regular foreign programming with live coverage of the Queen of England's annual address to Parliament. |

we had use of that equipment when we weren't doing that."[72]

In the early days there was considerable concern about costs. During the early call-in shows, C-SPAN used its three business lines for the phone calls. This "led to a number of rather interesting calls going out over the air."[73]

In employee Jana Fay's recollection, operating with a one-person technical staff for satellite uplink was pretty scary. One day she began getting calls that C-SPAN was not broadcasting. After repeated calls to the trailer at the uplink site, she began to fear that some sort of disaster had occurred. She jumped in her car and went out to the site. Her knocks on the door of the trailer went unanswered, and her fears for technician Don Houle's well-being began to rise. She imagined he had had a heart attack. Just as she was about to get help, the trailer door opened, and a groggy and chastised Houle appeared. He had fallen asleep, and he begged Jana not to tell Brian. She kept her word, only to find out that a penitent Houle had already told Brian. With more mock than real anger, Lamb then commented to her about not being told, "Now I know who my friends are."[74]

C-SPAN was not welcomed with open arms by the established media in Washington. Its young, nonunion staff seemed too eager to put in long hours. The established media's technical staff members, who might well be credited with establishing the "grunge look," were a bit disdainful of the C-SPAN staff, with its Lamb-enforced dress code. The established media were accustomed to zipping into a hearing, setting up the television lights to cover only the majority party members, and noisily packing their equipment after harvesting the initial sound bites. They could not understand a network that wanted to give full and unbiased coverage of a hearing gavel-to-gavel. At times C-SPAN crews found their cables mysteriously cut; more often, they were simply looked at as a transitory oddity that would not survive in the world of high-tech sound bites.

BUILDING AND MAINTAINING AN AUDIENCE

Just as C-SPAN began to hit its stride in terms of expanded programming and growing audience, it was faced with a major challenge. C-SPAN had begun broadcasting by sharing a satellite transponder with the Madison Square Garden Channel on the USA Network. It was a good match, since C-SPAN covered events during the day and Madison Square Garden covered evening events. But by 1982 an increasing number of channels were broadcasting twenty-four hours a day, and C-SPAN was forced to change satellite transponders. The shift often required cable operators to use another dish in order to keep C-SPAN on the air. Four hundred to five hundred cable systems, representing 2.5 million households, dropped C-SPAN after it discontinued its association with the USA Network. This was a significant drop, accounting for over 10 percent of C-SPAN's potential audience.

C-SPAN was forced to promote its service and cajole operators to keep its signal on the air. A major boost came from the C-SPAN audience. A basic rule of human nature is that people are more likely to be activated out of fear than out of hope. People who fear losing something they have become accustomed to are more likely to organize than people who want something they do not have. The activation of the C-SPAN audience bears this out. Two self-proclaimed "C-SPAN 'addicts,' " Shirley Rossi of Pueblo, Colorado, and William "Bud" Harris of Cherry Hill, New Jersey, formed "Friends of C-SPAN," a cooperative dedicated to keeping C-SPAN on the air."[75] They petitioned cable operators to reinstate C-SPAN. Although operating independent of C-SPAN, they obtained the names of viewers who had called to complain and created a mailing list for their newsletter. In response to their campaign, a number of cable operators reconsidered their decision to discontinue C-SPAN. As we will see later, this would not be the last time that C-SPAN would go through the throes of abandon and rescue.

For some citizens, C-SPAN was still not available at all. The provision of cable television is driven largely by economic considerations. Areas with a dense population of likely subscribers were the first to be cabled. One of the last major urban "islands" in which cable was unavailable was Washington, D.C. For C-SPAN this was symbolic, since its goal was to bring Washington to people outside the beltway. For citizens and political activists in Washington, the symbolism rang hollow. They wanted to be tuned in to know what the rest of the country was seeing. Necessity is the mother of invention. Mike Kelly, an English professor and head of the George Mason University Telecommunications Department, saw a need and endeavored to fill it. In 1981, the George Mason University Foundation created a microwave service for the C-SPAN signal, which could be picked up in hundreds of D.C. buildings equipped with a microwave dish. The Capitol Connection service, which cost about seven hundred dollars per year, was picked up by news bureaus, lobbyists, the national party organizations, executive agencies, and the White House.[76] When cable finally came to the District of Columbia in the late 1980s, Capitol Connection subscribers switched to a commercial cable service.

A POLITICAL THREAT

House members feared that television could be used against them in upcoming elections. Republican Whip Robert Michel (R-IL) took to the floor in October 1977 to warn his colleagues that the House should carefully consider the political impact of broadcasting the House proceedings. Little did Michel know that he would be one of the first victims of political use of the C-SPAN signal.[77] Even though the House resolution approving television coverage had included a provision that restricted the use of the televised proceedings ("No coverage made available under this resolution nor any recording thereof shall be used for any political purpose"), the restriction did not apply to challengers. During the 1982 campaign, Michel's opponent, G. Douglas Stephens, tried to make the point that Michel was insensitive to the economic troubles of his constituents. Stephens created a thirty-second commercial that included a short clip in which Michel said that some Social Security recipients were "fairly well-heeled." Michel barely won his seat back.[78]

Both parties began gathering an arsenal of damaging clips of opposition party members in a "video version of mutually assured destruction."[79] Each party knew that if it sanctioned the use of such clips, the opposition would up the ante by unleashing its own clips. There was little question that virtually anyone could be made to look foolish by a careful editing of comments. Thus both parties agreed that they would not initiate the "first strike." So far the agreement has held, with only minor infractions, largely perpetrated by challengers ignorant of the House rule and uninformed about the party agreement. Each time C-SPAN is alerted about a questionable use of its programming, General Counsel Bruce Collins sends a "cease and desist" letter telling the perpetrator of the congressional rules and asserting C-SPAN's right of ownership. In addition to not wanting to offend members of Congress, C-SPAN is concerned about the danger the misuse of its material might have on its public image. Many of the advertisements using C-SPAN material are very selective or use sophisticated graphic techniques to make a point. As Collins pointed out: "Our main concern is reputational. If we don't put our foot down, we become the free video production house of literally thousands of campaigns."[80]

Although the direct political use of C-SPAN footage is prohibited, the advantage of a member's on-air visibility is hard to measure. Staying within the law, members often inform their constituents and local news media when they are scheduled to speak on the floor.[81] Since C-SPAN allows the media to use up to three minutes of coverage without payment or approval, a member's one-minute speech can legally be used in a news broadcast.

As we will see in more detail in chapters 2 and 7, the political use of Special Orders speeches, an issue that gave Tip O'Neill heartburn during the first year of C-SPAN's operation, did not go away. Members of the Republican minority discovered such speeches to be potent weapons in an institution that limited their participation and in a mainstream media environment that tended to focus on the majority party.

Sorting Out the Questions of Access

As C-SPAN expanded its programming beyond simply transmitting floor proceedings, issues concerning the right to cover congressional events arose. As a result of the push for congressional openness during the 1970s, C-SPAN was allowed to cover most committees, but committee members could still ban coverage of specific sessions and types of sessions. Powerful committees like the House Rules Committee, which schedules legislation, were very hesitant to allow the cameras in. The Republican minority regularly complained that the Democrats closed such meetings out of fear of political embarrassment. Other committees, such as House Administration, used the excuse that their hearing rooms were too small to accommodate the cameras. Over the years, C-SPAN became more aggressive about commenting on denied access. When the House Telecommunications Subcommittee closed its markup session on cable regulation, C-SPAN bitterly commented: "It's ironic. A subcommittee hearing about television and they're not allowing cameras in."[82]

One of the major sticking points was access to markup sessions—the meetings in which the final details of legislation are agreed on. Many committees regularly barred the cameras from such sessions. Brian Lamb complained: "What I find interesting is that I've been personally lobbied by members of the House and Senate to cover any number of hearings. . . . It seems a little bit strange that when we get down to the most important part of the process, at a critical point, they won't let cameras in."[83]

Control over the House feed raised a related question concerning the substance of access. In 1988 when Representative Robert Dornan (R-CA) refused to relinquish the microphone at the end of his one-minute grant of time, the acting Speaker of the House ordered the sergeant at arms to cut off Dornan's microphone. The Republicans cried censorship when they realized that the C-SPAN signal was going out with video only. After heated debate, the House passed a resolution asserting that the "audio and visual broadcast coverage of House proceedings not be interrupted."

Conclusion

Over the years both C-SPAN and Congress have felt their way in an attempt to create a reasonable working relationship. By and large, the level of cooperation has been high; incidents of conflict have served as punctuation marks requiring further consideration. With limited cash and a lot of heart, C-SPAN created a national network and was well positioned to capitalize on its accomplishments. The future was not ensured, but the critical steps of creating a network and establishing a relationship with Congress had been accomplished.

CHAPTER TWO

Arrival and the Pressure for Expanded Coverage

From out of the corner of his eye, he sees the familiar scene. The elderly woman has just discreetly pointed him out to her husband in the airport lobby. A muffled conversation ensues, sprinkled with phrases like "It sure looks like him" and "If you are so sure, why don't you just ask him?" A few minutes later the woman timidly sidles up and asks, "Aren't you somebody?" She adds, "I mean, you look just like John Glenn." By then the older man has arrived to hear him say, "No, I'm Brian Lamb from C-SPAN." With a big smile the older man says, "Yea, I knew it, you're on *Headline News*, watch it all the time." Patiently Lamb responds: "No that's CNN. I am from C-SPAN, we cover Congress and public affairs." To which the older man responds, "Sure, we watch it all the time, great show."

For a CEO more interested in personal glory or for a network striving for ratings, such confused recognition would be frustrating. For Lamb and C-SPAN, even this amount of recognition is somewhat embarrassing but appreciated. A few years ago there would have been no

recognition at all.

Arrival in the Public Consciousness

The C-SPAN audience has never been large (see chapter 6), and C-SPAN staff members maintain a kind of "gee whiz" modesty about what they are doing. In the early years they were not even sure anyone was watching. When over thirty people called in during the first impromptu call-in program (see chapter 4), there was a true sense of amazement.

Believing you are important is less of a priority in politics than having others believe you are important. C-SPAN staff use a variety of benchmarks to point out how and when the network "arrived" as a nationally recognized force. Today they laugh about the days when C-SPAN was confused with the Spanish network. They remember that trying to book guests was "just horrendous." One employee noted, "We always had to explain who we were."[1] For a number of years, C-SPAN fought a quiet but frustrating battle to get the news media to use "C-SPAN," with all capital letters, as opposed to "C-Span" or some other variant. Callers often had problems finding the telephone number, since Washington directory assistance and telephone books did not know how to deal with the hyphen and often could not provide a phone number for C-SPAN.

Recognition of C-SPAN by the cable industry, which awarded the network its first of many Cable ACE awards in 1980 for "distinguished service," brought recognition from cable colleagues and resulted in positive publicity. Television writers began commenting on C-SPAN in their columns but always had to write out the name as the "Cable-Satellite Public Affairs Network" and had to explain its programming focus because most readers were not familiar with the network by its acronym. Real recognition did not come until the politicians began to see C-SPAN as important. For many members of the House, C-SPAN was invisible. It was not available in their offices, and their districts had limited cable penetration. President Ronald Reagan is credited with helping raise C-SPAN's visibility. At a White House Photographers' Dinner, the president put his fingers in his ears and stuck his tongue out, saying, "I always wanted to do that." C-SPAN had the only television cameras at the event, and the unpresidential gestures were carried by news media around the world. It became clear that Reagan was not only the subject of C-SPAN coverage but also a viewer. When the president called in during a Close-Up Foundation program, the media picked up the story.

There is a refreshing sense of excitement among the C-SPAN staff, who still get thrilled when they are mentioned. When Reagan called, people started running down the hall yelling, "The president is on the line!"

C-SPAN's big break came in 1984. Members of the Conservative Opportunity Society (COS)—a group of conservative insurgents within the Republican Party—used the C-SPAN coverage of Special Orders and one-minute speeches. Their goal was to bypass the traditional media, which largely ignored them, and make contact with the public. Under the leadership of more aggressive conservatives such as Newt Gingrich (R-GA), Vin Weber (R-MN), and Trent Lott (R-MS), they began a coordinated effort, inveighing against the spread of communism, the wastefulness of government programs, and the decline of the American family. They blamed the Democratic majority, often taking on colleagues by name.[2]

Based on a 1984 Congressional Research Service analysis, Republicans spoke twice as often as Democrats during the Special Orders period. As the minority, the Republicans found that such speeches provided a balance to the influence of the majority party during the rest of the legislative day. As Gingrich put it: "Special orders offer a voice to the minority. They're our time." Others pointed out that Special Orders were a good safety valve, reducing the frustration of members over the restrictions of normal House debate and serving as a good place to rehearse ideas for the actual debates.[3]

Initially, Speaker Tip O'Neill (D-MA) was unconcerned about the use of Special Orders by COS members. He saw the impact in a very narrow sense of lining up support for immediate votes and missed the longer-term intentions of the COS members. O'Neill commented: "I don't think they make votes in things like that. As a matter of fact our polls show the four members of the House that monopolized most of the time were an asset to the Democratic party because they showed a meanness, because they showed an unfairness. . . . Our polls show they helped us."[4]

After suffering in silence, the Democrats commissioned a study that concluded that the best Democratic response would be to begin using Special Orders speeches themselves. Bill Alexander (D-AR) began a coordinated effort to use Special Orders, but he found few takers among his Democratic Party colleagues.[5] It was the Republicans who mastered the technique and raised the ire of Speaker O'Neill in May 1984. As COS charges began to gain attention, O'Neill recognized his mistake in dismissing the utility of such speeches. In a moment of pique, he ordered the cameras to pan the empty chamber in the hopes of embarrassing the Republicans. Republican Minority Whip Trent Lott caught a glimpse of the unfamiliar scene from his House office while changing into his tuxedo for a GOP fund-raiser. As he raced to the House floor, hopping down the hall and trying to get his foot through his second pants leg, Lott got angrier and angrier. Robert Walker (R-PA), who was then speaking on the floor, knew nothing about the camera panning. Lott handed him a note telling him what had just happened. Without missing a beat, Walker called the move "one more example of the arrogance of power of the majority leadership." Lott then took over, fuming about the "underhanded, sneaky and politically motivated change" in procedures."[6] The "Camscam" affair eventually hurt both the Democratic Party and its leader, Tip O'Neill.

Tip O'Neill Takes Charge

Speaker Tip O'Neill retained full control of the House cameras, much to the frustration of the networks and the Republican minority. (Cartoon by William Garner, *Washington Times*, reprinted by permission.)

Many of the strongest charges against the House leadership came from Newt Gingrich. The day before the cameras were turned around, Gingrich had questioned the patriotism and voting record of O'Neill's political ally and Washington apartment-mate Eddie Boland (D-MA). The day after switching the cameras, O'Neill, in a rare speech on the floor, personally criticized Gingrich's comments: "You deliberately stood in that well before an empty House and challenged those people when you knew they would not be there. . . . It is the lowest thing that I have ever seen in my 32 years in the House."[7] A chorus of jeers and catcalls erupted from the Republican side during O'Neill's comments. His violation of House rules prohibiting personal attacks on colleagues led to the embarrassment of having his words stricken ("taken down") from the *Congressional Record*.

The term "Camscam" was coined by *Washington Post* reporter T. R. Reid to capture the conflict over panning the chambers. The media were not particularly enamored with Speaker O'Neill's decision to turn the cameras on the empty chamber and to attack the COS. One writer commented, "The large, boiler-shaped Irishman threw a video tantrum against some Republican whiz kids."[8]

Camscam sparked interest in watching the House in session and, indirectly, in watching C-SPAN. It helped put C-SPAN on the map. Since all news sources carried stories about the affair, the name C-SPAN became recognized. The experience and opinions of Susan Swain, currently a C-SPAN vice-president, were typical. She noted: "At last we didn't have to go into a big pitch about who we were every time we picked up the phone and called someone. It saved a lot of time and effort."[9] One congressional staff member verified the change by arguing that after Camscam, the awareness of C-SPAN among many legislators went from "What is it?" to "How can I get it?"[10]

The Battle for Senate Television

The U.S. Senate was engaged in a public debate over access to its official records. The outcome would have a dramatic effect on how much the press and the average citizen would know about what happened on the floor. Not everyone was sanguine about making the legislative process more transparent. Opponents argued that opening the doors "would promote oratorical pyrotechnics for the benefit of the gallery and press and would interfere with the expeditious performance of public business."[11]

Although this scene might well have occurred in the early 1980s, as senators debated the desirability of allowing television cameras to record floor debate, the debate in question had actually happened almost two hundred years earlier. The issue was whether the Senate should continue its closed-door policy or should open press and public galleries. There was strong precedent for secrecy. The Founders had written the Constitution under a strict code of secrecy. The Federalists, who dominated the political process of that day, believed that government "ought to be clothed in dignified aloofness, responsible to the people, but not constantly under their close scrutiny."[12]

In typical fashion, the Senate tested the potential impact of the change with an experiment. As the final judge of its own membership, the Senate had to decide whether Swiss-born Albert Gallatin of Pennsylvania was eligible to serve as a senator. Fearful of challenging a state's right to choose its own senator in secret, the Senate voted to open its doors for Gallatin's hearing on February 11, 1794.[13] When the Federalists closed the door on Gallatin's Senate career, they also voted 14 to 13 to shut their own doors again. A few days later, Senator Stephen Bradley of Vermont changed his vote and took along with him five new supporters. On February 20, 1794, by a 19–8 vote on a new resolution, the Senate voted to open its doors, but two years would pass before permanent galleries were constructed. The addition of public galleries had little effect on the Senate; most sessions were sparsely attended. There was little upsurge in either public favor or public derision, but going public "removed an irritant that had long alienated the Senate from the public."[14]

The parallels between this debate and the modern-day situation, with the Senate considering whether to expand its galleries through the use of television, form another example of how history repeats itself.[15] The arguments of the opponents in the 1980s had increased beyond a fear of playing to the galleries to include an expected discomfort from the heat of additional lighting and a worry that traditions would be threatened, thus undermining the constitutional purpose of the Senate.[16]

The final outcome in the 1980s depended on changing the perspective of key players such as Minority Leader Robert Byrd (D-WV), who earlier had opposed the addition of microphones to the chamber. Rather than jumping full force into the television age, the Senate again experimented, this time with a trial period. When the results showed relatively little effect on Senate procedures, an overwhelming majority approved the change. Similar to the experience of two centuries earlier, dramatic impact on Senate procedure was not immediately evident. Several years after the cameras began to roll, one of the prime opponents, Senator John Danforth (R-MO), called himself a "convert" and asserted: "The interesting thing about television in the Senate is that it has not changed life in the Senate that much. . . . I think that is good."[17] Final passage of the legislation allowing cameras into the Senate chambers involved a complex process of advocacy, timing, and compromise.

C-SPAN had been largely an indirect, bit player in the decision to allow television cameras into the House. But when the Senate began to seriously consider televising its proceedings in 1981, C-SPAN was a major player. To some degree, C-SPAN's growing audience and success in broadcasting the House served as the stimulus for Senate action. C-SPAN monitored senators' willingness to accept television and regularly published lists of supporters and opponents. During the four initiatives on Senate television between 1981 and 1985, C-SPAN regularly sent representatives from its staff and board of directors to testify at hearings. Senators who had served in the House after C-SPAN began televising used their experiences to argue, by analogy, both for and against Senate television.

The initial push for Senate television came during Republican control of the chamber in 1981. Senate Majority Leader Howard Baker (R-TN) attempted to put the move into historical context by arguing, "Television access is a simple extension of the public galleries in the House and Senate chambers."[18] The gallery analogy was apt, but the magnitude of the extension would be dramatic. The 611-seat Senate gallery, though often full during the spring tourist season in Washington and during historic debates, would hold only a minuscule portion of the potential television audience, even considering C-SPAN's early small audience. Although a strong supporter, Baker did not push televising the Senate to the top of the agenda, partly out of fear that he did not have the votes. In a chamber where one person can stop all activity through a filibuster, Baker backed off. He explained, "Television became a tool for someone to threaten me with."[19] Each time television coverage was discussed in 1981, Senator Russell Long (D-LA) began a filibuster. Charged with keeping the legislative process moving, Baker was afraid that opponents of Senate television or of any other legislation could hold him hostage with a filibuster if he pushed televising too strongly.

By 1985 Robert Byrd, then serving as minority leader, became the key player in the battle over Senate television. Majority Leader Baker was about to retire, and the incoming majority leader, Bob Dole (R-KN), supported televising the Senate, but it was not high on his priority list.[20] Dole's eventual support grew more out of the fact that his Republican predecessor had pushed so hard for the idea.

Byrd continued Baker's two-pronged set of arguments. The practical argument asserted that televising Senate proceedings was an idea whose time had come and that the Senate would look foolish or irrelevant if it failed to accommodate television's power in society. As Baker saw it, "If we don't open up the Senate to radio and television, I predict that in a few years . . . in the public mind at least, the House will be the dominant branch."[21]

A parallel philosophical argument allowed proponents to take the high moral ground by asserting that television would improve relations between the Senate and the public, thus strengthening representative democracy.[22] According to Baker, "A democracy thrives on public support, and public support thrives on open government."[23] With a touch of humor, Baker took his argument to *TV Guide*, a vehicle for informing television viewers. He asserted: "Otto von Bismarck, the 'Iron Chancellor' of Germany, is supposed to have said, 'If you like laws and sausages, you should never watch either one being made.' I say, 'Baloney.' "[24] Opponents were forced to face the criticism that they stood in the way of progress or that the Senate had something to hide.

Few would argue that Senator Byrd's love of Congress as an institution surpasses that of all other senators. He took to the floor in 1989 to present a series of meticulously researched speeches on the history of the Senate, speeches that were eventually published in book form.[25] Although he had initially opposed televising the Senate, Byrd is a realist who recognized the potential loss of the Senate's influence without television. He explained his switch: "We can't hold our own with the White House, and we can't hold our own with the other body if they have TV and we don't. Many people think Congress is only what they see on TV—Tip O'Neill and the House of Representatives—and it shouldn't be that way."[26] Byrd amplified this view by stating, "We finally came to the conclusion . . . that the Senate was rapidly becoming an invisible force . . . with the House of Representatives broadcasting its debates, with the President able with the snap of his finger to summon around him the television and print media . . . it was time for the United States Senate to televise its debates."[27]

During the debates over televising the Senate, Byrd received a dramatic lesson in the power of television during a trip back to West Virginia. Despite his long tenure as a senator and as a party leader, he was introduced to a local audience as Speaker of the House.[28] It was hard to imagine that he could be confused with the much more portly Tip O'Neill. Byrd realized that both he and the Senate had a visibility problem and that the Senate was at a severe disadvantage as long as the House was the only chamber on television.

Byrd was not alone in his concern. Senators were aware of Speaker O'Neill's boast that he was "on television more than any other politician in America." Both individual and institutional pride and jealously were evident. Senator Charles Mathias (R-MD) argued that senators were ready for the cameras because they were tired of "watching the House guys get all that coverage."[29]

Byrd was unwilling to venture out until he knew where the votes were. As a party leader, he was accustomed to counting votes. C-SPAN regularly polled senators on their openness to television coverage as part of its campaign to complement its House coverage. Over the years, the arrival of a younger generation of senators more comfortable before the camera had swelled the support to sixty-two solid commitments. Senator Byrd called Brian Lamb and Mike Michaelson over to his office and asked only one question, "Are these vote figures solid?"[30]

With an idea of where he stood, Byrd set about building the necessary coalition to pass the legislation. Under Senate rules, he knew that simply having a majority might not be enough. The right of unlimited debate (the filibuster) and other delaying tactics could allow one senator to tie up the entire chamber. As Senator Alan K. Simpson (R-WY) described the Senate process: "One person can tie this place in a knot. And two can do it even more beautifully."[31] In building coalitions in Congress, members hope to convert opponents into supporters and thus achieve a net gain of two votes (denying the opposition one vote and gaining one vote for their own side). Such a dramatic conversion is often difficult. An alternative strategy lies in toning down the opposition and obtaining a commitment that the opponent will not use all the strategies available. Recognizing the realities of coalition building and the potential problems of delaying tactics, Byrd pinpointed Senator Russell Long as a key stumbling block to Senate television.

Senator Long's opposition stemmed less from any concern about the public's right to know than from concern about the effect on the internal workings of the Senate. While proponents touted the potential opportunities of television, Long outlined the potential dangers. He was especially concerned about how television would change senators' incentives and opportunities for seeking publicity. Long took the position that the Senate's greatest strength was its committees rather than its floor deliberations. In his mind, the floor should be the place where one senator could staunchly defend deeply held beliefs. He would not accept the proponents' beliefs that floor debate was the legitimate forum for making wise policy decisions or for educating the public.[32] Long saw television as dragging senators away from their committee responsibilities to the bright lights shining on floor debate. He argued: "Every senator with an inflated ego or higher political aspirations would be taking to the floor to make eloquent speeches for the benefit of the voters back home. . . . I hope I never see the day when the floor of the Senate is a forum for senators to conduct their re-election or presidential campaigns at public expense."[33]

Long was particularly concerned that the arrival of television would require revising "the best rule" in the U.S. Senate, "the rule that protects the right of a single Senator to take the floor and hold it for a while if he thinks everybody else is wrong." Long noted, "The free debate in the Senate will have to go if the television comes in."[34]

After Senator Byrd became convinced that television should be allowed in the Senate chamber, he met with Long. Byrd recalled the conversation:

> "Russell, you and I are not going to be around here always. . . . Television is coming to the Senate . . . the American people are entitled to see us at our work . . . the Senate . . . is rapidly becoming the invisible force. The House is seen on C-SPAN. Everybody knows what the House is doing. Why shouldn't they see the Senate in which senators speak longer and in greater depth about the subject matter? . . . So, Russell, why don't you and I work to bring this about while we're here?" . . . So Russell Long was persuaded. He never was persuaded to vote for it. But his opposition to it lessened considerably, and it made it possible for television to come.[35]

Although the tempering of Long's opposition increased the likelihood of success, not all the opposition melted away. The Camscam brouhaha in the House was good television and made a name for C-SPAN, but it had a chilling effect on the Senate. Senators had been looking carefully at the House experience. One staff proponent in the Senate dourly predicted: "There goes TV in the Senate! If it were the heart, this would be the knife through it." Senator Paul Laxalt (R-NV) reacted to the House events by concluding that they would hurt the chances for Senate television: "I sensed yesterday in talking to several of my colleagues that we don't need that kind of problem."[36] Senator Long was described as "quite pleased" by the problems in the House.[37] Opposition continued, but the staffer and the senators underestimated other senators' interest in sharing the spotlight with the House.

Of major concern to senators was that colleagues would play to the camera. Senator Thad Cochran (R-MS) argued: "The camera is a natural attraction for a politician, and if a camera is here, we're going to be here. And we're going to say something, even if we have nothing to say." Senator Alan Simpson (R-WY) felt that the cameras would distort the priorities of individual lawmakers. He admitted: "I can see myself sitting at a committee hearing, doing my job. A staff member comes in and says, 'They're talking about blank issue over on the floor and you'd better get over there and get in on it or your constituents will never forgive you.' I just don't see how that is productive."[38]

Senator Russell Long fired a final salvo by arguing, "When the Senate is on television, we will see a substantial increase in expediency and we will see a substantial decline and erosion in statesmanship."[39] But the fight was gone from his opposition; he could see the handwriting on the wall. Long had made a commitment that he would not use his right to filibuster to kill Senate television, thus dramatically reducing his chances of success. His final strategy lay in bottling up the issue of Senate television in a battle over changes in the rules.

Senator Robert Byrd had seen the increasing support for Senate television as a vehicle for reforming Senate procedures. A common strategy in the legislative process is to allow a popular provision to carry along with it legislative provisions with less likelihood of success. Byrd felt that the balance between the rights of a small minority to thwart action and the necessity of the majority to act had tipped too far in the direction of minority rights. He proposed a package of reforms that would make filibusters and the introduction of nongermane amendments more difficult. This approach played into the hands of Senator Long and other opponents. If they could pile on enough rules changes, each with an identifiable block of senatorial opponents, they might win yet by creating a majority out of many minorities.

The Senate Rules Committee stripped all the rules changes on the legislation for Senate television in hopes of a clean decision on television without the distraction of rules changes. Senator Byrd faced a dilemma. His deal with Senator Long not to filibuster was premised on allowing certain rules changes to be brought up for a vote. Byrd and Majority Leader Dole met and reinstated the rules changes.[40] Byrd argued: "We have to get the votes or we won't pass anything. Sometimes we have to offer this Senator, that Senator, or a lot of Senators something that they want in order to get what we want . . . if there are certain senators who are dead set against television in the Senate but who would, nevertheless, vote for it in the event a particular rule or another rule is changed, or added or repealed, I hope we will let them have the opportunity to vote on their rules changes. That may be, in the final analysis, the only way we can get television."[41]

Senator Long knew that the one rule that might strip away potential supporters involved restricting nongermane amendments, a rule that he proceeded to champion with great vigor. In the final analysis, the strategy failed after an amendment to drop the rules changes succeeded by an overwhelming vote of 60 to 37. True to his word of allowing a vote on the rules, Senator Byrd opposed the amendment. Senator Long accepted the vote on the amendment as a good-faith attempt to vote on the rules and stuck with his commitment not to filibuster. In the end, the only rules change that remained was one reducing the number of hours allowed in a postfilibuster debate from one hundred to thirty.[42]

On February 27, 1986, the Senate adopted *Senate Resolution 28*, allowing cameras and microphones into the chamber for a trial period, by a vote of 67 to 21. Like the House, the Senate would control the cameras and would limit coverage to the senator holding the floor. Since senators speak from their desks rather than from a limited number of podiums, viewers were likely to get a broader view of the chamber. Closed-circuit television coverage began on May 1, and the feed was made available for the public (and C-SPAN) on June 2. At its launch C-SPANII, which was created to cover the Senate, reached 7.5 million households, considerably less than C-SPAN. Although there was no extra charge for C-SPANII, limited channel capacity and increased equipment costs slowed the availability of Senate programming to about half that of the House. By 1995, C-SPAN was available to 63 million households (approximately two-thirds, or 67 percent, of U.S. households) while C-SPANII reached only 39 million households (just over 40 percent of U.S. households).

The first day on the air reflected the senators' awareness of the television audience and the potential importance of their decision. Senator Albert Gore (D-TN) repeated his performance as the first televised speaker in the House, with a serious set of comments intoning the benefits of the "marriage of television and free debate" to benefit American democracy.[43] Senator Dole concluded: "Today we catch up with the 20th century. We have been the invisible half of the Congress for seven years."[44] Other speeches welcomed the cameras and spoke about the large potential audience. In a more lighthearted vein, Senator John Glenn (D-OH) mused about how senators

Is the Public Ready for Senate TV?

After its seven years of House coverage, C-SPAN's decision to televise Senate proceedings generated humorous reactions by cartoonists. (Cartoon by Lee Judge, *Kansas City Times*, reprinted by permission.)

had planned wardrobes and behavior for this day. He then proceeded to use a makeup brush to remove the shine from his forehead.[45] Senator Howell Heflin (D-AL) waxed poetic:

> Turn the spotlight over here;
> Focus the camera on my place;
> Pages, please don't come too near;
> Otherwise you might block my face.[46]

In some ways the fight over Senate television was symbolic of the Senate's approach to legislation and its role in the American political process. The fact that the Senate took seven more years than the House to accept television gives meaning to its oft-repeated role as a "cooling saucer" for new ideas.[47]

First Day of Senate Coverage

On the first day of Senate coverage, Brian Lamb, Paul Fitzpatrick, and Senator Robert Byrd met at the C-SPAN control room. (Photo courtesy of C-SPAN.)

After a six-week experiment with television during June and July 1986, the Senate agreed to suspend coverage for two weeks to review its expe-

Senator John Glenn with Mirror

On the first day of senate coverage, Senator John Glenn (D-OH) used a makeup brush to spoof the impact that he felt television coverage would have on the chamber and its members. (Photo courtesy of C-SPAN.)

riences. As the Senate moved toward a final vote on television, the initial conditions of the two-month test period came under question. With most senators pleased with the results, pressure developed to abrogate the two-week blackout period during which the final debate on Senate television would occur. Senate Majority Leader Dole felt uncomfortable with the confusion this would cause among the viewers and proposed a resolution continuing the coverage during the evaluation period. Considerable opposition to the resolution emerged. Fighting to shut out the cameras during the evaluation, Senator William Proxmire (D-WI) in his typical frankness commented, "We should have the opportunity to discuss this in a deliberate way without feeling that we are being watched by people who might feel that they were being shut out by the changes that we might want to make."[48]

Finally a compromise was reached: the cameras would be turned off for three days. During the debates on July 16–18, C-SPANII continued to broadcast the audio signal. The bulk of the debate revolved around fine-tuning rather than rescinding television coverage. Senators were wary that floor debate clips might be used in campaign commercials or in the development of a "TV Bloopers" videotape.[49]

The final 78–21 vote on July 29, 1986, favoring permanent television coverage in the Senate defies simple analysis. Opponents included both the telegenic freshman Dan Quayle (R-IN) and the Senate's longest-serving member, Democrat John Stennis (D-MS). On the final vote, younger and less experienced members were somewhat more likely to support allowing the cameras in. Detailed analysis indicates that most senators voted their perceived self-interest.[50] A handful of senators who had initially opposed the test period for television supported making the cameras permanent, whereas some initial supporters voted against the final resolution. Perhaps most important, the margin of victory for the three key votes on televising the Senate continued to grow, reflecting the degree to which the idea had taken root.

After the first year, most senators were pleased with the changes they had wrought. By this time Senator Byrd asserted, "TV in the Senate has been a success . . . [it] has allowed us to carry out the charge of informing the public in the age of television." Senate Republican Leader Robert Dole agreed: "Senate TV has delivered the people's democracy to their living rooms. That's exactly why we turned on the cameras and the lights one year ago and why television in the Senate is here to stay." However, many previous opponents were still not enamored with television. Senator Quenton Burdick (D-ND) argued that it had led to "longer speeches, increased visual aids and grandstanding." Senator William Proxmire noted: "Nobody is watching Senate TV. [It] may drive Sominex off the market, but it's not performing any useful function for our country." On the other hand, there were some converts. Senator Bennett Johnston (D-LA) concluded: "I think it has worked well. Some of the fears that I and others had have not materialized. I think Senate TV has been a success."[51]

To accommodate the cameras, the Senate made a few cosmetic changes. New paint, wall coverings, and curtains were chosen to provide a better television backdrop. A short wall was erected to stop the disruption of the flailing hands and feet of staff members walking behind senators who were speaking.[52] Overall, the physical adjustments were minor.

Allowing television coverage of Congress was not an isolated event. The change was part of a series of reforms associated with democratization and openness. Senator Robert Byrd argued that reforms such as televising congressional procedures, passing "sunshine" legislation to open committee markup sessions to the press and public, providing staff assistance to all committee members, and tightening the rules on filibusters all moved in the direction of "greater egalitarianism and openness in the legislative process."[53]

Arrival on the Broader Political Scene

Establishing a second network (C-SPANII) for the Senate, expanding the number of potential cable recipients, and encouraging the continued "use" of C-SPAN by politicians increased the visibility of C-SPAN and paved the way for more recognition. By the late 1980s, C-SPAN had become part of the popular culture: it was mentioned as a new political force and was spoofed on the *Gary Shandling Show*, *Saturday Night Live*, the *David Letterman Show*, and the *Tonight Show* (see box 4). News stories no longer added the once obligatory "(Cable-Satellite Public Affairs Network)" or the "channel that covers Congress" to every mention. By the early 1990s, C-SPAN became more interested in the attention it received from the media and started to distribute among its staff a "press router," which included all stories mentioning C-SPAN in the print media. By the mid-1990s, this had become a major undertaking, with over two hundred references per month.

While the cable industry continued to shower C-SPAN with awards for specific programming such as its election coverage (1981, 1992), and even for the C-SPAN bus (see chapter 3), more meaningful were the awards that came from dozens of organizations outside the cable fraternity. Despite the recognition, C-SPAN program planning is not award-driven. As the staff sees it: "We just try to do good work and the rewards will follow. We try not to let it go to our heads. The awards are temporary. We have to get up the next day and go on."[54]

During the 1980s, politicians regularly mentioned C-SPAN coverage in floor speeches, indicating that "the C-SPAN audience might want to know" or wondering if their "colleagues had any idea what this looks like to the C-SPAN audience." By the 1990s, politicians were regularly mentioning C-SPAN in news conferences and other events covered by the mainstream media. In announcing the 1994 Republican "Contract with America," Newt Gingrich laid down the challenge that anyone who questioned the Republican resolve only had to "tune into C-SPAN."[55]

One of C-SPAN's greatest frustrations is its inability to get the message

Box 4.
C-SPAN's Arrival in the Popular Culture

1988	C-SPAN is mentioned in Garry Trudeau's "Doonesbury" comic strip.
1988	C-SPAN program schedules are included in the *New York Times* television listings.
1988	Live C-SPAN footage is used on the *Gary Shandling Show* to spoof the network.
1988	C-SPAN plays a role in Robert Ludlum's spy-thriller *The Icarus Agenda*.
1989	Jay Leno, poking fun at C-SPAN on Johnny Carson's *Tonight Show*, comments, "The Home Shopping Network announced that it is going to merge with C-SPAN, for the convenience of buying a politician in the privacy of your own home."[a]
1989	C-SPAN is mentioned on ABC's *Who's the Boss*.
1989	C-SPAN is spoofed on *Saturday Night Live*.
1989	C-SPAN begins showing up on David Letterman's infamous "top-ten" lists. It holds the number 9 position in the list of "Top Ten Reasons Congress Deserves a Pay Hike" ("Because of C-SPAN, they all had to buy hair pieces"). Brian Lamb, rejecting an offer to appear on the *David Letterman Show* with humorous moments on the House and Senate floors, explains that C-SPAN would not use its unique position to poke fun at Congress.
1989	C-SPAN choice of music is the subject of a music critique in the *Wall Street Journal* (Alan Otten, "We Long for the Day They Play a Record of John Cage's 'Silence' ").[b]
1989	*Washingtonian Magazine* identifies Brian Lamb on its annual list of Washington's top fifty journalists.
1990	"What is C-SPAN?" is the correct question for the *Jeopardy* quiz-show answer, "This cable television network has been airing live gavel-to-gavel coverage of Congress for ten years."
1992	C-SPAN footage is featured on *Murphy Brown*.
1994	C-SPAN is an answer to the *New York Times* crossword puzzle inquiry, "Cable TV's C——").

Box 4 (continued).

1994 Virtually every column about newly elected House Speaker Newt Gingrich mentions the importance of C-SPAN in promoting his career. Gingrich mentions C-SPAN's role in his rise to power.

1995 *Newsweek* lists C-SPAN in its "Newt column" as the mouthpiece for the new majority.

[a]*C-SPAN UPDATE*, July 3, 1989.
[b]*Wall Street Journal*, June 22, 1989, p. B2.

across that it does not control the cameras in the House and Senate chambers. Members of Congress and the media continue to talk about the "C-SPAN" cameras when they are discussing the coverage of floor proceedings. Although the term has become a shorthand for many people who actually know better, the impression that C-SPAN already had control created confusion when C-SPAN sought broader access after the 1994 election.

Despite growing visibility, C-SPAN staff members remain awed both by their success and by those with whom they interact. Brian Lamb's *Booknotes* interview with Richard Nixon was a highlight for Lamb and his staff. Despite having worked in the Nixon White House, Lamb had never talked with the former president before interviewing him on the air. The interview was conducted at the C-SPAN studios in two one-hour shoots, with a nap and lunch for Nixon in between. Some staff members remained in the office so that they would not miss seeing the controversial former president. After the interview, the younger staff members stood in line to have their copies of his book autographed while the older staff held back.

The unwillingness of C-SPAN's staff to take success for granted continues. In early 1995, after former Vice-President Dan Quayle's release from the hospital, C-SPAN carried his press conference live. When he mentioned his appreciation for Brian Lamb and C-SPAN, and how he had watched it in the hospital, a cheer went up in the halls at C-SPAN. One of the C-SPAN insider jokes—having more than a touch of reality—notes, "More people discover us when they are sick or in the hospital." A few days later, when Senator Bob Dole complimented the educational value of the network during a Senate debate by calling it the "C-SPAN university," the same kind of exuberant cries rang out again. Unwilling to rest on its laurels, C-SPAN has continued to expand its programming, experiment with its format, and innovate (see box 5).

Box 5.
C-SPAN Milestones: The Second Decade

1989	Regular broadcasting of the British House of Commons begins.
1991	Closed-captioning of House debates begin.
1993	C-SPAN wins a Cable Golden Ace Award for its presidential campaign coverage and a Peabody Award for overall excellence.
1993	The C-SPAN school bus is launched.
1994	C-SPAN stimulates re-creations of the Lincoln-Douglas debates, winning critical acclaim for the broadcast.
1995	The C-SPAN school bus receives a Cable Golden Ace Award.
1995	C-SPAN requests expanded congressional access and coverage from the new Republican leadership.

A Contemporary Challenge

As we will see in chapter 4, much of C-SPAN's effort during the early 1990s focused on improving and expanding its programming and on encouraging affiliates to include both of its channels. C-SPAN had developed closer relations with its affiliates and had created an extensive educational program designed to teach college and high school teachers how to integrate public affairs programming into the classroom using C-SPAN (see chapter 3). Increased public visibility, improved quality of programming, and continuing diligence seemed to guarantee C-SPAN a bright future. Unexpectedly, an external challenge and an unexpected opportunity came along, forcing C-SPAN to redirect some of its efforts to maintaining and enhancing its work.

Unintended consequences are often some of the most devastating. The Cable Television Consumer Protection and Competition Act of 1992 sounds like an unadulterated good thing. Who could be against protecting consumers and fostering competition? In reality, the bill is a good example of the "law of unintended consequences." In an attempt to expand the range of available information and reduce costs to consumers, a "must carry" provision was included in the bill. The act required cable companies to carry local commercial channels being offered free. If the local channels did not want to give away their signals, they could negotiate retransmission contracts and charge a fee, or they could cut a package deal offering their signal free if the local cable company agreed to carry network-owned cable channels. The package-deal route was more palatable to most cable companies. ABC created ESPN-2, Fox developed fX, and NBC created an all-talk channel, "America's Talking." With cable systems typically limited to less than one hundred channels because of current technology, C-SPAN was caught in the squeeze. Whereas in the long run, new technology will expand the number of channels on cable systems to nearly five hundred, the potential short-term implications were significant.

During the debate, two strategies emerged. Some members of Congress offered to protect C-SPAN and make it one of the channels that must be carried. Brian Lamb, true to his philosophical commitment to an unfettered media, demurred at the offer: "I am totally opposed to having politicians require something be aired, even if it is my network."[56] Expressing a more practical concern, Lamb also asserted: "We did not want to ask any favors of Congress. With favors come paybacks. The Cable Bill messed us up. We got hit the hardest, but we still won't ask any favors."[57]

Lobbyists for the National Cable Television Association (NCTA) took a somewhat different approach, more typical of Washington politics. They used the potential for dropping C-SPAN to play the "Washington Monument Game." (In its classic form, the game asserts that if the Department of Interior is facing a budget cut, the department will threaten to close the Washington Monument, a highly visible action that would anger tourists wanting access to "their" monument.) Senator Christopher Dodd (D-CT) expressed frustration over the NCTA's use of the potential dropping of C-SPAN: "This isn't just a threat; it's almost blackmail."[58] Despite warnings, the Cable Act passed by an overwhelming majority of 280 to 128 in the House and 74 to 25 in the Senate, but President George Bush vetoed it. The legislation had been framed as guaranteeing inexpensive television viewing, and C-SPAN was caught in the wave of public support. Bush's veto was portrayed as his being "owned lock, stock and barrel" by the cable industry. The veto override was overwhelming, with a 74–25 vote in the Senate and a 308–114 vote in the House.[59] It was now up to the cable operators to determine the real impact on C-SPAN.

Since C-SPAN does not allow advertising and the revenue it would bring, there was a strong incentive for cable companies to go where the money was. Many cable systems dropped C-SPAN, forced it to share a channel, or relegated it to the top end of the dial, where it is hard to find. This may have been justified by the ratings. As TCI (one of the larger cable companies) Vice-President Bob Thompson argued, "We've done a lot of polling and C-SPAN is not always the most-watched network."[60] As one cable operator put it: "I don't have a lot of options. We have contract obligations to take fX, we have to take broadcast signals, we're channel locked."[61]

Over seven million potential viewers had their C-SPAN coverage removed or disrupted as cable operators followed the "must carry" dictates and chose more financially lucrative programming over C-SPAN. Most often, the disruption involved cutting back service rather than cutting it out completely. By 1995, only about half of the disruptions had been resolved. Although cutting one or both of the C-SPAN services is a problem, cutbacks in coverage and the shift of channels and times of broadcast for C-SPAN raise other difficulties. Since C-SPAN's balance in programming (see chapter 4) occurs over time, as opposed to within any one program, selected coverage vitiates much of C-SPAN's goal of broadly informing its viewers. As one television writer stated: "C-SPAN was designed as an 'equal time' channel, giving political balance on its daily 24-hour schedule. Now such balance and continuity has been destroyed. It is difficult, under the split channel system, for viewers to understand what C-SPAN does."[62]

Where C-SPAN was neither dropped nor forced to share a channel, there were other negative results. On several cable systems, C-SPAN—or, more likely, C-SPANII—was moved to one of the highest-numbered channels. Such placement reduces the number of viewers who might land on C-SPAN programming as they channel-surf through the major networks at the lower end of the available channels. Ideally, C-SPAN and C-SPANII would be adjacent to each other on low-numbered channels.

There was no clear intention on the part of Congress to give commercial broadcasters incentives and motivations to create new cable channels, but that was the result of the Cable Television Act. Brian Lamb expressed his frustration: "My attitude is one of controlled rage. The government writes a piece of legislation that sounded good, but it is the public service channel that is being hardest hit."[63] The impact on C-SPAN was clearly an unanticipated consequence. Brian Lamb noted, "Even though I testified that it was one of the worst pieces of legislation I have seen in twenty-nine years in Washington, and that it would hurt us, the members of Congress did not listen."[64] Even after the results of "must carry" became evident, Lamb still philosophically opposed plans to require C-SPAN carriage. In his words, "The Congress has done severe damage to this network, and there is no way to change that without doing us a special favor, which we don't want."[65]

In some observers' minds, targeting C-SPAN for actual abandonment by cable systems may have been part of a continuing strategy to reverse the law. David Moulton, staff director of the House Energy and Commerce Subcommittee on Telecommunications and Finance, explained, "There may be some Washington Monument syndrome here, where operators think they will get the most political response by picking on the most politically sensitive station."[66] And despite the partisan paranoia that often pervades American politics, none of the key players accepted the whispered argument that the Cable Act was a conspiracy by liberal Democrats to hamper C-SPAN, which had been so effectively used by conservative Republicans. The vote on the Cable Act belies any such conspiracy theory. Even the voting split on overriding President Bush's veto was not clearly partisan. In the House, the cable industry position and President Bush's position gained the support of 52 percent of the Republicans and 11 percent of the Democrats. In the Senate, partisanship was even less clear, with 34 percent of the Republicans and 8 percent of the Democrats supporting the veto. Legislators were swayed by arguments that the bill would cut consumer costs and increase programming options.

Many C-SPAN viewers did not take C-SPAN's loss lying down. Phone calls to cable companies burned with outrage, and letters-to-the-editors columns were filled with vitriolic attacks on the shortsightedness of decisions made by local cable operators. As one cable insider put it, "Everyone agrees on one thing about C-SPAN junkies: they know how to write letters."[67] In Waterloo, Iowa, Robert Snyder wrote, "By dropping C-SPAN, it is clear that TCI is more interested in serving its own bottom line than it is in serving the community with truly unique programming."[68]

In some areas, the protest was more demonstrative. In the American tradition of activism, the people of McAllen, Texas, reacted to their local cable operator's plan to cancel C-SPAN by organizing a protest group. "Friends of C-SPAN" flooded the cable operator with indignant letters. Eventually a deal was cut, with C-SPAN running during open times on two different channels.[69] Die-hard C-SPAN viewers in Eugene, Oregon, persuaded the local TCI manager to drop a pay-for-view channel. He pointed out: "It wasn't so much the number of people that complained, but the quality of their arguments. . . . C-SPAN viewers are very articulate."[70] He was "frankly amazed at how many people get up at 5 A.M. to watch C-SPAN. . . . The ones that watch it are extremely loyal."[71] In other areas, the results were not as favorable. Despite advertising in the local newspaper, Iowa TCI subscribers lost access when a poll of subscribers identified C-SPAN as the channel they were willing to drop.[72]

Even if cable companies did not completely cut out C-SPAN, reducing its broadcasts or forcing it to share channels was not acceptable to either C-SPAN viewers or C-SPAN itself. In the quotable—if vaguely familiar— words of C-SPAN's in-house lawyer Bruce Collins, "Carriage delayed is carriage denied."[73] Despite the fact that channel availability is a short-term problem that will soon be ameliorated by a dramatic expansion of cable capacity, dropping C-SPAN coverage was both a symbolic and an actual blow to C-SPAN and its staff. The passage of the Cable Act and its "must carry" provisions left C-SPAN bitter and baffled at the "incredibly punitive assault on cable." There was a feeling that the cable industry's "public-spiritedness" had "gone unappreciated by Congress."[74] Brian Lamb complained: "Congress has created two classes of communicators. If you come in over the traditional airwaves, you can demand to be carried. If you come in over satellite, you have no rights at all."[75]

C-SPAN joined cable operators in an attempt to get the courts to invalidate the "must carry" provisions. After losing the initial round, the case went to the Supreme Court. In an indecisive June 1994 ruling that failed to overturn the original decision, the justices held, in a 5–4 vote, that the government "had to present more evidence to justify the law's assumption that the retransmission requirement was necessary to preserve the economic viability of broadcast television." The case was then sent back to a special three-judge federal court, which had earlier upheld the law.[76] C-SPAN lawyer Bruce Collins expressed disappointment that the Court did not strike down "must carry," but he felt the case was important for its "defining statement about the First Amendment Rights of cable." He added, "In the old days, cable was considered ancillary to broadcasters."[77]

Licking its wounds, C-SPAN continued with more aggressive programming and launched the C-SPAN bus. The bus serves as a rolling studio for education and outreach to cable systems around the country (see chapter 3). It is clear the bus is one attempt to keep the pressure on cable operators to continue carrying the C-SPAN signal. Slowly C-SPAN has begun to get back some cable systems that had dropped its signal because of "must carry." Still, over three million potential cable subscribers are denied full access in 1995. As we will see in chapters 6 and 9, the C-SPAN battle for carriage and an audience is ongoing.

Conclusion

By the early 1990s, C-SPAN had reached its stride. It was recognized as a player in the new media environment and had moved from being solely a distributor of programs created by others to a significant producer of public affairs programming. Like the histories of most other organizations, the story of C-SPAN is not a simple trajectory from point A to point B. None of the early participants, Brian Lamb included, had a clear vision of the direction the organization would take, the opportunities it would capitalize on, and the reversals it would have to confront. C-SPAN's flexibility and its willingness to try new things organizationally, politically, and programmatically have strengthened its position. While attempting to remain true to its historical roots, C-SPAN has tried to creatively confront its day-to-day challenges.

CHAPTER THREE

The Corporate Culture

Management: A Board's-Eye View

The way in which Congress came to allow itself to be televised and the manner in which Brian Lamb positioned C-SPAN to capitalize on that event are matters of public record. Lesser known is the vital role the cable industry itself played in making the network possible and in guiding it through a series of crises, thus allowing C-SPAN to survive and to grow.

At the outset C-SPAN was an idea, not a business. "I never had a business course," Lamb recalled recently. "I didn't know a thing about business. I never had run a business worth a dime. . . . I didn't know the difference between a profit or loss or any of that stuff." He said, "The only thing I brought to the table was an enormous desire; nothing else was that important."[1] One early board member remembered: "It was unbelievable. We'd come in here and go through the financials . . . and you'd ask questions that you'd ask at any level of business, a $50,000 or a $50 billion annual revenue business, and sometimes the eyes would glaze over. 'What did you want to know that for?' " These were good people, he asserted, but "there wasn't a lot of experience" at the network.[2]

Lamb clearly needed a board to support him. Although he had worked for a cable publication, he was not an industry member. For C-SPAN to remain on the air, it needed the continued vocal support and advice that only industry leaders could provide him. It would be easier for board members to set an equitable fee structure for carriage than it would be for Lamb. An increase would hold more weight when approved by industry leaders. Board members could press for continued and expanded carriage in ways Lamb could not. He needed men and women who understood business. The establishment of a board meant that he could get thousands of dollars in consulting services free of charge, as well as have a pipeline to the cable industry's leaders. In the areas of finance, marketing, and satellite access, the board's help would prove vital to the growth and development of C-SPAN. Those in the industry who aided the network would leave their mark on the organization's culture. In chapter 1, we saw how vital a role Bob Rosencrans played in establishing the network. In 1996, the first board chair is still active in his support. It is time now to examine the roles of those who followed him.

THE COSTS OF A NEW SATELLITE

By 1981, C-SPAN had a well-established beachhead in suburban Virginia. Whereas Lamb's troops had made the move, industry leaders such as Rosencrans, Bob Titsch, Bob Schmidt, and Ken Gunter had made it possible. Rosencrans would be succeeded as board chair by John Saeman, of Daniels and Associates. Like many others, Saeman initially thought that the idea of starting a public affairs network was "a little farfetched." Although his firm had contributed ten thousand dollars to fund the project, Saeman at first doubted that "the cable industry would rally behind this cause and actually create the network."[3] Serving on the board made him a convert.

Rosencrans's priorities, Saeman remembered, "were to run the network as inexpensively as humanly possible."[4] The next step required spending money on equipment and facilities and on improving, or perhaps creating, C-SPAN's image in the cable industry. Under Saeman's leadership, the network would experience serious problems in doing so. First, and perhaps most important, by March 1982 C-SPAN was scheduled to lose its space on the transponder it shared with Madison Square Garden Sports. That spot would become the twenty-four-hour USA Network. Without the transponder space, there could be no C-SPAN.

Industry leaders met with Congressman Tim Wirth (D-CO), chair of the House Telecommunications Subcommittee, and with representatives of the Federal Communications Commission at a special conference at the Hyatt Hotel in Washington, D.C., to discuss options and opportunities.[5] C-SPAN eventually found space on the RCA satellite in 1982, which allowed the network to expand programming. The cost, however, rose from $250,000 to $1,200,000. Within months after occupying its new satellite space, C-SPAN went to a full twenty-four-hour day. Lamb credited RCA AmeriCom for its help in securing satellite space. "As everyone knows," he told a reporter in 1982, "there is a strong demand for transponder time and almost all of the new demand is coming from sports and entertainment services."[6] In December 1982, the network honored Andy Inglis, president of RCA AmeriCom, and its vice-president, Harold Rice, as "super citizens" for their efforts in ensuring that C-SPAN remained on the air.[7]

The switch in satellites created problems with coverage. C-SPAN feared that as many as a quarter of the cable systems then providing network programming would drop it. The reasons were multifold. Saeman realized that nearly two-thirds of the nation's cable systems had only a twelve-channel capacity.[8] To dedicate a full channel to C-SPAN—a network that produced no revenue—a cable system would have to delete something. C-SPAN could no longer share time with another network. Early "must carry" laws prevented the removal of local stations, so options were limited. Moreover, the cost for carrying C-SPAN would triple from one cent to three cents per subscriber per month. Some systems opted for the increasingly available advertiser-supported and entertainment-oriented programming. Systems that abandoned, or attempted to abandon, C-SPAN generally replaced it with "networks carrying cooking shows, endless reruns of 'Leave It to Beaver,' or a $10 extra a month pay movie service." Even larger systems like the thirty-five-channel system operated by NYT Cable, a subsidiary of the New York Times Company, canceled C-SPAN because the cable system did not think "there was any demand for it." The manager of the thirty-channel Syracuse News Channel Corporation, owned by the Newhouse newspaper chain, put it bluntly, "I will not give up valuable channel capacity for this kind of service, especially since the cost has tripled." Among the other systems removing C-SPAN was Communications Services, Inc., which was owned in part by Norman Lear, founder of People for the American Way. For one observer, such moves raised "serious doubts" about "the often repeated promises of cable-TV operators—a group that seems to have perfected the ability to speak out of both sides of the mouth simultaneously—that their medium can do absolute wonders for the home tube."[9]

The cancellations placed C-SPAN in a dilemma. Its board consisted entirely of cable company leaders. It was a nonprofit cooperative founded by the industry. It would be difficult to publicly attack one's funding source. Just before the transponder change took place, Lamb went on the air with an eight-hour call-in show to inform viewers about the shift in transponders and about the possible loss of coverage in some areas of the country. In doing so, he did not directly take on the industry. Instead he urged viewers to contact the cable systems to thank them for carrying the network and to suggest that they continue to do so. Although board member John Evans appeared with him, the tactic upset officials at some member companies. An angry representative of Cox Cable, for example, demanded that Lamb get off the air: "Tell him to stop churning up cable systems across the country." Anger, however, mostly flowed the other way on the air. Citizens in a number of communities began to mobilize because of the call-in. One Illinois viewer even contacted the major broadcast networks and urged them to pressure cable operators to keep C-SPAN. In the eighth hour of the call-in show, William "Bud" Harris of Cherry Hill, New Jersey, reached Lamb. He offered to form a nationwide "Friends of C-SPAN" organization.[10] The story of his success and that of other viewers in restoring C-SPAN in numerous communities is a testimony to the power of citizen involvement at the local level. It is chronicled elsewhere in this book (see chapters 1 and 6).

Cable system operators were more difficult to persuade than were loyal viewers. The blunt fact was that a number of market developments combined to squeeze C-SPAN off the air in many areas of the country. Under Saeman's leadership, however, the board began to take the case to the industry itself. Some operators balked at the rate increase and at the request to continue coverage. Saeman recalled applying "a fair amount of arm twisting with some of the more stubborn cable operators."[11]

As part of that fight Saeman sent a twenty-one-page pamphlet to each member of the board of the National Cable Television Association. "C-SPAN in Transition: Will It Survive 1982?" was conciliatory in tone, but its message was blunt. C-SPAN, Saeman explained, was "unique" to the industry. What it did and what it was doing could not be done "on a nationwide basis by any other communications outlet." It was the cable television industry's "best public policy voice." A page of quotes from House leaders made that clear. Majority Leader Jim Wright (D-TX) applauded C-SPAN and "the cable television industry" for "fulfilling their public service obligation." Minority Leader Bob Michel (R-IL) said, "I think [C-SPAN] alerts the public to the way the system works, and that's a public service on your part." The man for whom all politics was local, Speaker of the House Tip O'Neill (D-MA), chimed in. "I'm amazed at the C-SPAN response I have been getting from my district," he reported. "It's unbelievable how many people out there are tuning in the House debates." The pamphlet emphasized that coverage was threatened in districts represented by the Speaker, the minority leader, both party whips, six other key congressional leaders, and ten members of the House Telecommunications Subcommittee, including its chair, Tim Wirth. The pamphlet noted that one hundred representatives and thirty-two senators had appeared on C-SPAN call-ins in the last year. One page of the pamphlet brought Saeman's message home clearly. It contained only the words "CABLE RISKS LOSING HARD-WON GOOD WILL IF C-SPAN FALTERS."[12]

The efforts to hold the line in the industry and to eventually gain more support from it would be taken up as a major cause by the next board chair. The transponder switch had cost C-SPAN approximately two hundred systems. The one thousand remaining represented about 45 percent of the cable providers and 17 percent of all television homes. C-SPAN board members represented twenty of the top thirty cable systems, with a 51 percent penetration of the cable market.

FINDING A NEW HOME

Saeman's board had more to deal with than the transponder switch. He and others were convinced that if C-SPAN was to grow—and it had to grow if it wanted to survive—its headquarters needed to be larger and needed to be located closer to the action it covered. Lamb began the search for a better location within the District of Columbia. He found it at 400 North Capitol NW, a building within walking distance of Capitol Hill. The cost for square footage doubled as a result. Already accustomed to operating on a shoestring budget, Lamb initially sought to lease three thousand square feet, triple the amount of space he had in Virginia. Saeman toured the prospective site with Lamb; when the board chair heard the amount of space to be leased, he asked: "Can you get more? You're going to grow rapidly." Lamb took two thousand more feet but ordered that not all the space be occupied initially. "The minute we start opening up these rooms, they'll get filled," he said. In

1995, C-SPAN occupies close to fifty thousand square feet at the same location and could use more. Ever cautious, Lamb believed a forty-line phone system would be sufficient in the new location. After all, he had started with only three lines in Virginia. Today C-SPAN uses about 240 phone lines.[13]

C-SPAN certainly needed money to grow. In part, more dollars were tied to more system coverage. The move to North Capitol would cost an estimated $140,000. New broadcast cameras were ticketed at $50,000 each. New satellite costs made those items look like pocket change. In a two-year period, from March 1981 to March 1983, the operating expense budget grew from $715,000 to over $2,600,000.[14] A staff of fourteen arrived at the new Capitol headquarters. By September 1982, the number of employees reached eighty. Coverage increased from eight hours a day to twenty-four. By the end of 1982, C-SPAN had begun negotiating for a second transponder, which it hoped to use to cover the Senate. To defray expenses, the network began accepting advertising from AT&T and Merrill Lynch while considering proposals from four or five other organizations. Advertisements were to be accepted only from national companies. Product endorsements would not be allowed. The network hoped to raise one million dollars from advertising in 1983. The experiment with advertisements never produced close to the revenues expected, however. Within the corporation, some felt uneasy about using advertisements, which were a difficult sell because C-SPAN could not produce Nielsen ratings suggesting to buyers the size of the audience they might reach. Also in 1982, the network started the *C-SPAN Update*, which would eventually become a weekly paper and another potential source of revenue.

FINDING HELP FOR LAMB

Transponder changes, the need for revenues, and rapid growth led the board to suggest that C-SPAN needed organizational help. Lamb could not do it all. Someone had to be found to serve as second in command, someone who could push the as-yet-unwritten C-SPAN mission and supervise day-to-day operations. Lamb selected Mike Michaelson, the former head of the House Radio and TV Gallery, a man with thirty years of experience on the Hill. Michaelson's "Hill contacts" proved "invaluable," but organizing the early C-SPAN was like trying to train a puppy. "Whatever you grabbed," employee Jana Fay recalled, "you could hold on to."[15] Lamb himself had told C-SPAN staffers that there was a future at the network, but he added, "Anyone who drags or is unhappy, may find him or herself on the outside looking in." Rumors of a possible labor union led him to express intolerance "for someone in their early 20s making *demands* for this or that on C-SPAN."[16]

Michaelson could not train the puppy. He was more useful on the Hill and would play an important part in solving difficulties with the broadcast networks over coverage of Capitol Hill committee hearings and in opening doors for the telecasting of political conventions. He became executive vice-president, and Lamb brought in Bruce Collins, then a young staffer with the National Cable Television Association, to handle day-to-day operations and to be an on-air host. Typical of the hiring process in those days, Collins cannot recall just when he served part-time or at what point his appointment became full-time. As with the other hires he made in this period, Lamb hoped Collins would grow into the job. "All of a sudden [Collins] was supposed to be this leader," one staffer remembered. He was called "Director of Network Operations." Lamb, however, never clearly communicated Collins's role to others. Everyone continued to run around doing everything and anything.[17] Internal management would remain a problem for a while longer.

MARKETING AS A MISSION

One of Saeman's goals as board chair was "to make C-SPAN more aggressive on its own behalf" within the industry, but marketing in the early days was not a C-SPAN strength.[18] When Ed Allen became the third chair of the board, he made marketing his mission. Allen's firm was initially one of the smallest contributors to C-SPAN, but his efforts as chair would soon become worth far more than a hefty first donation. Like his predecessors, he was well respected in the industry. He eventually followed Saeman as board chair of the National Cable Television Association. Allen set two industry-wide marketing goals for C-SPAN. He made it his mission to see that each of the one hundred largest cable systems carried C-SPAN, and he campaigned to get the network on the air in each of the state capitals of the nation. He also had the added task of trying to deal with the cable systems that had been lost as a result of the fee raise and transponder switch. He took over the board during a period in which the courts were about to overturn the "must carry" rule. They did so in July 1985, but for Allen, C-SPAN was "the last remaining must carry," cable's "crown jewel."[19] He took out a full-page advertisement in *Cablevision*, calling on cable operators to carry C-SPAN. Nonconfrontational in style, Allen was a constant letter writer in the battle for industry support. Well-liked in the industry, an old-timer in a young field, he would be selected as "Cable Television Business Executive of the Year" in 1984.[20]

Allen's was a visible goal, a recognizable target. He would pursue it relentlessly. He initially concentrated on the chief executive officers of the top multiple-system companies. Without support from the top, he believed it was unlikely that local affiliates would readily add C-SPAN. At the local level, however, success and the bonuses that came with it were counted in revenues produced. Nonprofit programming was unattractive, especially if a system had only twelve channels available. A "must carry" message from the top, Allen believed, might solve the problem. Allen knew "most of the major players in the industry on a personal basis." A word from them might make C-SPAN a corporate priority. A number of them, however, "did not involve themselves in local carriage decisions." In some systems, carriage at the affiliate level was "generated up rather than from corporate headquarters." While still urging major multiple-system executives to bring C-SPAN to all of their affiliates, Allen concentrated on capturing the fifty state capitals for the network. C-SPAN's minuscule marketing staff would work on smaller systems and other affiliates as best it could.[21]

Allen's zeal paid off. By the end of his tenure as chair, C-SPAN had picked up five hundred affiliates, several hundred more that had been lost at the time of the transporter switch. Although the network reached about 50 percent of all cable homes, it was carried on only about one-fifth of the country's cable systems.[22] Within a year and a half of Allen's departure as board chair in the summer of 1984, ninety-four of the nation's top one hundred systems broadcast at least partial coverage of C-SPAN, and six more were scheduled to come on-line when their equipment was upgraded. Of the top one hundred, only Service Electric of Allentown, Pennsylvania, and Cable TV of Puget Sound, Washington, had no plans for carriage. By that time Club 100 was in operation. It consisted of multiple systems that offered C-SPAN to all of their viewers. Its six members made the network available to 2,870,000 homes. Five other companies had more than 90 percent coverage, adding another five million viewers. More subscribers would be on the way as the cable industry began to develop franchises in urban and suburban communities. Washington, D.C., for example, had yet to be wired. The forty-ninth capital—Pierre, North Dakota—was scheduled to come on-line in January 1986. Augusta, Maine, remained without C-SPAN; not until September 1989 would the service be provided to viewers there. The difficulty of the task that Allen undertook can be underscored by the fact that although his cable system, Western Communications, was a member of Club 100, his home, located in an area covered by another system, carried C-SPAN only from midnight to six in the morning.[23]

A FIVE-YEAR PLAN

During Allen's term, C-SPAN was poised for growth. System upgrades meant larger channel capacity, and the construction of modern systems in urban areas also promised a larger potential audience. The network had already established a niche for itself on cable. There was hope for coverage of the Senate on C-SPANII. Plans were under way to cover the 1984 nominating conventions in their entirety. Those in the industry considered CNN, which had become available in 1980, primarily a news synopsis and analysis service, different from C-SPAN's gavel-to-gavel, whole-event, well-rounded public affairs programming. A profit motive and its entailments differentiated the two as well. Furthermore, cable television programming had gone through a shakedown period and was now well established. All C-SPAN needed was a plan to grow, but in the early 1980s it did not have one in the corporate sense. It needed to get on a firm financial basis. It had done no long-range fiscal planning and did not even have its own in-house accountant until 1984.[24] Early on, the board began to push for the formulation of a five-year plan. To aid in the process, board member Jack Frazee, who headed the cable division for Centel, lent one of his staff, Rob Kennedy, an MBA, to the network.[25]

Some events, as we shall see, would demand attention before the five-year plan was formally adopted in 1984. Still, Lamb's vision needed a business sense, and the plan would provide it. In part it codified actions under way, in part it organized operations, and in part it made continuation of the vision possible. It would be revised and extended beyond the period it was designed to cover, 1984–88.

The five-year plan affected C-SPAN in a number of ways. To begin with, it formalized the network's goals and reinforced its objective to "produce, distribute and market a national public affairs programming service on a non-profit basis" premised on a "belief that viewers have the ability to reach conclusions through viewing the governmental process themselves." It mandated "the coverage of significant national and regional events and issues in as unbiased a manner as practical—as if the viewer had attended the event in person."[26]

The business strategy of the five-year plan linked four functions: programming, production, marketing, and organization. All depended on adequate funding. Without increased funding, little more could be accomplished. If C-SPAN's operational budget was to rise from $3.8 million in 1984 to $10.5 million in 1988, additional money had to be obtained.[27] The board members hoped that industry growth could provide some of the new funding, approximately $6 million by 1988. They neither wanted nor seriously contemplated a fee increase over the period of the plan. Because of "bottom line pressures" within the cable industry, they looked for other ways to increase revenues.

Planners expected that advertising would net about $145,000 annually, about 15 percent of the original projections. The thirty- and sixty-second advertisement spots, though institutional–rather than product-oriented, looked like commercial advertising on broadcast networks. C-SPAN would allow only corporate philosophies, not individual products, to be promoted, and advertisements would be shown only during breaks between programs, not during them.[28] Publicly Lamb expressed doubt that there would be much demand for commercial time on the network. Privately he and others were concerned that such advertisements might cause viewers and affiliate operators to become confused about sponsorship of the network. Furthermore, maintaining a lack of bias and keeping a balance were always tightrope acts. Advertisements might create the wrong impression.

A second small source of new revenues, the planners hoped, would be the *C-SPAN Update*, begun in 1982 to help viewers understand scheduling. In 1984 it grossed $33,000. It was projected to bring in $88,000 in 1988. (Instead, the *Update* had difficulty paying for itself over the period of the plan.) A third source for funds was expected to come from leasing the audiobands on C-SPAN's transponder. The planners estimated that this would add $500,000 a year to the coffers by 1988.[29]

The major new source of revenue was to come from the development of the C-SPAN Fund, a form of corporate underwriting especially attractive for tax reasons. The fifteen-second spots would simply feature a corporate logo and a voice-over stating, "The following program is made possible by a grant from . . ." No corporate sponsorship would be linked to a particular program on a regular basis. The plan estimated that more that eighty-four hundred spots would be available annually. The hope was to sell them in blocks of two hundred for $50,000 per unit. The goal was to sell forty units annually. Potential purchasers were the 250 largest U.S. companies, public and private foundations, and trade associations. Like the advertisements, however, these spots had the potential to dilute, in the public mind, cable's role in sponsoring C-SPAN. Lamb was "not uncomfortable" with the idea, which had first originated from an early board member; he just "wasn't in agreement with it." It would "send mixed messages," he thought.[30]

By 1989, twenty such groups were corporate underwriters, six of them telephone companies. At its height, underwriting brought in "about $400,000." The board suddenly realized that companies were getting "several hundred mentions a year" for $25,000 and, at the same time, stealing credit it believed belonged to the cable industry.[31] At a September 1989 meeting in New York City, the executive committee recommended termination of the fund.[32] Publicly, the main purpose of the fund had been "to increase awareness on the part of students, teachers and the politically active" that the service existed, even though their local cable systems might not have carried it. Supposedly, money gained from the fund was to be directed to those audiences who, once aware of C-SPAN, might pressure noncarrying systems to pick up C-SPAN. The announcement of the fund itself generated publicity but not the desired result.[33]

There were more immediate uses for the money than educational outreach, although C-SPAN's efforts in that direction would begin in earnest a few years later. While C-SPAN planned for the future, more immediately the network set out to cover the 1984 national nominating conventions and to follow them with *Grassroots '84*, a fourteen-city tour designed to give a local slant on the national campaign and to enhance cable's presence in local communities. C-SPAN did not acquire a mobile uplink until early in 1984. With industry support, the network was able to purchase it at what *Cablevision* called a "donated cut rate" of $211,000 from Scientific-Atlanta. The vehicle had been one of four originally bought by the Mexican government but not used. The mobile uplink, along with $15,000 from the cable operators in each of the cities visited, made *Grassroots '84* possible.[34]

SUCCESS AND A SERIES OF CRISES

In the summer of 1984 Jack Frazee, of Centel, followed Ed Allen into the board chair. Whereas Allen had been old-school cable, Frazee came from a company whose major assets were in the telephone business. The relationship between the telephone companies and the cable industry had not been good, especially in the early days. Frazee remembered an uneasiness surrounding

Centel's entry into the cable market. Either Ed Allen or John Saeman, he cannot recall which, had told him that if his company wanted to be involved in the cable television industry, it should support C-SPAN, and so he had joined the board. Now he was chair, and he was everything Ed Allen was not. Allen was a letter writer. Frazee could be blunt and direct in conversation. They did share one characteristic: they were both unabashed supporters of C-SPAN. Frazee was attracted to Lamb's "vision and his goal." Frazee believed in him, and much to Lamb's liking, the new chair believed the cable industry deserved "a wake-up call" when it came to carrying C-SPAN.[35]

Frazee would find that the realities concerning C-SPAN's operations threatened to make the five-year plan obsolete from the beginning. The plan called for "constant evaluation of equipment," for example. As equipment wore out or became obsolete, "opportunities to increase production quality" could be exploited, according to the planning document. In fact, the time for exploitation was at hand, not in a phased-in future. Basic items like cameras and videotape machines needed to be replaced immediately. The five-year plan also emphasized "that any move in improvement of facilities" required justification "based on efficient operation of the production department within financial resources, as opposed to simply the desire for more space." But larger quarters had to be occupied in a matter of months. So too, a new master control system was necessary shortly. More employees were needed, and compensation for current employees was inadequate. Success seemed more like a series of crises.[36]

Frazee was one of the most visible chairs C-SPAN had ever had. He was a hands-on person who did not like to lead from behind a desk. Frazee went with young C-SPAN crews to the convention in San Francisco. He was in Denver when *Grassroots '84* arrived there. He sat in on call-in shows and was moved and motivated by each experience. Frazee was a C-SPAN cheerleader, and he was a tough businessman.

Several things were immediately obvious. Industry growth and alternative revenue sources could not supply funds rapidly enough to meet the needs of C-SPAN. Expansion was further burdening the already strained internal management system. Although others had been given distinct management responsibilities, many employees, perhaps because of Lamb's zeal, still looked to him for day-to-day leadership. Lamb recollected that he had been "doing it all." By 1985 the board had a rate increase for cable systems under study. In the same year, at Lamb's urging, the board brought in Paul FitzPatrick as president and named Lamb chief executive officer. A personal friend of Lamb's since 1969, FitzPatrick had worked with Lamb at *Cablevision* magazine. Lamb recalled that he had asked FitzPatrick to come on board because C-SPAN "needed strong leadership."[37] FitzPatrick was "money conscious" and "very aggressive," much more "commercially oriented" than Lamb.[38] He was outgoing and well liked by the staff.[39]

It was FitzPatrick who would deliver the news of a rate surcharge approved by the board in October 1985. Bluntly he told the industry, "We found inefficient and inadequate equipment." In a sense C-SPAN had done too well with too little. First-run programming now composed 50 percent of the network's output. Coverage of non-House session events had increased dramatically. C-SPAN's new mobile capability added to equipment demands. Even the new mobile system would be "on its last wheels" within a year and would need to be "completely overhauled or replaced." The problem, of course, was not simply inadequate equipment and increased programming. FitzPatrick advised the cable operators that the network was "significantly understaffed" to meet its goals. It would need 138 employees to do the job by 1988. In the fall of 1985 C-SPAN employed 96. Furthermore "an extensive pay and benefits comparability study" revealed that "most C-SPAN employees merited improved compensation and benefits." Even their work space was inadequate, with ninety square feet per employee. C-SPAN needed better benefits for employees and more room. FitzPatrick told cable operators that C-SPAN had been "driven by conservative operating practices." He said, "We have always done our best to keep costs down." Frugality had its own costs, however. Without a surcharge, C-SPAN would "shortly experience significant revenue shortfalls." The one-cent "temporary" increase, once in place, would help build revenues by $1.6 million. The surcharge met minimal opposition from the industry.[40]

If Lamb was better at preaching the C-SPAN agenda, FitzPatrick would prove more efficient at taking up the collection and at organizing the church. Several long-term C-SPAN employees credit him with bringing meaningful structure to the network's internal organization. Heretofore, C-SPAN had been run largely by belief, faith, and commitment to an ideal. The eighteen months that FitzPatrick spent at the network turned C-SPAN from a week-to-week, event-to-event organization into one capable of more sustained operation. Somehow the change was engineered without losing all of the spirit that had initially typified those attracted to Lamb's vision.

Perhaps the energy both FitzPatrick and Frazee displayed was catching. In a sense FitzPatrick was the first Lamb peer whom C-SPAN employees saw daily. As good as those earlier assigned to managerial positions were (most are still valuable leaders at C-SPAN today), they did not have then the standing that FitzPatrick had. Lamb and FitzPatrick had different styles, and they differed too on the implementation of goals. FitzPatrick often had both eyes on the bottom line. Ideas not only had to be good but also had to bring in revenues. For him, a nonprofit organization did not exclude a businessman's attitude toward operations. Lamb, on the other hand, often kept both eyes on the mission and worried about anything that might taint it. Lamb saw employees as extensions of himself—dedicated, driven by an ideal. He looked for workers who were "on the same wave length he was."[41] FitzPatrick recognized the modern business world. Together, Lamb and FitzPatrick were needed in an important period of transition. C-SPAN was no longer a grand experiment.

The organizational structure at C-SPAN had not kept pace with growth. It seemed like there were "fifteen fiefdoms," one staffer recalled.[42] Before 1984 there was no "regular mechanism" to ensure coordinated action. New administrative units had been created, but "sometimes the operator's left hand didn't know what the office's right hand was doing."[43] Cooperation existed, but lines of responsibility and of authority were less than clear. It was as if a family business had expanded rapidly without realizing what growth might do to it. Some now see FitzPatrick's tenure as being marked by a "major reorganization" of C-SPAN; in a business sense, "organization" might be a better term.[44] There was much to be done.

Lamb turned his efforts to programming and policy. FitzPatrick handled finances and day-to-day operations. Titles and responsibilities were shifted. Bruce Collins moved from director of network operations to vice-president, corporate development. His duties also included legal matters. Brian Lockman became vice-president, network operations, with general oversight responsibilities for the engineering, field, and master control departments, including expenditures for related equipment. He also monitored network-production quality control and was responsible for developing an in-house training program. Susan Swain took on additional duties for "a very critical project": 1986 would become C-SPAN's "Education Year." Pam Fleming moved to her third position in five years at C-SPAN. She had been the network's first programming director, then director of special projects, and now, as director of personnel and administration, was to play a "more active role in particular in the performance and salary review area." Department heads were asked to develop yearly financial and operational objectives.[45] "If you don't have a plan," FitzPatrick said, "one day you could just wake up and say, 'We just spent a million dollars and where are we going?' "[46]

The lack of a well-developed personnel policy would cause problems for the board, Lamb, and FitzPatrick. Drawing up the five-year plan, the board members realized that the network had been built with "limited financial resources" and that this "limitation was most evident in the area of employee compensation and benefits." They understood that "C-SPAN's wages at all levels" were "below the competitive industry norm—particularly at the craft level." Standardized salary ranges had to be put in place. Benefits, "the other cornerstone upon which to build more competitive compensation," had to be improved. Life insurance, long-term disability, thrift plans, pension options, better health insurance, and "even bonuses" were on the table as part of a strategy to retain "key employees." Board members knew that a lack of attention to salary and benefits "had caused some turnover." Still they believed time was on their side. They could move incrementally and be fiscally responsible because, they believed, "management employees" were "dedicated to the ideals of C-SPAN" and had "established this dedication throughout the organization." Furthermore, "vacancies at the craft level" had been filled "by enthusiastic individuals" who had "willingly learned C-SPAN production and programming values." The five-year plan emphasized retaining key employees and providing "a gradual increase in the competitive level of salaries and benefits" for all.[47]

From the beginning and with board encouragement, C-SPAN had "operated on the cheap." Given minimal funding and his own personal inclinations, Lamb had been "parsimonious." He often sought part-time employees and used a two-week trial run for some employees who would eventually become full-time.[48] He himself never sought a raise. "He lived, ate, and breathed C-SPAN," Tom Wheeler, former president of the National Cable Television Association, recalled. "If ever there was total dedication," Lamb had it.[49]

Although employees looked to him for vision and for guidance, for some staff members C-SPAN was not their whole life. Resources, equipment, and people were stretched to the limit and beyond. By September 1984, signs of unrest were in the air, and C-SPAN's *Grassroots '84*, a major part of its first venture into the field, was about to go on the road. Crews were scheduled to cover fourteen cities across the country in two months. Meanwhile board members boasted that theirs was "the only cable company" that had "constantly operated in the black" and that its yearly budget was "the equivalent of what a network spends for a couple of evenings of prime time programming."[50] In-house, however, complaints large and small began to be heard, about everything from dress codes to the midwestern makeup of the staff, from long hours to salary and benefits. For the moment, the chance to learn, to gain experience, to participate in C-SPAN's mission lost its meaning. One employee attempted to get an industry publication to do a story exposing "the 'slave labor' environment at C-SPAN."[51] Some staff members left the network. Part of the dissatisfaction undoubtedly came with rapid growth. The workplace culture had changed. The close proximity in which C-SPAN technicians worked with employees from other networks also fueled the dissatisfaction. C-SPAN employees were younger and less experienced. Their pay, working conditions, and benefits were no match for those offered by ABC, NBC, and CBS.

As the network geared up to work on employee benefits in the spring of 1985, technicians talked of forming a labor union. In April, the National Association of Broadcast Engineers and Technicians (NABET) petitioned the National Labor Relations Board to conduct an election among production employees in the hope of organizing a portion of the network staff. Their talks with what they believed to be at least half of the possible voters left Harry Coyle, NABET'S director of organization, feeling pretty "confident" about the outcome.[52] While a network representative suggested that a successful strike would force industry costs to go up or cause employees to be laid off for budgetary reasons, Lamb began what one long-time employee recalled as "a personal crusade."[53] Wise to the ways of business, FitzPatrick warned supervisors, "Don't promise, don't interrogate, don't threaten, don't spy."[54]

It was not the thought of a union that bothered Lamb so much as a feeling that some employees doubted "his sincerity." Frazee advised: "Meet with them, tell them what we're able to pay and why, open your books up to them. Don't fight it; listen, talk to them." Frazee went to Washington, D.C., himself and practiced what he preached.[55] Lamb began to meet employees one on one and in small groups. His message was simple: "We screwed up; we won't do it again."[56] On May 14, 1985, he wrote to employees, "If there is one thing I've learned since *Grassroots '84* and especially during the last few weeks, it is that the whole company would benefit from better communication." He admitted that some units did not "fully appreciate or understand" what others were doing. He recognized that there was "across-the-board unsatisfied curiosity" about the network itself. Newer employees did not know "how this place got started." All were curious about "the prospects for survival" and about what plans, if any, were being made for the future. None knew about the actions of the board in late March or of plans for the fiscal year beginning in April. Better benefits were already in the offing. For Lamb, communication was the solution. He began holding small luncheon meetings with employees drawn from several departments in the company. "The primary goal," he said, was "to share information."[57] The now-famous Lamb lunches continue today.

While Lamb talked, FitzPatrick, with the help of key staffers and with board approval, worked on wage scales and benefits. The joint effort proved successful. The vote was 3 to 1 against a union. Bruce Collins, now legal counsel, believes the strike threat made C-SPAN a better company. It caused the network to set work rules and to improve salaries, and it forced the board to realize that C-SPAN had to be capitalized differently. The network had survived another crisis, just as it was about to go on the road again for *States of the Nation '85*.

MARKETING ONCE AGAIN

While C-SPAN worked on getting its house in order, efforts continued on the marketing front. "Perhaps no other function of C-SPAN will undergo such a major change and play such a critical role in the network's success—as marketing," the five-year plan asserted. In October 1983 the marketing and public relations staff consisted of three people. Susan Swain, who had been hired as an evening producer, *was* the public relations department. Brian Gruber and an assistant handled marketing for the whole country. C-SPAN was on eleven hundred cable systems. More staff members were assigned to the network's newspaper, *C-SPAN Update*, than to marketing.[58]

Early efforts at making C-SPAN available to the public had primarily been a board function. Members took on the task of contacting others in the industry whose systems did not carry C-SPAN and persuading them to do so. The industry was their primary focus, and they could contact leaders directly. Going to the public required an indirect route. Lamb was often on the road working on the same task. By the end of the five-year plan, the board hoped to increase the marketing staff to eight, but public relations would remain the responsibility of one person. Plans included the possibility of setting up regional marketing centers. The object was to build awareness of the network among operators who already carried it so that they, in turn, could more "effectively promote C-SPAN to their subscribers." A second course of action directly targeted viewers themselves.[59]

Marketing was then, and continues to be, a problem for C-SPAN. In a sense

the whole notion of marketing went against the grain of the network's founding philosophy. How was C-SPAN—determined not to produce star-system
journalism—to sell itself? If some of its programming, as one board member
put it, was as "interesting as paint drying on a wall," how could the network attract attention? Would the industry at the affiliate level spend its energies and
dollars touting a programming format that cost, rather than made, money?
With C-SPAN itself unable to predict much of what would air when, and for
how long, how would systems be able to promote programs that might be of
interest? In a sense, marketing was a strange concept to Brian Lamb. He did
not like it. A board member recalled, "We had to be very careful at meetings
about using terms like packaging and marketing, terms that were consistent to
a going business concern." Such terms "didn't fly with Lamb or his troops."[60]
In Lamb's world, programming would always take precedence over either
marketing or press relations. Late into the 1980s, the attitude was still the
same: "If the press manager leaves, we'll be fine, we'll make do." Growth
in the 1980s was more a product of the expansion of the cable industry and
of technological enhancements than of sophisticated marketing schemes.

At the board's behest, the network held a "C-SPAN Invitational" in the
nation's capital for cable marketing leaders. Jack Frazee sponsored a similar
event in Chicago, and a third meeting was held in San Francisco. The gatherings were a combination of meet-and-greet and show-and-tell. "We gave it
our best shot," Lamb recalled. Direct results were slow in coming, however.
Lamb noted: "We realize that C-SPAN will always be the last thing an operator sells. All we can do is do the best we can, work with the people who
do support us and they include some of the best people in the industry—and
hope the word of mouth will convince others to come along."[61]

Still, the small marketing staff found that local marketing managers were
often unaware of a corporate commitment to carry C-SPAN. Some did not
realize that under the C-SPAN fee structure, many multiple-system operators
paid for the programming even when a local affiliate chose not to carry it.[62]
Attempts to aggressively market by directly contacting community leaders
sometimes upset local operators, one of whom claimed C-SPAN stood for
"Cause Some Pain and Nuisance."[63]

The marketing efforts of individual members of the board and the executive committee were invaluable. C-SPAN was not their primary business, but
some devoted extraordinary effort on its behalf. For example, Gary Bryson,
of ATC, involved his directors of marketing and public affairs as well as
the vice-president of consumer research in developing plans to market C-

SPAN itself; more important, he urged them to use the network as part of their own effort to market cable in general. Gene Schneider, of United Cable, later served as board chair; he simply committed his company "to carry C-SPAN—period." John Evans, then of Arlington Cable Partners, devoted a significant portion of his time to chairing the executive committee's marketing development group.[64] Others helped to make C-SPAN more visible by lending staff and facilities and sometimes providing money to help C-SPAN carry the political conventions live in 1984 and to support *Grassroots '84* and *States of the Nation '85* programming. For many in the industry, however, carriage was a big enough contribution. Viewers could find C-SPAN if they wanted it.

Several other schemes also increased visibility. C-SPAN used advertising in trade magazines to keep the network's name before the industry. The return, however, was difficult to measure. The *C-SPAN Update*, designed to provide viewers with schedule information and to highlight certain programs, was still seen both as a potential revenue source and as a marketing tool. It would not succeed at either. The delivery system was often too slow, the schedule information was too sketchy, and system operators seemed unwilling to distribute it to key officials in their area. In 1986, C-SPAN began yet another form of indirect marketing when it sought foundation support to develop its "C-SPAN in the Classroom" program. It was hoped that this project would increase awareness and usage of C-SPAN in classrooms at all levels, from grade schools to universities. The next generation would grow up knowing about and appreciating C-SPAN. Furthermore, this program might link local cable systems more closely to the schools, providing a public relations benefit to the former and useful information to the latter. In 1993 the board created the C-SPAN Educational Foundation to extend the network's outreach into the schools. "C-SPAN in the Classroom" did, and still does, fulfill some of its original goals. By April 1995, nearly 11,500 teachers had affiliated with C-SPAN's program for teachers. One hundred and twenty faculty development grants, totaling $60,000, had been awarded. Nearly six hundred college faculty had participated in seminars sponsored by C-SPAN. The C-SPAN school bus carried the network's message across the country. Once the network began to use on-line services, C-SPAN distributed lesson plans based on its coverage. For the 1995–96 season, C-SPAN expanded its "Equipment for Education" grant program for teachers. The two educators submitting the best proposals were invited to attend either the Democratic or the Republican National Convention, at the network's expense. Thirty other winners received a TV/VCR for their schools. The network continued to fund its High School Teacher Fellowship, which brought an educator to the nation's capital for four weeks in the summer to assist C-SPAN's education staff in developing lesson plans for national distribution. As part of its "First Vote 96" campaign, the network also offered $100,000 in scholarships to high school seniors. Among the other awards to be earned was a visit of the C-SPAN school bus to fifty winners' schools.

If increased network visibility both with and by cable systems operators was the key to the success for the first five-year marketing plan, that effort met with minimal success. Recalling the effort, Pam Fleming, now vice-president for marketing, remembered that the network would send out representatives to as many systems as possible. The work, however, produced "no success stories." Fleming said: "Where was the print? Where was the hard copy? We had nothing."[65] An inability to predict programming, and thus to provide accurate scheduling information, meant that viewers could not turn to their daily papers to find out what might be on the network. Scheduling information that scrolled on local cable systems screens proved to be too general to be useful. The press that C-SPAN did obtain came not from advance notice from system operators but rather from unique programming of major events. C-SPAN simply had a problem with finding a way to promote itself consistent with its philosophy. A minor case in point strikingly illustrates this and is emblematic of C-SPAN's public relations dilemma. Late in 1985, a young viewer wrote to the *Update* editor and said that she had "never read the newspaper" until she started watching C-SPAN. Now the *Update* was to be one of her Christmas presents. She asked for a picture of Brian Lamb. "We're sorry," the editor replied, "but because C-SPAN does not want to promote its moderators as media personalities, we do not send out photographs."[66] To this day, callers to the *Washington Journal* requesting biographical information are met with the polite response, "Thank you for your call."

SATELLITE PROBLEMS AGAIN

Board leadership changed hands in May 1986 when Jim Whitson, of Sammons Communications of Dallas, Texas, took the helm. The executive committee authorized an additional nine seats as the board expanded once again to involve more industry leaders in C-SPAN. During Whitson's term, the long-awaited C-SPANII would finally go on the air as the Senate moved to open its chambers to coverage. Televising the Senate would once again mean taking the case to the industry; additional board members would help. Another nonrevenue channel had to be sold to the operators. A new five-year plan was approved, covering the period from April 1988 to March 1992, and with it additional requirements for personnel and equipment, an upgraded benefits package, and plans "to increase C-SPAN's 'beyond the beltway' programming."[67] C-SPANII would increase the budget by 5 percent. In the same period the network would lose 10 percent of its revenues with the failure of Studioline, which leased audio transponder space on its satellite carrying the original C-SPAN.[68] Because of a technical malfunction the network would also lose its signal on the transponder on the second satellite carrying C-SPANII. And it would lose Paul FitzPatrick to the New York-based Request Television.[69]

Before he left, FitzPatrick helped engineer what Brian Lamb called "the satellite deal of the century," providing protected space for both C-SPAN channels on Galaxy III until 1994. C-SPAN had lost its first satellite space because it had occupied the least expensive and most preemptive spot. It had continued to use the cheapest, and hence most vulnerable, transponders thereafter. Now a satellite failure threatened the appearance of C-SPANII. Fortunately for C-SPAN, Attorney Henry Goldberg, a longtime friend of Brian Lamb's, also was the attorney for Eddy Hartenstein, then with a company called Equatorial, which was based in northern California. Hartenstein's company, it seems, owned a number of transponder spots on Galaxy III. Looking for space, FitzPatrick searched out others and put together a package deal that Lamb today believes was "probably the most important single financial happening" in the network's history.[70] It provided C-SPAN with two fully protected transponders. If either transponder, or the entire satellite, failed, a backup was guaranteed.[71] This solved a problem that had brought the board's executive committee together five times in 1986, more times than it had ever met before in a single year and more times than it ever met again in that space of time. C-SPAN switched to Galaxy III from Satcom IV on October 31, 1987; C-SPANII followed in December of that year. The satellite also carried MTV, VH1, Nickelodeon, the Weather Channel, and Viewers Choice I and II. The Galaxy III arrangement did more than solve a problem; it allowed C-SPAN to set up a contingency fund in case of future emergencies and to begin long-term planning.

The marketing of C-SPANII, the loss of a transponder, and the departure of the president dominated board efforts. C-SPAN planned to provide the new channel free of charge to operators, with a proviso. They could not carry the Senate without first carrying the House, nor could they switch between the two on a single channel at will.[72] Put simply, the board urged systems to carry as much C-SPAN as possible. If for some reason operators found the Senate to be more attractive than the House, they would first have to carry the House proceedings. In some quarters this was "déjà vu all over again." Telecommunications, Inc., the nation's largest cable operation, typified the problem. Although it served 3.8 million basic subscribers, many of its affiliates were small cable systems serving rural communities. The issue was a pragmatic one. As John Sie, the system's vice-president, put it, "How can we best utilize what limited channel capacity we have?" He noted, "In areas where we have the channel capacity, we'll put them both on." William Strange, vice-president of Sammons Communications, claimed that regardless of channel capacity, "the tendency of the operator" was "to have it pretty much full." He added, "It's hard to take one channel off and put another on." By May 1986, operators had committed to only 2.6 million subscribers. Eighty-eight systems had signed on.[73]

As C-SPAN had done earlier, when it had first gone to twenty-four-hour coverage, the network gathered industry leaders in Washington, D.C., this time to "give them a first hand understanding about the Senate" and about what was "about to happen." For years C-SPAN had courted Senator Robert Byrd (D-WV). Now that the Senate was about to follow the House on television, Senator Byrd would be used to court the industry. As the upper body of Congress was about to begin trial-run coverage before a final vote in July 1986, Byrd told the audience, "We need your enthusiastic cooperation." He argued, "By dedicating a channel to C-SPAN's coverage of the Senate, you can make this a true test."[74] Perhaps the fact that the Senate was about to discuss tax reform attracted some operators' interest. More likely, board members had delivered. Whatever the reason, 305 systems, 24 percent of the cable industry, broadcast C-SPANII to over ten million homes by December 1986. Still, only half of American homes were wired for cable at that point.[75]

The numbers would be a good beginning for the network. As always, it had proceeded cautiously. "One scintilla of a possibility" existed that the Senate might vote against coverage, a C-SPAN representative told a reporter. Thus he explained, "One thing we do not want to do is to install a full-blown master control center."[76] The system the network would use was built in-house by Richard Fleeson, who had been with C-SPAN since its days in Crystal City.

One other action is important to the C-SPAN story. In 1986, the network gained a historical sense. It began discussions with Purdue University,

Lamb's alma mater, to set up a video archival facility. There all C-SPAN programming would be taped and made available, at a minimum cost, to teachers for classroom use. The archives began operations in the fall of 1987. Under the direction of Professor Robert Browning, the archives became a valued repository for those who brought C-SPAN into the classroom.

FitzPatrick's eighteen months at C-SPAN proved invaluable. At his departure, Lamb retained the title of chief executive officer. From the board's perspective, the key industry player would now be the chair of the executive committee. With board approval, Lamb established a new management structure in-house. He created a four-person executive management committee to guide the day-to-day operations. The three existing vice-presidents—Susan Swain, Brian Lockman, and Bruce Collins—were joined by a fourth, Rob Kennedy. Lamb brought Kennedy back to the organization to oversee its business operations. Kennedy had been at C-SPAN years before "on loan" from Centel, courtesy of Jack Frazee, and had been the "principal author of C-SPAN's first five-year plan." He had also been involved in updating the plan over the years.[77] In the long run, his arrival was vital to the network. In the short term, management by a four-person committee reporting directly to Lamb proved to be too cumbersome for efficient operations.

C-SPAN AT TEN

As it reached its tenth anniversary, C-SPAN could look back with pride on its accomplishments. The network celebrated its success with an attempt to once again answer the question, "Who watches C-SPAN?" It would do so by publishing *America's Town Hall*, a compilation of interviews with viewers from all walks of life. Like so many other things at C-SPAN, the book was a collaborative effort, involving more than thirty C-SPAN staff members. The task meant "accompanying" Georgia Congressman Newt Gingrich on one of his 6:00 A.M. walks around the Capitol, in order to get his story. Staffer Mary Holley "had to work through the intricacies of the Missouri state correctional system to talk with Russell Epperson, an inmate in the Moberly prison." The celebrities interviewed ranged from Ronald Reagan to Phil Donahue to Frank Zappa, but the book also introduced fellow viewers to housewives, parents of members of Congress, and local political activists.[78]

Writing in the foreword, Jeff Greenfield, of ABC News, said that for him, C-SPAN was "like having a $12 million research operation at [his] beck and call." He claimed: "The real pearls of great price come when C-SPAN picks up its cameras and winds up at a coffee klatch in Iowa, or a lobster dinner in New Hampshire, where a hopeful presidential candidate is speaking. Or when it covers a conversation between a political strategist and a group of reporters and editors over coffee and doughnuts. Or where it shows up at one of the hundreds of seminars in Washington on *Media Coverage of Politics*." C-SPAN, he argued, had brought politics to "the greater public." He concluded: "If the men and women seeking to lead us actually come to discover that the country is listening to what they say, it just may result in a more serious level of public discourse. And Lord knows, that is long overdue."[79] These were the kinds of comments that made Lamb proud. But Appendix 1 of *America's Town Hall*, "The C-SPAN Audience," the latest in a series of surveys of viewer demographics, was of more interest to the board members. It was at their urging that such studies were done.

One of the programming highlights of the year was a four-hour retrospective of the network, complete with viewer call-ins. Both Lamb and Susan Swain seemed slightly embarrassed by the praise they heard. The program did, however, help the network respond to advice given to Lamb by the in-house "10th anniversary group." The last comment in the September 14, 1988, memo stated, "A final point: most of us agree that the cable connection should be front and center in all of our efforts."[80]

The anniversary year was also a cause for celebration among board members, for reasons beyond the external praise the network had received. Once again, C-SPAN shifted its management style. Gone was the cumbersome four-person executive management committee. The board restructured the lines of authority in the company to make it a sounder enterprise in business terms and to streamline procedures. The improvements brought a sense of relief to at least one long-term board member. One of the reasons to join the board, he said, was "to share in the glory of the network." Industry leaders had other interests and problems to worry about. C-SPAN had always been like "a low maintenance child," he remembered, but now it could "manage itself."[81]

In 1989 the board updated its bylaws, both clarifying membership issues and reaffirming C-SPAN's mission. Its desire to leave programming decisions to the network was made clear once again. The number of board members had grown to the high thirties from its original twenty-two in response both to the nature of the industry and to a desire to involve as many of its leaders as possible. The new bylaws designated seats for "the chief executive officer" or "the chief operating officer" of the nation's "fifteen television operating companies with the largest number of subscribers." Fifteen more seats would be elected by the board. "Three associate directors" would be chosen by ballot. These directors would come from "companies that manufacture, distribute or supply equipment, material or programming services to or in connection with cable television systems."[82] For potential members to be eligible for membership under either of the first two categories, their company had to carry C-SPAN at least at the 70 percent level. Several seats were also set aside for former board chairs.

The C-SPAN Board

Brian Lamb is surrounded by members of the C-SPAN board of directors in the 1980s. *Clockwise from lower left:* John Saeman, Jim Whitson, Gene Schneider, Bob Rosencrans, Amos Hostetter, Ed Allen, and John (Jack) Frazee. (Photo courtesy of C-SPAN.)

MOVING FORWARD

With the revision of internal management procedures and the clarification of board membership issues, a series of industry leaders stepped in to head the executive committee. Their styles would be different and so would their contributions. Although all were elected, they were Lamb's choices, each selected to help with a different problem. For example, Gene Schneider, of United Cable TV Corporation, seldom said anything, but what he did say was what Lamb liked to hear: "All my systems will carry C-SPAN twenty-four hours a day, seven days a week." The same was true for C-SPANII.[83] Amos Hostetter, of Continental Cablevision, stressed marketing and education. He was responsible for doubling C-SPAN's marketing budget, providing funding for the C-SPAN school bus, about which we shall learn shortly. Unlike many of his predecessors, Hostetter was not a top-down manager. He let his system's operators make programming decisions. He hired good people and let them do their job, but those who worked for him knew of his devotion to C-SPAN, and they carried the network.[84] John Evans, then president of Hauser Communications, was the driving force behind C-SPAN 2000, a blueprint for the future now being implemented by the network. Among the subjects it discussed was how to deal with technological innovation. It also stressed both increasing international programming and marketing C-SPAN around the world. Jim Gray took the leadership of the executive committee at a time when his company, Time-Warner Cable, had not been an active supporter of C-SPAN. He promised to move the company to 100 percent coverage, and he did. Jim Robbins, of Cox Cable, guided the committee in the face of rapid consolidation in the industry. In his last meeting as chair, the board approved funds for a second C-SPAN bus and authorized the establishment of C-SPAN bureaus outside the beltway. In the summer of 1995 Tom Baxter, of Comcast Cable Communications, assumed leadership of the executive committee. Because of industry consolidation, the board had an unusually large number of new members. Joining him on the executive committee was Brendan Clouston, the chief operating officer of Telecommunications, Inc., the nation's largest cable system.

In the late 1980s and throughout the 1990s, the board was called on for advice on a number of matters, chief among them being the "must carry" legislation. It needs to be pointed out here that the general industry's response to the legislation and C-SPAN's response were different. "Must carry" caused a loss of viewers for C-SPAN. For some board members, the legislation meant a decline in revenues. Several in the industry were disappointed that their support of C-SPAN had not provided better leverage with Congress. "Must carry" posed, and still poses, a problem beyond affiliate losses. Although C-SPAN firmly believed in staying out of the way of Congress, it was now confronted with legislation that directly affected the network. "I find myself in an unusual role today," Lamb said as he began his testimony before the House Telecommunications Subcommittee in June 1991. "At C-SPAN, we are observers of the legislative process, not participants," he explained. Lamb's message was simple. "If Congress chose to reinstate must carry requirements," Lamb argued, the action would "convey second class citizenship to program services like C-SPAN." Lamb concluded: "I am not here today to ask for any legislative favors for C-SPAN. Just continue to give us equal access to cable systems. Keep the environment open and give EVERY channel an equal and fair opportunity in the market place."[85] At the board level, faithful C-SPAN supporters urged systems not to drop C-SPAN or, if they had to drop it because of the law, to return the network to the air as soon as possible.

Although the matter pales before the problems created by "must carry" legislation, the story of C-SPAN and the 1992 Olympics is a fitting testimony to the loyalty of its board members to the network's mission. To pull off a pay-for-view triplecast of the 1992 Olympics, NBC and its partner, Cablevision Systems Corporation, needed additional cable-channel capacity. They sought out C-SPAN as a likely prospect. Preemption of C-SPAN programming would cost cable system operators nothing and would bring them revenues. The executive committee chair, Amos Hostetter, rejected the idea immediately. "Who decided the House and Senate are less important than fencing?" he asked.[86] Still NBC persisted. On May 1, 1991, Marty Lafferty, NBC's vice-president in charge of the pay-per-view project, wrote to Hostetter. He described the "total six-month public relations program which would include extensive C-SPAN publicity on NBC." The commercial network's proposal, he explained, represented "an extremely significant contribution" to C-SPAN "from a financial perspective." To overcome Hostetter's reluctance, Lafferty suggested that the U.S. Olympic Committee and its backers would petition "the leadership of the House of Representatives" and ask them to "request C-SPAN I to serve as a home for one of the three TRIPLE-CAST channels on behalf of the U.S. Olympic movement and as Congress' visible show of support for our American athletes."[87]

Hostetter exploded. C-SPAN not only would not agree to cooperate in the venture but also would hold NBC, and Lafferty himself, "personally responsible for any continuing suggestion" that C-SPAN channels should be used to carry Olympics programming. No inducement would budge Hostetter. He noted, "Your persistence is disrespectful to C-SPAN and its viewers [and is] damaging to our reputation and good standing as a public affairs network."[88] Lamb told a reporter, "You couldn't pay us enough money to interrupt [coverage]."[89] C-SPAN's prime channel would carry its own style of Olympics. While the nation watched track-and-field events, C-SPAN's cameras went, once again, to the National Governors' Association meetings.

As the year 2000 approaches, C-SPAN will obviously face more difficult decisions than the one presented by NBC. Carriage problems certainly will persist in the short run. The emergence of DirecTv and PrimeStar, using home satellite dishes to bypass cable, and the possibility of telephone companies bringing their own video signals to the marketplace will undoubtedly have an impact on C-SPAN and possibly on its board composition. Already, consolidation within the cable industry has had an effect. These and other concerns will be addressed in the final chapter of this book.

CABLE'S CROWN JEWEL

The story of the role of the cable industry in the development of C-SPAN is a remarkable one in may ways. Without the industry, C-SPAN would probably not exist in its current form. This is the tale of industry leaders willing to give their time, talent, and money and of others less willing to follow them. No doubt the industry leaders who served on the C-SPAN board brought a sense of business acumen to a network that needed it. Furthermore, they were never involved with the content of C-SPAN's programming. Their dedication extends beyond board service. Some have even left C-SPAN in their wills. Others have helped to endow a scholarship in honor of Lamb's father. One has made a six-figure gift to C-SPAN's Educational Fund. An idea, one that only a few people thought would work, was made possible by the cable industry and especially by those within it who became C-SPAN devotees. As striking as the story is from the industry perspective, however, there is another way to follow C-SPAN's development. This view too reveals more about the personality of a network than about network personalities.

Looking from the Inside: The Employees

ORIGIN TALES

Most of the top decision makers at C-SPAN in the mid-1990s joined the network in its first three years. Susan Swain was initially hired as an evening producer when the network was moving to a sixteen-hour schedule. She is now one of two executive vice-presidents. The vice-president for programming, Terry Murphy, started out behind a camera. Bruce Collins joined the network part-time in 1980. He is now vice-president and legal counsel. One of the original four employees in Crystal City, Jana Fay today serves as vice-president for finance. Kathy Murphy, who heads programming operations, began as a telephone operator. Pam Fleming, who came to the network as program director in 1981, is now vice-president for marketing. Together with a few others, such as chief engineer Richard Fleeson, they constitute the network's cultural memory. Unlike early board members, they can tell the inside story of the network that Brian Lamb built. These tales serve not only as archetypal examples of what life was like at the beginning but also as markers of behavior in the present. Although such stories were never intended as value tales, they have taken on that status. The morals they reveal are now codified both in C-SPAN's mission statement and in its list of employee expectations. For early employees, the stories are lived experiences. For employees who joined the network later, they take on mythic qualities. Myth in this sense is not a lie but rather the result of a combination of retold stories, with several variants, used to represent the corporate culture.

Not surprisingly, success stories dominate the corporate culture at C-SPAN. Multiple tellings convey a similar message about the qualities necessary for success both for the corporation and for the individuals working within it. The stories told about the early days have identifiable roots in American culture. The tales sometimes contain lines that sound as if they were borrowed from *Mr. Smith Goes to Washington*. Occasionally they bear traces of the camaraderie that can be found in the upbeat 1930s Hollywood musicals. Often one sees signs of hard work and community effort similar to that necessary for a nineteenth-century barn raising. One can catch glimpses of a hard-scrabble farm, willing hands, long hours, determination, pluck, and luck. Less frequently one hears tales of an idea that is too big and moving too rapidly to be well managed. These, however, belong to another set of stories. Taken together, the tales still, unconsciously perhaps, drive C-SPAN. As Jana Fay said in early 1995, the people who are in the department-head meetings now "are the people who have been [at C-SPAN] forever, who understand the mission."[90] Anecdotes they remember are not simply accounts of what was; the stories contain "what is" and "what will be" for C-SPAN.

The memories are as midwestern in tone as is the founder. They reflect nineteenth-century, small "d" democratic values and virtues. In the beginning, everybody did everything, and they did it for next to nothing. Former Vice-President Brian Lockman recalled a company so strapped for funds that if an employee needed an electric cord, he or she brought it from home.[91] Richard Fleeson was barely twenty when he came to C-SPAN from a television-repair shop. "I remember seeing Richard the first day I walked into the place for an interview," Brett Betsill remembered. Everyone pointed to Fleeson and said, "That's the guy who built the technical end of C-SPAN." Fleeson knew how everything worked. "He had all the schematics in his head." When something went wrong, Betsill noted, "the first thing" anyone did was to call him at home. If Fleeson could not talk the person through the problem, he would be down at the studios "a half hour later," no matter the time of night. "Things were pretty much jury-rigged all over the place." Any piece of equipment was "welcomed with open arms." "Richard would find a way to hook it up and make it run and make an improvement in our operation."[92] Fleeson was "indefatigable." He logged over two hundred hours in overtime and comp time during the two months of *Grassroots '84*.[93]

Everyone indeed did do everything. Jana Fay, C-SPAN's first employee, was sometimes called on to operate the camera during early call-in shows. Although some employees today think she started as a technician, she had no technical experience. When the House remained in session until after midnight, engineers stayed on the job regardless of when they had first come in. They simply "had to be there." The eight-hour day or the five-day week was unknown in the beginning. C-SPAN, Fay remembered, relied on what amounted to "slave labor."[94] Everyone pitched in. When the corporate board came to Washington for a meeting, all hands—from Brian Lamb down—worked into the night to make sure the facilities were spotless. Bruce Collins recalled scrubbing the baseboards.[95]

"Hard work always comes to mind," Vice-President Pam Fleming remembered. "The only thing that permeates everything [about the early years] is the fact that the hours were very difficult."[96] Her desk doubled as studio. From there she read "live schedule updates." One could never be certain of when the House would finish its business or a committee would end its work. It was best to stay close to the desk at certain times. Like others, she did many things. Gail Rubin, who worked with her then, recalled that they were "the programming department." Rubin noted that the two women "scheduled call-in guests, created program schedules, and logged information in on each program" necessary for re-airing and "found out about events to cover." Rubin also "ran a camera, answered telephones, copied tapes and undertook a self paced education on that creature called the U. S. Congress."[97]

The three-button phone system hardly qualified C-SPAN as a state-of-the-art network. Six- and seven-day work weeks come to mind when the first employees reflect on the early days of the network. There was, nonetheless, an air of hope and excitement in the small northern Virginia studio where everyone could hear everyone else's phone conversations and wave across the room to get attention.

Above all there was opportunity. Fleming, for example, had taught high school English for a number of years before coming to C-SPAN. In between she had served for six months as a press secretary for a congressman who thought her job was "to keep the press away from him." Her sole experience with the visual media consisted of coursework toward a master's degree in radio and television. "I knew a little about production," she said. "I was shocked to get the job here." Brian Lamb hired her as a producer. Before coming to C-SPAN, she never believed that she would "come up with ideas and actually see them go on air."[98] Bruce Collins, now vice-president and legal counsel, joined C-SPAN six years before completing his law degree. "The job was more interesting" than school. "I was practicing law without a license from day one."[99] He was also in charge of day-to-day operations for a while and served as a call-in host, as he still does today. He was the prototype part-time hire, initially hosting a one-hour call-in on Thanksgiving Day in 1980.

That was the C-SPAN way early on: little experience, much pressure, and a great opportunity. Well into the mid-1980s an employee-of-the-month would be described as having had "no solid television experience" but a "great" attitude. "On his resume he wrote, 'no job too small.' That really appealed to us." A history major who joined the network the same year remarked: "I'm starting on the ground floor of my career, and I'm able to do a lot of fascinating and challenging things here. There's so much to learn—it just never ends."[100]

Those who survived and are now in leadership positions would not trade the experience "for a million dollars." Jana Fay remembered: "It was like Camelot. It was a campaign. . . . You came in, it didn't matter, you just did whatever it took." During the first on-air fire drill, Pam Fleming thought: "Oh, my God, can we really keep people here? . . . Nobody left . . . we all had to be there for the show." She recalled: "There was always pride in working here. I always knew that something was going to happen at the network."[101] That same excitement is still found in the accounts of newer employees returning home from a tour on the C-SPAN bus. They learn that the network makes a difference and that they are a part of the experience.

Another aspect of the early and later tales is important if one wants to understand C-SPAN. Early on, Brian Lamb became a "visionary" for his employees. Many speak of him as a "father figure." Fleming recalled: "We always looked up to Brian. . . . Brian hired people on the same wave length. We were like an extension of Brian."[102] Others speak fondly of his loyalty to those who were loyal to him in the beginning. Younger employees still vie to tell him stories of their experiences on the road.

What do all of these stories mean about C-SPAN as a corporate entity? First they reveal an ethic, a set of values. The phrases "hard work," "effort," "loyalty," "everyone did everything," and "it was all worth it" appear too frequently to be ignored. The accounts illustrate an adherence to a mission. Few of the early tales emphasize the importance of what C-SPAN covered; in this case, actions speak louder than words. Although the emotions that brought commitment to a mission are difficult to pass on, the stories of advancement are known to every new employee, and the tales of the beginning find their way into each new employee-training session. Living examples of the values formed in the beginning still walk the halls at 400 North Capitol NW.

THE MISSION STATEMENT

Not surprisingly, when employees speak of the early days, few describe it as a network with a formal mission or mention the decisions that led to programming choices. Early C-SPAN was like a nineteenth-century backwoods fundamentalist church. Faith was more important than doctrine; hard work and deeds were more vital than codified rules. The choir did not need temperamental soloists; it needed loud voices. Lamb's dedication and drive were mission statement enough. A mission existed, but employees simply intuited it or learned it on the job rather than read it in a manual. Sometime in the mid-1980s, Jana Fay recalled, C-SPAN finally prepared a formal mission statement.[103] Even so, the codification of "corporate values" did not occur until the early 1990s. Formalizing the mission statement, identifying corporate values, and making rules based on both signified a coming of age, a maturing of the network. These actions also signaled a generational shift that necessitated rule making in the organization.

One of the earliest attempts at codifying the C-SPAN mission appeared in a letter that Brian Lamb wrote to Speaker Tip O'Neill on March 5, 1982, just as the network was about to reach its third anniversary. Bruce Collins, then director of operations, distributed a copy to "All Personnel" under the heading "C-SPAN's Statement of Purpose." It was, he told them, "a concise statement" of what they were trying to do at C-SPAN.[104] Lamb told the Speaker, "An unchanging fundamental rule at C-SPAN is that all public business be presented from gavel-to-gavel and without editorial comment." Such an approach, he asserted, was "essential" if C-SPAN was "to fulfill its public service goal of non-partisan, unbiased programming." This approach would also guide the network's coverage of committee hearings, making C-SPAN "the only media outlet that fully and accurately" showed "the American people the nuts and bolts legislative work . . . conducted off the floor of the chamber." An interest in presenting "public business" in long form also led the network to present speeches made before the National Press Club and to cover the National Governors' Conference and over ninety more special events, forums, and speeches. Lamb told the Speaker that C-SPAN cameras also attempted "to demystify the rule-making process . . . by devoting programming time to such bodies" as the Federal Communications Commission and the Consumer Product Safety Commission. In addition, C-SPAN gave its viewers "an inside look at the news media" by visiting the newsrooms of major newspapers to show the public how the news became the news.

The "most satisfying program," Lamb said, was the daily live call-in program, which was "rapidly becoming a sought after forum by Washington policy makers." He explained to the Speaker, "We invite our viewers to call-in their questions and comments to a Congressman, Senator, or Administration official who is a guest in our studio." In the words of Representative Dick Cheney (R-WY), the call-in provided "a sort of instant feedback system." Congresswoman Barbara Mikulski (D-MD) believed the format allowed one to answer questions more thoroughly than "on commercial television." So far, Lamb reported, 264 guests had "answered over 4600 questions from a thousand communities in all 50 states." Of the "141 Congressional guests, 73 were Democrats and 68 were Republicans."[105]

The mission statement grew, and perhaps many of those involved did not even notice the growth. Reviewing the six-page appendix of programming accompanying the O'Neill letter, Collins told his fellow workers, "Just seeing what we have already accomplished is pretty impressive, and something to be proud of." He said, "Our work seems much more worthwhile when you realize that C-SPAN is absolutely unique, that not only does no one else do what we do, but also no one else CAN do what we do." The formal mission statement, which Collins would later help draft, could not capture that same pride and emotion.[106]

PHILOSOPHY, GUIDELINES, AND PROCEDURES

Today new employees receive a thirty-three-page "Philosophy, Guidelines, and Procedures" manual.[107] It, by necessity, reads prescriptively. Those new to C-SPAN are told that the network's programming philosophy is unique. C-SPAN, they are warned as much as advised, "is a different kind of television—television designed not for entertainment and escape but rather for education and enlightenment." It "does not stand between the viewer and the event, interpreting, dissecting or clarifying." C-SPAN makes it possible for citizens "to see public events and decision makers themselves and to judge for themselves." The network, new employees are told, "takes no position on issues." The "founding principle" of the network was "to present as much as what is happening in Washington and across the country as its television cameras can capture." Call-in programs allow guests and viewers an opportunity to "learn from each other." C-SPAN is "a live civics lesson," "a tool for democracy," and "the New England town hall of the modern age."

The document represents the 1990s workplace, an environment different from that of 1979. Who then thought of overtime or holiday pay, shift differential, meal allowance, a smoke-free workplace, lunchroom etiquette, sexual harassment, disciplinary procedures, performance reviews, grievance procedures, personal dress and grooming, or hazardous weather guidelines? Growth and the sociopolitical climate allowed just over two pages of the document to address the network philosophy; nearly thirty pages dealt with work rules. The times also mandated a legal warning at the outset: "This manual is an advisory guide concerning the present guidelines for the management of C-SPAN. Neither this manual nor any other policy or form shall be construed as a contract concerning employment with C-SPAN." Coming of age would have its costs for a corporate culture.

PERFORMANCE EXPECTATIONS

Begun as a network in which, for most employees, a term like "corporate culture" would have been buried in a long-forgotten, perhaps never read, college textbook, the modern C-SPAN put into place "corporate values and performance expectations" to guide evaluations and promotions.[108] The idea that "you just did everything it took" would have to be codified. If the mission statement clarified the network's programming philosophy and the "Philosophy, Guidelines, and Procedures" manual laid down the ground rules for working at C-SPAN, evaluation procedures would have to be put into place.

The new terms became "fair and equitable management," "professional environment for achieving and learning," "competitive compensation," and "diverse workplace and programming." New employees had to be told "to make a commitment" to network "principles."

Those joining the C-SPAN of the 1990s were advised that the company valued "employees who watched the network at work and at home," who wanted to "learn about public and current affairs," who understood "C-SPAN's role in the industry," and who showed a "dedication to integrating their knowledge into their daily work." They were cautioned to be responsible "for commitment to quality" while being "cost and resource efficient." They had to be "adaptable and open to changing priorities" while responding with "professionalism" to any changes that occurred. Showing "enthusiasm for the job and the mission" and "working extra hours" if needed were signs of "leadership potential." A commitment to teamwork was highly valued. Of course, even more was expected of those in supervisory roles.

The evaluation standards are in many ways a response to growth, to respectability, to coming of age, and to changing expectations among those who later came to the network. They are also a response to an organization larger than the original family, one not only with fourth cousins but also with grandnephews and grandnieces. No longer did everyone know everyone else at 400 North Capitol NW. One might have come to work dressed in jeans and no socks in the early days, but respectability demanded a dress code for the 1990s. Early employees, those imbued with the original mission, were appalled when new employees chose soap operas over C-SPAN on the lunchroom television. Formal standards would put a stop to that behavior.

JUST CALL ME BRIAN

Although rapid growth and the normal turnover among recent graduates flocking to Washington, D.C., for their first jobs necessitated codifying the rules, a strong sense of family remains at C-SPAN. Brian Lamb is responsible for much of this attitude. He "manages by walking around." A look at his daily appointment book is deceiving. Lamb roams the halls, addressing everyone by their first name and treating everyone equally. It takes only a few moments of following him around to recognize that he is interested in those who work at C-SPAN and in what they have to say. The talks range from family to trips on the C-SPAN bus to a recent program seen or a book read. Those at C-SPAN have learned not to be surprised when he speaks excitedly about an overnight program on C-SPANII featuring President Mary Robinson of Ireland discussing the diaspora. He has a thirst for information, and he likes it best when those with whom he talks are just as thirsty. Lamb often plays a version of political-historical *Jeopardy* with some employees. Interns first give him an "oh yeah" look when he tells them, "Call me Brian." That look turns to incredulity when, for example, in the midst of his talk with them, Teresa Easley, a human resources assistant, pokes her head in the door and says: "That's Brian. Don't go to lunch with him."[109]

Lamb's lunches are an institution at C-SPAN. Started in response to the possibility of a strike in the early 1980s, they are yet another way to check the pulse of the network and to involve those working for the organization. The talk is open, free, and wide-ranging, albeit somewhat intimidating to newer employees who are still learning the C-SPAN way.[110] After assuming major management responsibilities, Senior Vice-Presidents Susan Swain and Rob Kennedy extended the meal opportunities to breakfast meetings. Although both follow C-SPAN's "open door" policy, neither are as visible on a day-to-day basis as is Lamb. Swain's duties as a host on the *Washington Journal* do, however, put her in face-to-face contact with programming operations personnel on a regular basis. Both she and Kennedy have also been on the road with the C-SPAN bus.

"C-SPAN Week," the new employees' orientation, does attempt to do more than simply lay out the network's philosophy and organizational structure. A few of the origin stories are heard there, but shared experiences mean more to new employees than tales of the past. Shared experiences can come early at C-SPAN because it practices a kind of horizontal management style. Although many ideas originate from the top, they are developed, perfected, and occasionally vetoed by a series of committees formed to review them. In key matters, those committees are likely to consist of people who know the C-SPAN way, but that is not always the case.

Two principles are at work when committees are appointed at C-SPAN to study major programming options. Members from outside the programming department are included to prevent tunnel vision. The mix across departments also fosters a better understanding of how a decision can affect the whole network. Institutionally, such committee staffing promotes a sense of total involvement and a better understanding of the range of tasks that help C-SPAN run. Major special events like conventions or the Lincoln-Douglas debates reenactment can involve large numbers of employees in planning and in execution. Out of the 190 employees at C-SPAN at the time of the reenactment, for example, 118 traveled to Illinois to make the debates work.[111] Teamwork is a way of life at C-SPAN. In terms of viewership, the Lincoln-Douglas debates may not have been a major success, but they did wonders for the corporate culture. Everyone did everything to make the project work. In a smaller and more lasting way, the C-SPAN bus accomplishes a similar goal. It too brings together groups of individuals from a range of levels and a variety of departments.

C-SPAN is a company of committees. "We laugh about Congress having a Committee on Committees," Lew Ketcham remarked. But he added, "If you can't find somebody [at C-SPAN], you can pretty much bet they are in a meeting."[112] Togetherness, of course, has its costs. A decision filtered through committees takes time, and involvement costs time and money. But involvement is the C-SPAN way. What else would one expect from a network that has been called "America's town hall"? In a company in which the chief executive officer expects all employees to call him by his first name, horizontal management allows everyone to meet everyone else.

THE STAFF

C-SPAN is more than a mission statement or a code of corporate values and employee expectations. Like the employees at any other organization, the people at C-SPAN make the place. C-SPAN's "no star" philosophy downplays discussion of personal lives, however. Callers who request such background information are most often met with a polite "thank you for your call," and that is all. They may be surprised to learn that Steve Scully comes from a family of sixteen children, including five sets of twins. They may find it intriguing that Brian Lamb likes nonfat yogurt, that he buys tapes at Wal-Mart, that he has visited the gravesites of all of the deceased presidents, or that he wore shorts when C-SPAN visited Vietnam. Perhaps they would like to know that Bruce Collins comes from Apalachin, New York, smokes expensive cigars, and is a banjo picker. Would it interest them to discover that Susan Swain likes to sail during her free time? Is the name of the in-house band important? Such information, C-SPAN believes, gets in the way of, rather than helps fulfill, its mission. Viewers who are sure that bias exists at the network want to know about the political proclivities of those at C-SPAN. Many employees involved in programming or on the air are Capitol Hill veterans from both sides of the aisle. Working at C-SPAN, however, is like entering a monastery: one leaves the past behind. Those at the network are frequently warned not even to let a political opinion slip in the halls of 400 North Capitol NW.[113]

Demographic data about the company is easier to obtain.[114] In 1989, two-thirds of the employees were under thirty years old. By 1995, the average employee was thirty-two. Over the past several years the average length of service has risen from 3.1 to 5.3 years. Just under one-third of the forty-nine staff members pictured in the 1989 tenth anniversary book remain at C-SPAN.[115] In 1995, slightly over 10 percent of the employees at the network had been there for ten or more years, whereas 15 percent of the staff were in their first year.

C-SPAN sees itself as a company of the future. "I've always thought we have a wonderfully diverse company," Lamb told a group of students in April 1995. "Women here are as equal as you'll find anywhere in society." he said. "They get exactly the same amount of money as a man doing the same job."[116] Nine of C-SPAN's top managers are women. Over half of the employees who sit regularly on the editorial board are female. Not quite half of the total number of employees are women. C-SPAN has neither a gender gap nor a glass ceiling. One-quarter of its employees are classified as minorities. They compose 13 percent of "officials and managers," 18 percent of the "professional staff," 34 percent of the technicians, and 37 percent of the office and clerical staff. "You don't set out to build a company saying, 'I want to have 50 percent men, 50 percent women, and 13 percent African Americans simply because that reflects the population of the country,'" Susan Swain told an interviewer.[117] Talent, a real interest in understanding politics and journalism, and dedication to the mission count at C-SPAN.

HIRING

Gender equity and racial diversity were not a high priority when C-SPAN began operations. There simply was work that needed to be done. Lamb approached the task with an idea in mind. He told a reporter that as "a matter of policy," he did not want to hire "a finished product." He wanted employees to "grow with C-SPAN, not arrive with preconceived ideas." He explained, "I'm doing this because I want to provide the news as it happens and not to filter it through any kind of editorial operation." He did not want the network to "bear the imprimatur of professional journalists."[118] Looking back in 1995, Lamb said that in the early days, it was almost impossible to get "journalism types" to do the C-SPAN "thing." He noted, "I looked for people who didn't fight me."[119]

Getting Lamb to even consider hiring in the early days was a problem. Pam Fleming remembered pressing him for weeks to add an evening producer, since the network was about to make its move to sixteen hours in late 1981. Around Christmas of that year, her husband, who was a legislative assistant on the Hill, called her. "I just interviewed the neatest woman," he said. She had "great presence" but no legislative experience, so there was no way he could employ her. Maybe C-SPAN could use her. Her previous experience in public affairs for Raytheon and for the Up With People organization did not quite fit the producer's slot either, but Fleming interviewed Susan Swain anyway. She liked Swain immediately. Lamb was reluctant to interview anyone. Fleming sent Swain to Bruce Collins, then director of operations. He and Fleming both agreed the network needed evening help and finally persuaded Lamb to talk to Swain. He interviewed Swain four times, Fleming recalled, before he offered her the job. "That was the way it was," Fleming said. "It was so difficult to get anybody hired."[120] Swain remembered it differently. She recalled that she had a six-hour interview with Lamb, Collins, and Executive Vice-President Mike Michaelson and then started work the next day.[121] In any case, the money was tight, and Lamb was tight with the money he had. Swain certainly "grew" with C-SPAN, as did many others.

Intelligence and a willingness to work meant more than experience in the early days at C-SPAN. Even in the 1990s, there are still "you can come in and learn" jobs at C-SPAN. Lamb "feels strongly" about having such positions available.[122] There is still room for personal growth and for advancement, especially since the network has plans to add three more channels in the future, and a rapidly changing telecommunications world will obviously cause C-SPAN to develop its marketing and sales units. The word at C-SPAN now, however, is experience.

In April 1995, the network sought a field producer to assist with the enhanced coverage of the Senate. He or she would work with the coordinating producer of House and Senate coverage and with the producer charged with Senate programming. The applicant would need five years of related experience, preferably on Capitol Hill. The network hoped that this experience would include television and news production. The ideal candidate would be able to demonstrate a "thorough understanding of, and consistent adherence to, C-SPAN's corporate values, policies and procedures." He or she should also have a "strong knowledge of, and interest in, Congressional procedures, public policy issues and players." Another plus was "experience with editorial decision making." Night and weekend work was part of the job.[123] The days in which someone from master control could simply call the House or Senate cloakroom to find out what amendments were up were long gone.[124]

What attracts applicants to a network that programs for eyeballs, not dollars? Certainly it isn't money. Lamb himself makes next to nothing compared with commercial network evening broadcasters. The stories of how three individuals whom readers often see on the air came to C-SPAN may explain why working there is attractive. Lew Ketcham, Steve Scully, and Connie Brod arrived at the network with media and Capitol Hill experience. Whom they worked for on the Hill does not tell you what they believe, but their media experiences hold the key to C-SPAN's attraction. Ketcham had worked as an executive producer for an NBC affiliate in Norfolk, Virginia. "Frankly," he said, "I was tired of chasing ambulances, and that is pretty much what local news became to me." There were times, he recalled, when crews would come back from covering a fire "really pumped" because they had beaten the trucks to the scene. He noted, "We got plenty of flames." That was news. A rapid response to a fatal injury brought on a typical response: "It was great; we got the body." Ketcham said: "I got to thinking 'Why are we doing this? What's the point?' Telling people about the latest body count didn't seem to be serving the public good."[125] Ketcham left and went to C-SPAN.

Scully too came to the network by way of an NBC affiliate. His interest was covering politics. For the Rochester, New York, station where he worked, ratings came first. The station brought in what Scully called "a dress and drama" consulting firm to improve the ratings. The firm taught reporters how to smile and how to do exciting stories. Close shots of "the faces of people crying" at a murder scene were important. When ratings week came around, the station reporters would have to "go down to check out the hookers." The news director did not seem interested when House Speaker Tom Foley (D-WA) came to town to support a local candidate. When Scully was invited to the White House for a regional reporters' press conference, the station managers said he would have to pay his own way. They wanted their anchor to attend instead. At the conference, Scully asked a question that appeared on CNN's coverage of the event, and suddenly the local station told its viewers on the evening news, "Our reporter Steve Scully . . ." Scully later said: "You couldn't pay me to go back to local news. I used to fight to get fifteen extra seconds on the evening news. Here they say if you want an extra hour, go for it."[126]

"I swore I'd never go back to regular journalism and I wasn't doing it in

New York, I was doing it in Kansas," Connie Brod asserted. Brod now heads the team that produces the *Washington Journal*. She too had worked on the Hill. In Kansas she was with a radio network. "I still hate the ninety-second stories," she stated. "Even when you were balanced, you were only balanced in black and white." Of these three employees, she most clearly articulates the network's mission. She has learned to approach issues not with the question "What is my opinion?" but rather with "What is the range of choices?" Of course, Ketcham and Scully speak of the mission, but Brod references most of her answers to it. For her, "the emotion behind the mission is so strong" that it drives everything. "Many people who come here can do the job, but never *get* the network," she told a new employee. "If you don't *get* the network, and if you don't *get* the mission," she said, "[you're] not going to work" out at C-SPAN.[127]

It is clear that experience is a major factor for those joining C-SPAN beyond the entry level today. A "great presence" alone will not do it, nor will a willingness to do anything and everything. It is true also that experience alone will not ensure employment. The mission, for C-SPAN, is not simply a public document; it is a way of thinking and hence acting. Experience is not simply knowing how to do things and getting done them; it is also understanding what television news is and what it can be.

The C-SPAN School Bus: Bridging Two Cultures

It may seem strange that a forty-five-foot, 44,000-pound yellow bus that spends about three hundred days on the road a year would prove emblematic of the two cultural strains that make C-SPAN what it is today. Believing in fiscal responsibility and marketing, the board got both in the C-SPAN school bus. C-SPAN could not hire away a commercial network anchor for what it paid for the bus, even if it chose to do so. Certainly no network star could be maintained for about $200,000 a year.[128]

Undoubtedly the school bus was a marketing tool. It strengthened ties with local cable systems, promoting both them and the network. By the end of October 1995, 277 cable affiliates had sponsored stops in forty-eight states. Almost every stop generated local news coverage. The bus, however, proved to be much more than a marketing gimmick. It allowed the network to introduce C-SPAN's philosophy and programming to nearly thirty thousand teachers and students as it traveled across the country. At each stop, network representatives used video equipment, computers, and CD-ROM technology

The C-SPAN Bus Concept

Brian Lamb introduces the bus concept to the board of directors in 1993 as board member John Evans and C-SPAN Vice President Susan Swain look on. (Photo courtesy of C-SPAN.)

to show teachers how to incorporate C-SPAN programs into "video lesson plans." They taught teachers how to use on-line information services to access C-SPAN program schedules and how to research the library of past programs housed in the Purdue University Public Affairs Archives.

The bus staff served as a production crew, providing a variety of programming for the network throughout the tour. The bus was also instrumental in network community building. While its crews always included a manager and a marketing representative, the remainder of the staff consisted of volunteers from across the company. Vice-presidents and lower-echelon employees worked side by side to tell the C-SPAN story. The C-SPAN spirit even spread to the bus driver, who often used the CB radio to explain the network to passing truckers.

The multiple functions performed by the bus and its crews earned C-SPAN both the Cable Golden Ace and the Golden Beacon Awards from the industry. Everywhere, the bus was a success. Typically, the *Los Angeles Times* editorialized: "C-SPAN's yellow school bus may seem hokey, a throwback to Norman Rockwell in an age of Beavis and Butt-head. But the traveling studio helps to demystify TV by allowing visitors to see firsthand what goes on before and behind the camera at C-SPAN. That is educational in and of itself—a lesson that will help viewers assess TV programming with a more critical eye. The yellow bus adventure is about government and so much more."[129]

Over an inch of bus-related press clippings greeted board members when they arrived for their summer 1995 meeting in Washington, D.C. With funds in reserve for C-SPANIII and with an industry not yet prepared to receive the new channel, they approved funding for a second bus. Both marketing and educational programs benefited from the decision. So too did *Road to the White House* and the *Washington Journal*. Accompanied by a satellite uplink, the bus became a moving studio. The school buses indeed bridged the two cultures at C-SPAN: the board's entrepreneurial urge and the staff's dedication to extending coverage.

The Mission and the Pragmatics of Programming

When asked how programming choices are made at C-SPAN, Vice-President for Programming Terry Murphy usually starts his explanation with a favorite network phrase, "It's not rocket science."[130] The selection of what to air, however, is clearly an exercise in editorial judgment. Each weekday at 3:00 P.M., members of the assignment desk team, several field producers, a representative from programming operations, and one from the *Washington Journal* gather with Murphy to decide where to send crews the next day. C-SPAN almost always airs what its cameras cover. The group may be faced with as many as fifty choices selected from a larger list compiled by Carolyn West, who follows Congress for the assignment desk, and by Craig Brownstein, who tracks off-Hill events. Final selections are constrained by the commitment to cover Congress in session and by the availability of field crews.

Brian Lamb is never present at shoot meetings, but his spirit is evident in the room. As Murphy put it: "Brian had a vision for the company. He taught us well enough what that vision is that we are able to make day-to-day decisions on our own."[131] Lamb's critiques of such decisions are more likely to be about opportunities missed than about the events selected to air.

Based on observation over time, it is clear that some who regularly attend the shoot meetings express points of view that might identify their personal politics or their preferences for the coverage of certain types of events. Others, however, seem to be so mission-driven that even Murphy claims not to know their politics. Many gathered in the room have been at the network since 1982 and before. They are well aware of each other's personal proclivities and are comfortable with each other. Challenging what some might see as preconceived notions is not a problem. Those looking for impassioned outbursts or strong accusations will not find them by attending a shoot meeting.

Perhaps one of the reasons for the lack of heat at shoot meetings is that well before the group meets, the next day's live programming has already been selected by the assignment desk so that it can be promoted on the air. Another seems to be that selection guidelines are mission-driven. Congressional committees usually make the cut based on relevance, importance, informational value, and balance in terms of who is testifying before them. As a rule of thumb, if other media outlets are carrying a hearing live and in its entirety, C-SPAN will choose to tape the event and air it later. Forums stand a better chance of being selected if they involve public policy, have national implications, are balanced, and include participants who are recognized experts. Many such events, of course, are designed to promote a particular point of view. The availability of programs representing an opposing view figures into the decision-making process. C-SPAN assignment desk personnel realize that some think tanks and special-interest groups are better at the public relations game than others. The assignment desk is always looking for groups that may not know how to present their programs to C-SPAN. People or events providing diversity to the public dialogue stand a good chance of getting on the air. Political candidates are more likely to gain additional coverage if they address audiences with viewpoints different from their own. C-SPAN decision makers seem to favor outside-the-beltway events whenever coverage is logistically possible. These rough guidelines serve as the major influence in any argument about whether or not an event should be covered.

The mission statement demands that programming selection address the questions of bias and balance. Data that might satisfy a political scientist is available in the roughest form. Of course, numbers are kept. The assignment desk produces monthly figures regarding issues covered and positions taken. *Washington Journal* staff supply data on guests invited. Those involved in *The Road to the White House* produce a to-the-minute breakdown of time allocated to various candidates. Numbers are even kept on the political leanings of those who appear on *Booknotes*. The data is discussed at weekly programming meetings. Just where any event or participant falls on the political spectrum is, however, a rough judgment call, one made by the individual charged with preparing the data. What is the dividing line between a centerist and those to the right and left of that position? The data is only a rough indicator of balance or its absence. With event-driven programming, balance is a long-term process, sometimes achievable only long after its absence is noticed. So long as both sides of an opposing viewpoint complain about not receiving equal coverage, a kind of crude balance has been maintained.

Faith in and commitment to the mission appear to dominate decision making at C-SPAN. This mission is a product of the way in which C-SPAN grew as an organization with the help of its board, with Brian Lamb's vision, and with employees who chose to work within its constraints. The mission led to the development of what might be called signature programming. The next two chapters are devoted to the development and sophistication of this format.

CHAPTER FOUR

Signature Programs and Their Significance

At C-SPAN, programming is like a rope, a series of intertwined strands of formats woven together to give the network its identity and its strength. By examining some of these strands independently, one can gain a sense of how both the mission statement and the programming philosophy that springs from it come into play at the network, how ideas spur actions, and how actions shape and limit ideas. To do so requires a developmental approach. How an institution grows tells much about what it is.

Four programming strands are fundamental to C-SPAN. They are recognizable signature components of the network. Two, coverage of Congress and the call-in format, are embedded in the mission. Two others, following presidential campaigns and interviewing book authors (*Booknotes*), flow logically from it. By tracing their development, one can get a better sense of what C-SPAN is, what it does, how it does this, and to what end. In this chapter we will concentrate on the growth and development of C-SPAN's coverage of Congress and on the call-in format. In the next chapter we will turn to politics and prose, focusing on the way in which the network came to cover presidential campaigns and discussing the place of *Booknotes* in C-SPAN programming.

The decision to emphasize signature programming has its costs. This focus means downplaying the network's attempt to cover the presidency as an institution. It neglects C-SPAN's efforts to bring the judicial process to the people. It deemphasizes the network hours spent covering politics at a state level. It favors discussion of presidential campaigns over discussion of campaigns for state office or for Congress. It does not detail C-SPAN's efforts to bring international programming to viewers. It bypasses the treatment of special events, which C-SPAN covers so well. All of the above have been, and are still being considered as, the focus for C-SPANIII, IV, and V. On C-SPAN and C-SPANII, they supplement signature programs.

The First Strand: Covering the Hill

CARRYING CONGRESS

When the House of Representatives first went on the air in March 1979, the programming would prove to be no match for competitors such as *The Young and the Restless* or *The $20,000 Pyramid*. Television, former FCC Chair Newton Minow's "Vast Wasteland," was primarily an entertainment medium. *Three's Company*, *Mork and Mindy*, and *Laverne and Shirley* were America's favorites that March. All were better known than most of the members of Congress and more popular than the evening news. Few Americans may have seen Representative Al Gore (D-TN) make his first televised speech from the floor of the House that day, but this did not matter to C-SPAN officials. They had made the first crack in the Berlin Wall of commercial network news dominance. Viewers could now see their legislators in action without interruptions or commentary.

From a programming standpoint, carrying Congress live made great sense for C-SPAN. It fulfilled the network's mission and was cost-effective. It required only a handful of people. Congress controlled the signal, the cameras, and the shots selected. Like the early cable industry itself, C-SPAN thus began as a transmission service. The desire to bring citizens to the House and the Senate and to let them see and hear the action without journalistic interference undoubtedly cleared the path to coverage. It also reduced the investment the cable industry had to make and produced a better return.

Although the House and Senate floor discussions now compose roughly 15 percent of C-SPAN's programming, they have become a network trademark. Like all actions, the commitment to follow the floor whenever in session would have its consequences. To citizens unaccustomed to long-form, real-time television, the proceedings often seem dull, as confusing as a cricket match for a nation raised on baseball. Few caught up in the process can follow all day, every day. Working America is left with evening reruns of a portion of the coverage or with late-night sessions. Especially in the 104th Congress, this meant seeing Special Orders, in which Representative Bob Dornan (R-CA) seemed to *be* Congress. It meant listening to an often poorly scripted exchange between new Republican legislators tracking the successes of the "Contract with America" and feigning alarm over the excesses of big government. It meant hearing Democrats describe Speaker Newt Gingrich (R-GA) as the grinch who stole Medicare from the elderly.

Designed to bring the people to Congress, gavel-to-gavel coverage is more likely to be overheard in the offices of representatives and senators, in newspaper offices, and in think tank headquarters. Even there, for much of the day, or even for many days, the coverage may sound like the political equivalent of Muzak. The benefits for the viewers, however, are measurable, as we shall see in more detail later. Reporters can remain at their desks and follow legislation. For legislators, having C-SPAN in their offices allows them to return to the floor to change their vote or to challenge a speech heard during Special Orders.[1] It is clear from the number of times legislators refer to the C-SPAN audience and from the charts and graphs they use designed for the cameras, not the chambers, that they believe an important part of America is watching.

From a programming standpoint, dedication to gavel-to-gavel coverage can be, and often is, a nightmare. The House and the Senate set the rhythm, and C-SPAN orchestrates programming around it. The length of the sessions determines how much additional programming the network can air and when. During the calmer times before the 104th Congress, C-SPAN could reasonably count on airing the event of the day at 8:00 P.M. E.S.T. The Gingrich-led House, however, striving to symbolically emulate Franklin Roosevelt's first one hundred days, often ran well into the evening. Special Orders consumed even more programming time. Under House rules, these could run until midnight. Even before the 104th Congress began its marathon sessions in January 1995, C-SPAN had already decided to end its evening call-in programs because of the uncertainty of consistently getting them on the air.

As is often the network's style, in 1989 C-SPAN executives chose forty-three employees, divided into focus groups, to evaluate programming. Among the items considered was what to do with Special Orders. A minority argued for eliminating them, or at least for taping and then airing them on the overnight schedule. This case had three bases: (1) Special Orders presented scheduling problems; (2) they were not "part of legislative business"; and (3) some employees resented the fact that representatives were using the network "for free air time."[2] At least two Speakers of the House had privately supported elimination or taping.[3] Those defending airing Special Orders live pointed out that these speeches gave "the minority party the chance to speak on any issue." Since the majority party sets the agenda, Special Orders were important. Furthermore, these speeches sometimes "made news." Some of them were even "interesting." Taping and re-airing would not work because the speeches would have to be aired in their entirety or the network might be accused of "favoritism." Their unpredictable length "would make it difficult to guarantee that they would be shown gavel-to-gavel without making evening scheduling difficult."[4] As a focus group unto himself, Brian Lamb thought C-SPAN was "lucky that they exist." Discussing C-SPAN coverage of the House, he noted, "We want people to say that we do the best job possible."[5] It did not bother Lamb if members of Congress used C-SPAN. "I wish more members would use us," he said. "We're here to let members speak directly to their constituents, without being filtered by the media."[6] C-SPAN would continue gavel-to-gavel coverage, including Special Orders.

Getting Congress to abandon Special Orders would be like trying to persuade a grade-schooler not to go out to recess. C-SPAN would live with the programming problems that covering the House and the Senate live entailed. This coverage could mean fewer committee sessions carried and fewer special events. It could mean that the programming operations department would have to juggle schedules and that the assignment desk would have to make tougher decisions, but full coverage was C-SPAN's mission. That coverage would give rise to a number of myths about C-SPAN's programming philosophy and even its ownership.

A surprising number of viewers believe C-SPAN is funded by the government itself. When the Republican leadership decided to remove entirely, or at least decrease, funding for the Public Broadcasting Service, some callers worried about what would happen to C-SPAN without federal support. In part, their calls were a telling comment on the extent to which C-SPAN had informed the viewing country about who sponsored the network, and in part, the calls indicated how much viewers watched. For years C-SPAN had carried the promo "Funded Entirely by the Cable Television Industry." The belief that C-SPAN was funded by the government has led some people to assert that the network is the pawn of the powers that be.

The second myth is a variant of the first. It revolves not around sponsorship but rather around who owns and controls the cameras that bring House and Senate coverage—in short, who editorially selects the shots. This myth is promoted partly by carelessness and a lack of precision by politicians themselves and by representatives of the media. Speakers on the floor are forever referring to C-SPAN carriage as if the networks were responsible for the pictures shown. As late as 1994, Brian Lamb cited survey data indicating that "as many as half of the Senators" and "as many as two-thirds of the House members" did not know that the cameras were "controlled by government employees" under procedures "established" by the Senate Rules Committee and the Speaker's office.[7] Imprecision in the press fosters this misunderstanding. When Speaker Gingrich directed the House-controlled cameras to pan the floor in March 1995, a headline in the *Washington Post* ambiguously read, "Under Speaker's Order, C-SPAN Pans the House." Eight paragraphs into the article, reporter Howard Kurtz attempted to clarify the matter. "C-SPAN's cameras are controlled by Congress," he wrote, "not by the network." The confusion might not make much difference for many. If, however, the cameras really belonged to C-SPAN, why then did the network request the leadership of the 104th Congress to "allow C-SPAN cameras to cover House and Senate floor debates"? The answer is simple. At that time, congressional rules prohibited the government-owned cameras from panning the chambers unless the leadership ordered otherwise. No reaction shots were allowed. C-SPAN wanted to place its own cameras on the floors of both houses to cover debates in its own established style.[8]

A third myth has its basis in the way in which the House and the Senate have traditionally used the cameras. The constant focus on a speaker led many outside the organization, and some within C-SPAN, to believe that the shot selection was representative of the network's commitment to emphasize content over camera angles.[9] It was, in their minds, what differentiated C-SPAN from the commercial networks.

All of these myths create expectations about C-SPAN's programming and are sometimes used to evaluate its performance.

More Carriage to Provide Coverage: Airing Committee Hearings

With no control of the floor cameras, until the mid-1990s C-SPAN simply carried the House and Senate signals to America using scrawls across the bottom of the screen to identify the legislation under discussion and the speakers addressing it. Commitment to a long-form, whole-event, no-interpretation format led some, like veteran correspondent Daniel Schorr, to rule out C-SPAN as a journalistic enterprise. It did not cover events; that is, it did not place them in context and interpret them.[10] Almost as soon as the network was technically able to do so, it began to provide context in the broadest sense. In January 1981 it started gavel-to gavel coverage of congressional hearings. By 1983 it had experimented with a call-in show conducted from committee chambers immediately following a hearing.[11] Although the latter would prove impractical from a scheduling standpoint in the long run, committee hearings themselves did become a part of C-SPAN's programming.

C-SPAN Covers Committees

In July 1985, C-SPAN's Bruce Collins (left) and Mike Michaelson testify before the Senate Rules Committee, where they argued against granting commercial networks exclusive pool-coverage rights for committee hearings. (Photo courtesy of C-SPAN.)

In the beginning, televising committee hearings caused several problems. C-SPAN was "the new kid on the block," and the broadcast networks did not take kindly to it. "Some network personnel . . . tried to sabotage coverage by invoking union rules and intimidating young C-SPAN technicians," C-SPAN executive Mike Michaelson remembered. "They didn't think we were a bona fide news organization."[12] Committees with no breaking news and no star witnesses were not a problem, but when sessions promised to provide a feed for the evening news, the trouble began between sound bite and long form. Lighting the room and placing the cameras made a difference. C-SPAN staff members were viewed as replacement players. At best the network crews saw them as amateurs, at worst as scabs. Network crews would light the room, get the shots they wanted, turn off the lights, and leave C-SPAN cameras in the dark. They were known to move C-SPAN equipment, C-SPAN employee Garney Gary remembered.[13] In the halls outside the chamber, the space necessary for supporting equipment was limited. Like the big boys in the neighborhood, commercial network personnel assumed that what worked best for them was theirs. Often sitting cheek to jowl with the young C-SPAN employees, they would fan discontent about the new network's working conditions and pay scales. Those at the earliest hearings often could not recall specific incidents by place, time, or network, but they do remember the tension. Nor can they recall the turnaround day when network officials realized that a hearing clip that had originated with C-SPAN would do as well as one produced by their own crews and at less cost. But that day did arrive. Whereas C-SPAN previously had to fight for the right to carry hearings to provide context to the floor debates, the commercial networks now count on it to do so. The broadcast networks attend only high-profile hearings. Each day Ellen Schweiger, head of C-SPAN's assignment desk, gets calls from the commercial networks about what events C-SPAN will cover.[14]

The networks were, however, only part of the equation working against committee coverage. Under the rules in effect before the 104th Congress, committee chairs could prevent their sessions from being telecast. In fact it took a "unanimous consent agreement" by the committee members before C-SPAN could begin to cover proceedings. Objections ran from the heat of the lights to members "playing to the cameras" to the sensitivity of the hearings. Although coverage was usually unrestricted, as late as 1992 C-SPAN was barred from committee rooms four times between March and June. One of those incidents involved the editing and amending of a controversial cable-regulation bill. In each case, print journalists were allowed to follow the action. One hearing was in fact a press conference.[15] Although the 104th Congress opened up most committee sessions to C-SPAN, carriage still is not ensured because major committees often meet on the same day. With limited capacity to air the hearings because of commitment to gavel-to-gavel coverage of the floor, C-SPAN has few choices. Committee sessions often appear in the overnight schedule and on weekends.

From a programming standpoint, using committee hearings to provide context for floor debate can be a daunting task. In assignment desk meetings, participants have to choose which committees to follow and for how long and at the same time have to worry about balance. Hearings on agricultural subsidies are all-important in some sections of the country; they may not play well in Manhattan. The task for C-SPAN is to reflect what Congress does, not simply to meet the interests of some of its audience. By choosing which committees to cover, the network in effect makes an editorial judgment, one that it believes is guided by a desire to inform the audience. Should the network follow the money? If so, budget-related hearings are primary. Should it select hearings that better help viewers understand the entire legislative process? This question guided the network's choice to follow a House committee to a hearing in Montana in the spring of 1995, which showed that Congress does move outside the beltway. Every day those at the assignment desk have to decide. Is it better to follow the Senate or the House? When both bodies have committees examining the same subject concurrently, should the two committees be carried on both networks so that viewers can see the differences? Balance and fairness are the keys.

Using hearings to provide context for floor legislation by showing committees in action raised yet another concern. C-SPAN eschews oral commentary, but cameras comment by what they show and do not show. Like an empty chamber, a sparsely populated committee room sends a message, one that may easily be misinterpreted. For those who do not understand what a legislator's day might be like, an empty chair is a sign of failure to perform a job. What should the C-SPAN networks show? Two principles exist in tension at the network. First, carriage—by principle if not by written rule—has always been driven by content. In the literal sense, content implies focusing the camera on the committee member who is speaking or on the witness who is testifying and leaving it there. Content suggests few or no wide-angle shots, little if any panning, and no reaction shots. In short, it means covering committees just as the House and the Senate cover their floor debates. By a second principle, however, C-SPAN wants to show events as if the viewer were in the room. This principle calls for shots that establish the scene before proceedings begin—showing the entrances to the room, panning the dais, and visually taking viewers to their seats. Context in this sense favors wide-angle shots at the outset, shots that establish the scene. It calls for medium close-ups once the proceedings begin. The object is to follow what is being said and, at the same time, to convey the atmosphere of the room. As Camscam revealed, the selection of shots carries meaning. Shots can convey or take away power. Improperly employed, they can turn the meaningful into the meaningless and the unimportant into the important. They can take the drama out of the dramatic or make the mundane into first-rate theater.

Not surprisingly, C-SPAN uses its cameras with caution and conducts ongoing workshops for camera operators. The object at hearings and elsewhere is to respond to the query, "If interested viewers attended the session, what would they see?" An in-house document on directing hearings reads, "If something or someone in the audience becomes the temporary focus of attention, show it but don't over emphasize it." The document suggests that shots directing viewers away from a speaker should be "proportional to how much attention the majority of the people in the room give it." A hearing should never look like "an entertainment event."[16] For C-SPAN, the "best picture" takes on a different meaning than it does for commercial networks. However, camera angles that are as free from bias as possible do not necessarily make a presentation unbiased. Viewers who do not realize that Congress does many things at once, not one thing at a time, may see the room and determine, because of the absence of many committee members, that the legislators are not doing their job. C-SPAN's unwillingness to comment on the comings and goings, its silence, may in fact allow viewers to misunderstand and to misjudge their legislators. True to its no-commentary, long-form mission, C-SPAN would attempt to correct that impression in other ways and at other times, as we shall see.

"Fire bell in the night" hearings—such as Iran-Contra, the Keating Five, the Bork nomination, the Hill-Thomas clash, Waco, Whitewater, and Ruby Ridge—attract the major media, which often present these hearings almost in long form. The presence of the major media presents another challenge to C-SPAN's programming philosophy. The crowded conditions that multiple media outlets create often necessitate the use of pool cameras, which can wreak havoc with "the viewers in the room" concept. The footage of the Oliver North hearings has become a part of C-SPAN's training program on the bias of camera angles. The costs for the commercial networks to compete with C-SPAN can run high, however. When they devoted five days to Clarence Thomas's confirmation hearings in October 1991, "weekday and prime time preemptions collectively cost the networks between $15 million and $20 million in lost advertising revenue."[17] That amount would have more than paid for C-SPAN's operating expense budget for the entire year of 1991.[18] Since many of the hearings that C-SPAN brings its viewers, including ones on controversial subjects, occur while the House and the Senate are in session, the network cannot air them live, as the commercial networks do when they decide on coverage. C-SPAN's re-airing of those sessions actually makes them available to more people, who can thus see the action after work hours or on the weekends.

Beyond the Floor and the Committee Room

The puzzle pieces of how and why Congress works are scattered all over C-SPAN and C-SPANII. Program choices are made to help citizens understand their government and the issues that confront it. Viewers, however, have to be patient and have to work for an understanding. Rather than attempt to explain legislation, C-SPAN relies on call-in segments and on forums and conferences sponsored by academic institutions and think tanks. The latter gatherings help place issues and positions in a larger context. Because many forums represent a particular point of view, those charged with selecting programming constantly look for balance. A Republican-leaning discussion of federal mandates is sure, in time, to be followed by one expressing a liberal slant. Those at C-SPAN, of course, cannot choose a conference agenda; they can only recognize it and hope to counterbalance it in the future. C-SPAN programmers have to know sponsors and to cultivate those less skilled at getting their programming on the air.

Forums and conferences do, however, let viewers get a better understanding of the players, the issues, and the rules of politics. Especially before partisan groups, viewers can learn things they might never encounter elsewhere. With C-SPAN, they can stay at home and be inside the beltway for a Cato Institute forum one day and be at Harvard for a conference the next. Topics run the gamut. Viewers are not surprised to see a forum on "The Effect of Drought on Agriculture" or one on "Evaluating the Performance of Congressmen." They have the opportunity to take a seat at a discussion on "Budget Reduction and Its Implications" and at one on "United States–Vietnam Relations." Over a year, as many as one hundred such opportunities are available. As is always the case, not every event captures a large audience. Even some of the gatherings telecast are sparsely attended. As Peter Ross Range wrote, "An afternoon of C-SPAN can feel like a life sentence in the gulag of windowless Sheraton banquet rooms, surrounded by a punishment brigade of earnest men and women in dark suits." He concluded, "It's torture by white tablecloth."[19] Sophisticated meeting planners understand the rhythm of C-SPAN programming and have learned to sponsor events at slack times and to determine which speakers and topics are more likely to be aired. For viewers, it is another matter. Often they find the puzzle piece by luck alone.

To further place legislative debate in context, in the 1980s C-SPAN began programming and producing events that take viewers behind the scenes on Capitol Hill. *Inside Congress* is typical of such programming. Like most other C-SPAN programming, it is not host-driven but rather consists of a series of events in "video verité" designed to reveal an "insider's view of how the actual transition of power begins" at the start of each session. In 1994 it ran weekly from November 20 to the beginning of the 104th Congress. Viewers met several newly elected senators and representatives, attended a gathering of the freshman transition teams, and watched the office lottery for space on the Hill. They were guests when several members of Congress moved into new quarters. Viewers were able to follow a GOP group exploring the Star Warehouse, a property rented by the federal government at a cost of $600,000 to store out-of-date agriculture yearbooks. As the 103rd Congress was about to begin, they could follow Carrie Meek (D-FLA) into the office of Chair of the Rules Committee Joe Moakley (D-MA). There Meek sought a committee assignment normally reserved for more senior members. That same year Representative Peter Hoekstra (R-MI) opened his first staff meeting to C-SPAN cameras. Among the agenda items was picking an office and a parking space. The vignette got him "all kinds of calls from across the country." People, he found, were really interested. Neither he nor C-SPAN could know at the time that in 1995, Hoekstra would lead a task force on opening up Congress to television.[20] In his hands would rest the fate of a C-SPAN proposal for expanding public access.

Occasionally C-SPAN is able to follow a representative home to talk to his or her constituents. Such was the case with Billy Tauzin, a Democrat from Louisiana. There, before a public forum, he discussed the problem of being a conservative in a liberal party and publicly mused about switching affiliations. In the spring of 1995, Republicans dangled the nomination for the soon-to-be-vacant Senate seat before him if he would change parties. Tauzin finally switched parties, and C-SPAN covered the announcement. C-SPAN also allows viewers to learn more about process. It has aired everything from a seminar for Hill staffers on parliamentary procedure to an hour-and-half program focusing on the Rules Committee of the House. When the opportunity arose, as it did in March through May 1994, C-SPAN devoted its Monday evening call-in program to the appropriations process. Major item by major item, Appropriations Committee members, staffers, journalists, and students of the process answered viewers' questions. Now that the evening call-in has fallen victim to an energetic Congress, one can watch the *Washington Journal* seven days a week to pick up similar information. By focusing on process and policy, C-SPAN has created its own "no network commentary" style. The struggle to follow this style, through long-term, over-time programming, is daunting. The large mural that C-SPAN presents, however, paints a clearer picture than the miniature that commercial networks provide.

Humanizing Congress

The angry citizens of the 1990s are nothing if not vocal. For them, senators and representatives, though usually not those from their own states or districts, have lost touch with reality. In their eyes, lobbyists or a search for fame, or both, have turned legislators into prime subjects for a casting call for a remake of *The Night of the Living Dead*. From the beginning C-SPAN has tried to show the public that legislators are real people too. On October 12, 1982, the network made "television history" with a nine-hour special, "A Day in the Life of Congressman Richard Bolling" (D-MO). Soon to retire after thirty-four years of service, Bolling headed the all-powerful Rules Committee. A year later C-SPAN followed Minority Leader Robert Michel (R-IL) for a day. The shows were "contexting" at its most meticulous. Bolling appeared to carry his laundry into his office in what looked like a pillow case. Michel sang along with the radio as he drove to work. It would be difficult to sound-bite an audience for eight hours, and neither program did. Over the period of a day, the viewer came to see them as real people—and likable ones at that—just doing their job. The glitz and glamour, the pomp and posturing, had no place in a full workday. As the day wore on, both men seemed to forget the cameras. It was as close as most Americans can get to being in Congress without running for office and winning. Few citizens put eight straight hours into anything except work; fewer still had the time to watch. But those able to see a part of the day saw something very different about their legislators and the job the men undertook.

Shorter in length, and usually scheduled for holidays, the series *An American Profile* proved to be a more practical way to help citizens learn more about the men and women elected to serve them. An hour in length, and conducted in the C-SPAN style of interviewing, the profiles have a way of opening up people. They become conversations, not interviews. Although politicians seldom let their guard down completely, the profiles may come as close as is possible in a media age. At best the programs become conversations among friends; they have a level of intimacy that political advertising seeks but never achieves. In such cases the events and motives that caused the politician to appear are revealed not in policy pronouncements but rather in small anecdotes and remembrances. The talk ranges from how the guest became interested in politics to who influenced him or her when young, from general views on politics and society to what life was like when he or she was growing up. Small anecdotes often reveal how and why those profiled think and act as they do. Why is capital investment so important to Jack Kemp? Listen to him describe how his father and mother "took all of their savings, and everything they could scrape and borrow," to begin a delivery service and make it work. Hear how they closed down a motorcycle distributorship with declining sales and, in the process, traded a cycle for a truck. "I learned my first economic lesson from that trade," Kemp recalled. "You can deliver more packages in a truck than a motorcycle." He said pointedly, "The truck was a capital investment."[21] The tales can become even more personal. Lew Ketcham recalled talking with Kwame Mfume (D-MD) about the representative's formative years. Mfume detailed a life "headed down the road to prison," a path not taken because of a religious experience. In print the story loses its force; on the air it left Ketcham "slack jawed."[22]

To cynics the Kemp and Mfume tales sound like narratives delivered on the campaign trail. But there is something to the C-SPAN style. Before his appearance on *An American Profile*, Ben Bradlee, then executive editor of the *Washington Post*, thought: "Here we go again. Another interview." Almost halfway through he admitted, "We're doing questions that I haven't done before . . . generally you get someone who hasn't done much homework and who's got about five minutes to talk to you." Journalists usually hit the big issues, he asserted, and then "sayonara, they're out of here."[23] In long form, C-SPAN profiles help viewers understand legislators they have learned to hate. The profiles help tell America's story.

Long-Form "Contexting" and an Impatient Society

"This is the place where you can get it raw, and see for yourself," Brian Lamb once told a reporter. "Listen, I'm a classic 'C' student from Purdue University in Lafayette, Indiana," he said. "I'm not a brain surgeon. If you watch it, you can figure it out."[24] Life today, however, does not run on real time when it comes to politics. Real time rides on a horse and buggy; the speed of politics is measured in terms borrowed from physics. C-SPAN is one step at a time in a world of quantum leaps. Not even Brian Lamb can see all of the C-SPAN programming about Congress, let alone this and the ancillary programming too. The philosophy of C-SPAN, however, is simple. If people cannot see it all, at least the network will make it possible for them to see what they could not before. C-SPAN is painfully aware that "the network of record" still cannot show everything. One cannot imagine "A Day in the Life of a Lobbyist," for example. Those in Congress cannot see it all even with dozens of eyes provided by staff. Still, for C-SPAN, the more opportunity there is to see Congress and all of the actions that make Congress work, the better it is for the viewers.

Sometimes, simply seeing is not enough. One has to watch more often than just occasionally to figure out what takes place in Congress. Without understanding the rules of the game, it is difficult to judge the quality of the players. C-SPAN's programming formats provide a large context, perhaps too large for viewers in an age of channel-surfing. The network next turned its efforts to helping viewers understand process and procedure on the floor of the House and the Senate. No longer would some viewers feel that they had entered a movie that was already halfway over when they switched to a C-SPAN channel.

"CONTEXTING" CONGRESSIONAL COVERAGE: THE NEW C-SPAN

C-SPAN is wary of interpreting anything for anybody in any way. As 1994 drew to a close the network began to make plans for its programming of the House and the Senate to become more viewer-friendly. Rather than less information about legislation, it would provide more, and in a more timely fashion. The network seems to be continually conducting a self-study of its programming. An in-house series of focus groups, involving nearly a quarter of the company, met in the summer of 1989. Each of the groups mentioned the need to provide "more context for programming" as a high priority.[25] That same idea would appear again in the midcourse report of the committee evaluating programming plans for "C-SPAN 2000." One of the goals of its Core Product Committee was "to continually offer the viewer more context."[26]

In a memo to that group, Mike Michaelson put forward a number of ideas that had appeared over time in numerous earlier discussions both in committee and in the hallways. Pointing to a recent report sponsored by Centel Corporation, and based on a "lengthy discussion with members of Congress," Michaelson urged change. The report found that "a growing number of Americans were distressingly uninformed about the ways Congress and government work." C-SPAN, he believed, could help lessen that ignorance. Of the dozen suggestions he put forward, one-half concerned the way the network covered floor debates and committee hearings. Ten minutes before each body convened daily, he suggested, a C-SPAN representative should explain the upcoming legislative agenda. An on-air person would be more effective in doing so than simply an off-screen voiceover, he argued. Visibility added "more dramatic emphasis." During quorum calls, C-SPAN should interview legislators about the bill on the floor. It should also consider special programming that would trace a piece of legislation from inception through a committee markup and on to the floor. How the Rules Committee treated the legislation, how it was handled in the other body of Congress, and what its eventual fate was would all be part of the program. Another program, perhaps a call-in, could offer an in-depth explanation of the legislative process. Here congressional parliamentarians might be useful guests. Michaelson also made the case for C-SPAN to begin a variant of the old *Ev and Charlie Show*. To do so, the network would have to invite the leaders of both parties in both the House and the Senate to participate in a weekly discussion of legislation. Finally he urged the network to begin its own legislative week-in-review program, even if this deviated from C-SPAN's "original" format.[27]

The report of the C-SPAN 2000 Committee proved far less ambitious. It supported continual gavel-to-gavel coverage and carried forward two of Michaelson's suggestions in modified form. The group recommended the use of "an on-screen host" to provide a ten-minute summary of the legislative day before every meeting of Congress. It also favored the use of a host during quorum calls and votes. The host's task would be to interview those involved with the legislation. "This," they believed, "would be preferable to having the explanation all come from C-SPAN personnel."[28] The suggestions were hardly new. Writers and callers to the network had been making them for some time.

The line between providing context and offering comment would rule out C-SPAN personnel from explaining anything to anybody. "Lamb remains dead set against adding any commentary to C-SPAN's congressional coverage, because he believes that some slanting would inevitably creep in," Lee Winfrey told her *Philadelphia Inquirer* readers.[29] The network took the recommendations of the C-SPAN 2000 Committee under advisement and moved slowly and cautiously toward their implementation. But despite the 1989 focus group suggestions and those put forward by the C-SPAN 2000 report, the network looked much the same in 1994 as it had earlier, when it had first started carrying the House and Senate proceedings.

The January 1st Groups and the 104th Congress

Perhaps because of the effects of "must carry" legislation on viewership, perhaps because of the changing nature of technology, or perhaps because of some other reason, in the fall of 1994 C-SPAN began to worry more about visibility. As Lamb put it, the network would begin to compete "for the eyeball not the dollar."[30] For the third time in six years, C-SPAN took a serious look at its programming. Before the votes were tallied for the 104th Congress, the "January 1st" groups were in operation. Teams of five staff members each were selected from across the company to prepare recommendations for change. Ruth Kane, of in-house programming, headed the House group. She was joined by four others: Jim Mills, a producer hired to cover Congress; Maurice Haynes, from programming operations; Eloise Collingwood, of viewer services; and Joanne Wheeler, from education and marketing services. Mark Farkas, a field producer, would lead the four other members of the Senate group: Kristy Schantz, then a producer for the *Sunday Journal*; Chad Pergram, from in-house programming; Valerie Matthews, from programming operations; and Joe Kipikas, of marketing services. A third group would examine coverage of committees.

The two groups charged with floor coverage again recommended special C-SPAN programming before the beginning of the legislative day. They again suggested a weekend-magazine-style congressional almanac program. They encouraged promoting on-line services for more information about legislation under discussion. Both called for improved graphics. Once more they made suggestions for better use of the period devoted to quorum calls and voting.[31] Of the two reports, that of the Senate group read for the "eyeballs." It sought to provide what Senior Vice President Susan Swain called "KILLER APPLICATIONS," and it looked to the future.[32]

"There is a difference between *continuous* coverage and *never ending* coverage, particularly if the event you are covering looks the same from one day to the next," Farkas's group asserted. "Such is the case with our U.S. Senate coverage." There the issues might change, but "the visual imagery does not." The report cited a page of examples of ways to alter the imagery, the most radical of which called for squeezing the screen during the debate itself to provide information about the speaker and the history of the legislation under discussion. "Viewers will want to know who these people are, what the cost of legislation they are discussing is, where the appropriated money will go." In the case of something like the discussion of the California Desert Protection Act, the group proposed squeezing the screen to provide video of the sites affected. The suggestion was a radical one for a network committed, from the beginning, to "uninterrupted live gavel-to-gavel coverage."

When it came to ideas about how to handle quorum call programming, the group was on safer ground, but it was every bit as direct. "Everyone we canvassed in the company," it reported, "felt that the worst time to watch the U.S. Senate (and hence the most probable time for viewers to change the channels) was during quorum call time." The report noted, "From the classical music to the cryptic informational the Senate puts up on the screen to the deadness of the viewed activity, [the Quorum Call] is the desert of our Senate coverage." The group listed over a dozen suggestions about how to make that time period more useful and visually pleasing to viewers.

Like its counterpart House group, the Senate conferees called for more outside-the-beltway video to enhance coverage of Congress. As part of its proposed *Weekend Capitol Re-Cap Show*, the group suggested using the C-SPAN school bus as a production vehicle. The crews should bring "the man on the street" into the studio; doing so would make issues not only more "contextual but *visual* as well." The report noted, "By producing such a show as this, we would fill a great void here at our network: lack of outside the beltway, people without coats and ties, viewpoints on issues." Call-ins might also do this, but these shows were visually limiting. The group turned again to the California Desert Protection Act of 1994 as an example. "Here was a continuous piece of legislation debated and passed in the Senate. Dianne Feinstein's political life depended on its passage." What did C-SPAN viewers see? Floor coverage. What else did they hear? "Possibly some call-in shows focusing on it." The group claimed, "Beyond talk, they received no contextual information." Why not make use of cameras available in California? Let viewers hear from and see those affected by the legislation "in the desert." Show "real people, living with the real impact of Washington

D.C." The committee asserted: "The visual impact will draw viewers in. The discussion, if good, will keep them there."

Finally the Senate group suggested that C-SPAN begin, in whatever ways possible, to enter into the interactive age that new technologies would bring. C-SPAN should promote on-line services, for example. It should prepare for the day when callers too could be seen on-screen, when viewers, by splitting screens, could monitor more than one program, when computers and televisions would be used together. Interactive programming could get C-SPAN outside the beltway, away from a "Washington look and perspective." The group added that the network should use the resources it already had to do so.[33]

Reacting in Slow Time: C-SPAN Responds to the Reports

As is their style, C-SPAN officials moved cautiously on the January 1st reports. Some suggestions would be modified and moved into the format of the *Washington Journal*, planned to begin in the first month of 1995. A few would be met in part by actions of the 104th Congress. More would be integrated in a new look for House coverage that would start when the new Congress met. Those suggestions that contained the possibility of commentary or that appeared to take away from uninterrupted coverage of the floor met with opposition from staffers born and bred in the C-SPAN tradition. Lamb was on public record on the matter. C-SPAN would never try to "liven things up by editing or adding production values, even if it had the money."[34] Internal discussions were tempered by what was, for C-SPAN, a bold proposal, the suggestion to congressional leaders that the network take control of floor coverage. (See chapter 9).

C-SPAN would not wait for Congress. It would make do with what it had, both in terms of the new opportunities open to it and in terms of the limited resources on hand. Innovation in the programming of Senate sessions would have to wait until summer; C-SPAN began with the House. Hired to be C-SPAN's Capitol Hill producer, Jim Mills was loaned to programming operations because C-SPAN needed a "righthand man" on the Hill to implement changes. Paul Orgel was temporarily assigned the task of handling House operations back at 400 North Capitol.[35] "On loan" and "temporary" seem to be common words in the C-SPAN vocabulary. The two men, plus a one-camera crew, were charged with changing how C-SPAN covered the House floor debates. From a world in which classical music was queued up, ready to play during a vote or a quorum call, C-SPAN now filled part of that time with its own programming. In some ways, the situation was like Crystal City again. Mills could get the House feed only at his booth in the press gallery. There was no computer linkup to a C-SPAN database. Mills could not access *Congressional Quarterly* on-line, for example, for information. And he could not pick up available audio feeds from his desk.

Long-form C-SPAN had covered the action whole; it was never forced to react. Now C-SPAN programmed for breaks that did not occur on schedule. Gavel-to-gavel coverage remained, but the silences it provided were filled. Reacting, even without C-SPAN commentary, would prove to be a different world. Orgel and Mills talked to each other by cellular phone "forty to fifty times a day," floating different scenarios for the breaks in action. Mills, whose background was in radio, was on "a learning curve," trying to develop journalistic and legislative contacts for interviews when "the bell rang" and legislators were called back to the floor. The most innovative ideas for the uses of boxes and split screens would have to wait until later. Given the resources available, C-SPAN, in Susan Swain's words, "would have to crawl before it walked."

On a day-to-day basis C-SPAN was more accustomed to booking guests well in advance, at scheduled times. Occasionally, filling a slot on the *Washington Journal* could go down to the wire, but obtaining guests for breaks in legislative action would accelerate the pace. Reporters on the Hill had stories to file, deadlines to meet, and interviews of their own to conduct. Quorum calls proved an easier time to lure them to the cameras than during a voting break. C-SPAN had to develop a series of "evergreen" interviews, re-airable and already prepared talks with experts on procedural matters. These would explain how the amendment process worked or what was needed to get a rule for a bill. Members of the Congressional Research Service proved invaluable here. C-SPAN programmers also had to learn the ebb and flow of legislation. Which bills would bring which reporters to the press gallery? Who were the experts on specific legislation? Was someone from *Bloomberg Business News* more likely to cogently explain tort reform than a *Washington Times* reporter who had the Hill as his or her beat? Mills would roam the press gallery, asking reporters the classic question, "What are you working on?" Who could provide the information that C-SPAN viewers needed? "A lot of what I do," Mills said, "is evangelism work." He tried "to woo people into wanting to help [the] viewers." For him the search for potential interviewees was like "playing without the ball."

Members of Congress have been cooperative for spot interviews. Mills learned early that scheduling their appearances through press secretaries was a time-consuming task. It proved easier to pick them off on their way to a quorum call or a vote or to reach them by a message to the floor. The more interested and involved the member was in a piece of legislation, the more knowledgeable the interviewee was likely to be and the more partisan. Though short-term balance was a worry, the long term was a greater concern. Here too C-SPAN keeps records to guide future actions. In the short run, the network uses several strategies. If, for example, C-SPAN decided on a live interview of someone favoring a piece of legislation, Orgel would search the debate covered for a clip of a member speaking in opposition. When interviewees became too partisan on the air, Mills would remind the audience that someone on the other side had appeared earlier. Throughout the process, those providing Hill coverage looked for material that might also air on the *Washington Journal*.

Mills preferred live interviews using what he called "Kodak moment" questions, but one can never be sure of getting a live interview when needed. To supplement the "evergreen" clips available, he interviewed legislators and members of the press to have the clips on hand to air during a break in action. They might be used in ten minutes or in three hours depending on the flow of legislation. Such interviews were less pressure filled. They did not require a phone call to the programming operations department while Mills walked a legislator to the camera. No thirty-second warning was necessary. They did, however, present difficulties. Those appearing had to be reminded that the conversation would be played later. A "just five minutes ago" comment would not work well if the interview was played two hours later.

It may not appear that what viewers finally see is a product of work weeks that can run sixty to eighty hours. C-SPAN is on hand while the House is in legislative session. House coverage now employs a rapid-response team to give viewers more information. The process looks simple. When a vote begins on an amendment, for example, the tally will be squeezed down in size for a part of the count and a box will appear in which an interview will be shown. Typically, when the vote began on an amendment to a Coast Guard funding bill, viewers first saw the vote totals alone. Within three minutes, the totals were reduced in size and an "evergreen" interview with Walter Oleszek, of the Congressional Research Service, appeared on-screen. He explained the amendment process used in the House. Next, the House feed of the voting took over the whole screen. Within minutes the picture was compressed again, to show a live interview with Elizabeth Parlan, of the *Congressional Monitor*. Off-camera as always, in C-SPAN style, Mills asked her why so few members had voted at that point. The answer would help viewers understand that committees were meeting at the same time. That same day, for example, budget discussions were taking place. Mills asked Parlan, "What other stories are you following?" Viewers learned that a wetlands bill would be up the next day but that the budget was the big story. The interview ended, and once again the voting tally consumed the whole screen.

The additional context provided by the new House coverage may seem minor to many, but it involved hours of discussion. The imaginative suggestions of task force members were treated conservatively by C-SPAN's management in creating the new format. Time was spent worrying about how to maintain balance, about how to avoid commentary on C-SPAN's part, and about how to selectively use open phones during the breaks. Should viewers be heard then? Some worried that the short interviews were a move away from the long form and toward sound bites. Others watched and attempted to judge what the product looked like. By April 1995, C-SPAN would begin the search for producers to bring a similar format to Senate coverage. In the summer Paul Orgel was placed in charge of both House and Senate coverage, and Chad Pergram was assigned to produce Senate coverage. Just how effective were the changes in providing viewers with more pertinent information? C-SPAN did not know; it could only listen to callers and read viewers' letters. No out-of-house focus groups were set up to measure impact. Few surveys were commissioned.

A NETWORK OF RECORD

In bringing gavel-to-gavel coverage and the ancillary programming that supported it, C-SPAN hoped Americans would see the action in Congress for themselves and not simply base their judgments on what others told or showed them. Starting first with floor coverage, it provided viewers with a grand national civics lesson with no editing, commentary, or analysis. To balance points of view, C-SPAN relied both on long-form coverage of related congressional events and on call-in programming. To give viewers more information, it profiled Capitol Hill leaders and showed the day-to-day routine followed by legislators. It aired programming explaining legislative process. In an age of acceleration and sensationalism of the news, a time of star journalism, C-SPAN slowed events down to real time, attempted to place them in context, and then stayed out of the way.

One way to assess what C-SPAN has given Americans is to think about their access to information before its founding. They could read the *Congressional Record* after the fact, if they could find it. They could trust the data the media condensed and interpreted for them. They could believe the newsletters franked home to them by those they elected. They could listen to special-interest groups who claimed to speak for them. They could travel to Washington, D.C., and observe for themselves. They could rely on what they had learned in high school and college government courses, if they remembered it, to figure out some meaning. C-SPAN added another choice.

In time C-SPAN would come to call itself "the network of record," in large part because of its coverage of Congress. Of course that record is not complete, and C-SPAN knows this. Should the House and the Senate comply with the network's requests for camera access to the floor debates, the record will be more complete. A proposed third channel, likely to be initially designated for more committee coverage, is waiting in the wings. The network may not have opened all the closed doors on the Hill, but since its inception in 1979, it has cracked open quite a few. In covering Congress, C-SPAN has come a long way since it began as a one-room operation; in the process, it has brought the country closer to seeing what Congress actually does.

The Second Strand: America Talking

Barely eighteen months after C-SPAN embarked on its first mission—to bring Washington to America, alive and unedited—the network undertook the second half of the job. For the first time, on October 7, 1980, it brought the people's voice to Washington. Viewers would have a chance to talk to policymakers. Still located in cramped quarters in Arlington, Virginia, the network hoped to piggyback a call-in format onto a speech carried live from the National Press Club. For ten months C-SPAN had provided citizens from across the country a seat there. Now, using the George Mason University microwave transmitter, cameras provided by Forrest Boyd of the International Media Service, a small back room and phones belonging to the National Press Club, and the goodwill of the featured speaker and four guests, C-SPAN attempted to begin a national dialogue.

The twenty-year-old chief engineer, Richard Fleeson, patched together what he had begged and borrowed. In terms of television, it was a crystal-set operation. Staffers even used blue National Press Club tablecloths as part of the backdrop. Whether anyone would watch or even call was problematic. The program had not been promoted, and viewers would have to pay long-distance rates to be heard. Still the chance was there. The previous Thanksgiving, on the spur of the moment, during the airing of a speech by Henry Kissinger before the Republican Governors' Association in Texas, C-SPAN had flashed its Arlington number on the screen, requesting viewer comments. It had received over thirty calls in a matter of minutes.[36] Still, the featured speaker this time, Charles Farris, the chair of the Federal Communications Commission, was no Henry Kissinger. Federal communications policy could not rival foreign affairs for attention. As C-SPAN staff readied the set, Farris waited patiently to be interviewed. In the wings were Don West of *Broadcasting* magazine, *Cablevision's* Pat Gushman, Dawson "Tack" Nail of *Television Digest*, and Professor Mike Kelly of George Mason University, a board member of the Corporation for Public Broadcasting. The four were scheduled to follow Farris to answer questions about the issues that had been raised. Even if the phones did not ring, an extended roundtable discussion provided a fallback position.

Before the program could begin, the two cameras had to be relocated from the ballroom of the club to the East Lounge. The inexperienced crew untangled cables and switched outlets as rapidly as they could. With minutes to go, the producer, Gail Picker, "a seasoned Capitol Hill hand with little TV production experience," was ready to begin, but when the set lights were turned on, all power promptly went off. After a frantic struggle to jury-rig the equipment, the power came back on. Next several studio lights failed. Again the crew scrambled to make things work. With the lights finally back on, Brian Lamb began the interview with Commissioner Farris. Given the situation, the interview went well. The four guests then took Farris's place at the table. After a short discussion, they were prepared to answer viewer calls. Picker changed hats and became the phone operator. Things looked promising; all three phone lines quickly lit up. The experiment, it seemed, would work. Unfortunately, however, the hastily patched-together phone system was too close to the set. The first call produced horrible feedback and a garbled message. The staff struggled to repatch the phones and move them farther from the set. The system ended up far enough away to eliminate the feedback but was still too close to the set. Picker's voice could be heard on the air. Finally from Yankton, South Dakota, came the first intelligible call. Americans listening heard a phrase still regularly repeated today: "Thank you for C-SPAN."[37]

First C-SPAN Call-in Program

Guests at C-SPAN's first call-in program, broadcast from the National Press Club on October 7, 1980, are pictured with Brian Lamb. *From left to right:* Michael Kelly, then a board member of the Corporation for Public Broadcasting; Pat Gushman of *Cablevision* magazine; Don West of *Broadcasting* magazine; Dawson "Tack" Nail of *Television Digest;* Lamb. (Photo courtesy of C-SPAN.)

Brian Lockman, later a C-SPAN vice-president, had done everything that evening from preparing the set to moving equipment, solving crises, and directing the cameras. For him the production was "a nightmare." The relatively short time on the air "seemed like an eternity." For Brian Lamb it was "the biggest roll of the dice C-SPAN had taken in its brief history, and it paid off." Lockman marveled, "There were really people watching, [and] they knew their stuff."[38] The format would become one of the network's trademarks. Within four years C-SPAN was running three call-in shows daily. The call-in format would also be used after major events aired on the network. By the end of the decade, over fifteen thousand citizens a year had taken advantage of the opportunity to talk with politicians, policymakers, and journalists.

In the early days, however, C-SPAN was very much a low-budget, shoestring operation, a fact that occasionally showed on the call-in programming. Once, a camera "literally fell apart" in the cameraperson's arms, with ten minutes left in the program. Three days after the federal ruling that broke up AT&T's phone monopoly, C-SPAN invited three representatives of the company's new competitors to discuss the implications of the ruling. As the host took the first call, the phone lines went dead, remaining that way for forty-five minutes. Fifteen minutes into another call-in, a studio light exploded. The host recalled "a loud boom, a bright flash"; the set was "plastered with flying glass." With a red-hot coil simmering on the table in front of him, Representative Tom Downey (D-NY) calmly turned to the host and said, "This is exactly why I have introduced the Safe Television Act of 1985!" Other, less traumatic events also made programming difficult. Broken air conditioners produced sweating guests under heated lights. "Live call-ins" in the early years "were still somewhat experimental and looked it," Brent Betsill remembered.[39]

When the equipment was not a problem, getting guests could be. Pam Fleming, now vice-president for marketing, came on board as a producer in 1981. One of her many duties was booking guests. Since the network had little name recognition on the Hill, that could be a problem. Fortunately for C-SPAN, some members in the House, such as Jim Wright (D-TX), and others in the Senate, such as Paul Simon (D-IL), understood what the network was attempting to do. Their willingness to appear on the network helped. Booking guests was also made easier by the arrival of Mike Michaelson, who joined C-SPAN at about the same time that Fleming did; he was hired to serve as second in command and run the network's day-to-day operations. His thirty years in the House gallery gave the network name recognition and an expanded Rolodex.

From the outset C-SPAN concentrated on two things, the guests and the callers. Occasionally guests could be a problem. Bruce Collins, now legal counsel and veteran host, remembered the congressman who had to go to the bathroom midprogram. After slipping Collins a note to that effect, he misread the host's expression, "got up, undid his microphone, waved good-bye, and walked off the set." During the five minutes until he returned, Collins killed time by encouraging a caller to go into great detail about the origins of his own political point of view. Connie Brod too had a disappearing guest, an expert on national transportation policy. During the program he suddenly seemed to lose energy, turned pale, and began to stare at the table. Then came the note, "I need to stand away and take an allergy pill." She remarked, "Sometimes you're thankful for verbose callers." Luckily the caller on the line was determined to go into great detail about the California transportation infrastructure, and she let him. Just as the recovering guest returned to the table from the anteroom where he had been listening, the caller said, "So what does your guest have to say to that?" The guest replied smoothly, and the program continued.[40]

There are other stories: guests who had a drink too many, beepers that went off midprogram and summoned the guest back to the Hill for a vote. In another instance, a representative of the League of Women Voters was invited to talk about the presidential debate that her organization had sponsored. Five minutes into the program, she admitted that "she didn't get a chance to watch it." Technological problems still occurred on rare occasions. The international access code was not listed on the screen during a call-in viewed by satellite in Europe during the 1988 Democratic Convention. Callers who forgot to use the code reached a couple in Holland, who eventually had to take their phone off the hook.[41]

The early stories reveal much about the network, undercapitalized by mainstream media standards and understaffed but dedicated to a mission. The stories about call-in glitches make one point but miss a more important one. Despite the technological difficulties of the early years, and the continuing foibles of guests, the format has brought Americans important and sometimes vital information that is not easily accessed by the average citizen. From the outset C-SPAN sought "to provide the audience, through the call-in program, direct access to elected officials, other decision makers and journalists on a frequent and open basis."[42] In doing so, it has frequently done more. *America's Town Hall*, an account of the experiences of 104 viewers, captures the American spirit. So too did the calls on the evening of the 1983 "March on Washington." Sometimes callers reflect "a national catharsis," as they did the evening the Persian Gulf War started: C-SPAN "opened the phones all night and got hundreds of calls." The questions addressed to Charles Keating, during the savings-and-loan crisis, are a testament to the financial savvy of the country.[43] Through the angry, the poignant, and the pointed, Walt Whitman might still hear America singing—on C-SPAN.

TALK AND THE MISSION

The call-in format rests at the heart of the C-SPAN mission. The philosophy that engendered it reveals an attitude about the capacity and the capability of the average citizen. The risk that C-SPAN takes is a risk born in the fundamental nature of our society. The network will not "stand between the viewer and the event, interrupting, dissecting or clarifying."[44] Viewers are trusted to judge for themselves. The faith that empowers that view belongs to the nineteenth century of Ralph Waldo Emerson and Theodore Parker. For them the true national literature of the country was declamatory. Our willingness to talk and to talk freely set us apart from other nations. In that talk, Parker noted, "many things come disguised as truths." But he claimed, "By and by their bray discovers them."[45] By providing a conduit for as many voices as possible, each respected for what was said and not for who said it, C-SPAN would become a source for democratic life. Put simply, C-SPAN would present as much of what was happening in Washington and across the country as its television cameras could capture. In doing so, it would provide average citizens a chance to question policymakers. The network hoped that Americans could then make more informed choices and that guests and viewers could "learn from each other."[46]

The philosophy and faith embodied in the mission led to a set of principles that would govern how talk would take place on C-SPAN. Key to the process would be the host, whose job it was to promote "informed discussion of issues between guests and callers."[47] From the beginning, hosts were to stay out of the way of the dialogue. Every call-in program began with the camera focused on the guest. Instead of "Hi, welcome to the Larry King Show," for example, C-SPAN hosts start with a question for the guest. The host's job is not to give opinions; he or she should not even appear to have any. Calls would not be screened, nor would they be managed in any way that might promote one point of view over another. To maximize the number of voices heard, the network instituted a thirty-day call-in policy, prohibiting more than one call per viewer during that period. In typical C-SPAN fashion, viewers were to place themselves on the honor system. Care would be taken to ensure balance over time through the selection of guests representing a wide range of opinions. Callers would have a cafeteria of ideas from which to choose. Through the interview process in a call-in show, the host would ask questions designed to clarify what was at issue, to define key concepts and terms, and to establish the credentials of the guests. Many of the early call-in programs were devoted to a single issue. Whatever the subject, the host's job was to help viewers more fully understand, to offer them an opportunity to learn, and to establish a reputation for fairness. The manner in which the host conducted interviews was designed to set the tone for the discussion that followed.

The reputation for fairness is an extremely fragile one. By the early 1990s, what had become a preferred way of talking for hosts had been codified in several house documents. John Splaine, a professor of education at the University of Maryland, had been brought in as a consultant to ensure that the C-SPAN way of questioning was followed consistently. Splaine had first come to the network's attention because he frequently contacted Brian Lamb whenever he thought he detected bias on the network. The shelves of Splaine's C-SPAN office are filled with videotapes bearing titles such as, "Susan Not Asking an Open-Ended Question." He conducts seminars for hosts and staff on the questioning process.[48]

The single-sheet in-house documents, which Splaine helped to prepare, bear titles such as, "A Summary of Guidelines for Formulating Effective Questions for Call-In Programs" and "Tips for Conducting Interviews 'C-SPAN Style.'" That style ensures that there will be "no C-SPAN stars" or media celebrities. No show would bear the host's name, and if the show was well done, any moderator could substitute for another without a noticeable difference. "We don't allow our hosts to appear at a regular time" on the call-in programs, Lamb told an interviewer. "You never hear me say, 'I'll be back with you on Monday.' You won't have the slightest idea who is going to be hosting on Monday. What we want the audience to do is to watch and hear the guests and the callers."[49]

The hosts are there to facilitate the passage of information and no more. Since they are trained to avoid the appearance of clubbiness, they never address a guest by his or her first name. Distance and decorum are important. Hosts are advised never to "try to impress the audience" with their own knowledge. They are told never to set up questions with their own theories. When they ask about theories or opinions, they always reference the statements to "a creditable source." Instead of asking, "Don't you agree with me?" a host might say, "An editorial in the *Wall Street Journal* asserts . . ." Hosts are taught to never answer a question from a caller unless it is about C-SPAN. They will often pass off comments about the network with a simple "Thank you for your opinion." Any indication of political preference expressed verbally or nonverbally is out of order. So good are they at this that one caller addressed his comments to Brian Lamb by calling him "O Great Poker Face!" The hosts' job is to help inform the people and to help the people to inform their guests. The camera shots aid in downplaying the host as a participant in that dialogue. When clips from C-SPAN programming are used to invite discussion, one is unlikely to see the interviewer. Certainly the viewer will never hear network patter like, "Recently our Steve Scully caught up with Senator Dole." C-SPAN is a no-names operation.

Striving for anonymity during call-ins, the hosts help viewers understand issues in context not by providing summaries of events but rather, through the kinds of questions they ask, by inviting the guests to do so. No commercial network newsperson would ask a senator, "Just what does GATT stand for?" Or, "How does a conference committee work?" C-SPAN hosts can be counted on to do so. The audience might not know the answers; someone might need the information. Just as they attempt to help callers get a better understanding of the issues, so too do the hosts try to provide relevant information about the guests. Again they do so by asking questions. Viewers quickly learn, if they listen, that those inside the beltway are just as likely to have come from middle America as they are to have been born, bred, and trained either in New York or in the nation's capital.

From the beginning, C-SPAN's on-air style was conversational, not confrontational. Some saw the style as a sign of fairness and of respect, perhaps even as a model for callers' behavior. The tone provided guests with an opportunity to explain, defend, and amplify positions. Open-ended questioning allowed them to get beyond sound-bite responses and boiler-plate answers. Some at C-SPAN believe that the gentler question and the longer time allowed for an answer might actually lower a guest's defenses and reveal information beyond the pat answer. The interest in a guest's career path and background might also help viewers get a better feel for the guest's actions or positions on issues. Unlike commercial network public affairs programs on Sunday mornings, no host would ever ask a "gotcha" question; no host would pit his or her own personality against that of a guest. Dialogue was all-important. Information counted far more than points scored. Discussion between guests and callers was the name of the game. Success was measured by what was learned, not by the increasing visibility of the host.

No style, however, is without its costs. The C-SPAN way meant that hosts would sit by while misinformation, disinformation, error, and sometimes outright untruths were mixed in with the information necessary for viewers to begin to make informed decisions. Hosts let it all pass by, unfiltered, in the hope that subsequent callers would be sagacious enough to correct the error. They themselves, in C-SPAN's tortoise-over-hare approach, knew that in time every spin would eventually be reversed on the air. Viewers who watched and who read enough would recognize the seesaw of politics and would discern the balance point. This would, however, require patience, patience to see and hear enough, and many people seemed to prefer microwavable political talk. Of course astute callers, if they could get through, had the chance to gain clarifications or correct errors. Still, few Sam Donaldsons live in middle America.

Few guests get away with "using" C-SPAN call-ins. Whereas major politicians and political pundits may have become comfortable with alternative media formulas and formats, self-serving answers are easily recognizable. Some guests, Brian Lamb believes, are "their own worst enemies." He explained, "When they give a boiler-plate-cassette-in-the-head answer to a political question, it's no wonder the audience glazes over . . . in the long run it is one of the reasons why the audience, the American people, are so angry . . . just tell it to me straight."[50] Even though the talk on C-SPAN is unfiltered in the strict sense, a kind of filter does exist. Viewers interpret and respond to guests. The best guests respond, in turn, by attempting to understand perspectives other than their own. They try to adjust ideas to people and people to ideas. They listen as well as talk. In an unusual way, performance on a call-in reveals character and character traits.

C-SPAN, as the call-in format reveals, is a different kind of journalism. Its hosts know that politicians at all levels want to deliver unfiltered messages, messages that often serve the politicians themselves as much as, or more than, they serve the viewers being courted. They realize, as NBC's Tim Russert pointed out, that in 1992, George Bush "did sixteen five minute local TV interviews in one day." The "it's great having you here . . . how's the weather today . . . what's on your schedule today, and how are you doing" formats may have promoted both Bush and the local channel, but they did "little to inform America." They know too that by early 1995, Bill Clinton began to favor New Hampshire media outlets. The commercial networks' "gotcha" questions provide a voice different from C-SPAN's, in order "to get an honest answer to a real question about real problems affecting real people."[51] Their attempts to point out discrepancies and inconsistencies are seen by many as a sign of a left-leaning antenna, especially in a period in which talk radio has become a substitute for news. Both the networks and talk radio can be heard; what often is lost is the voice they interrupt and the opportunity to respond to it. Politicians serve as punching bags on the evening news; they function as strawmen on the radio airwaves. C-SPAN hosts simply will not confront guests as a matter of policy. Hosts provide balance by party, by news medium, and even by geographic area when possible and have the statistics to prove they have done so. Their approach and that of those at the commercial networks both have costs.

THOSE ANGRY CALLERS

If C-SPAN hosts will not challenge politicians, policymakers, or journalists, neither will they take on their callers. They respect their viewers as much as they do their guests. The closest a host comes to a rebuke is likely to be, "Where did you get that information?" One should remember that in the early years, "there was a mystique about the call-in . . . period." Lamb recalled, "People didn't care who was on, it was a call-in show."[52] Bruce Collins described the early callers "as delightful, curious; they knew what they didn't know."[53] The callers were a source of energy for a fledgling network. Jana Fay remembered that in the beginning, when Brian Lamb was frustrated about the future of C-SPAN, he would say, "We are going on air; I'm going to talk to the viewers." She noted, "He would sit there for three hours and his whole mood would change; he would suddenly be recharged and happy again." Lamb liked "talking to people" and he liked "finding out about them," she claimed.[54] He still does; their voices are an important element of the mix.

From the very beginning, the "thank God for C-SPAN" callers did not always dominate the lines. Those seeking information were often matched by those who were angry because they thought they had detected bias. Increasingly, a part of America seems to be "talking itself into anarchy."[55] Bruce Collins believes their outcries are a sign of societal change. "A lot of people," he claimed, "are consumers of a new form of information, Limbaugh-Donahue infotainment." The callers of the mid-1990s, he finds, are often "meaner and less informed."[56] When they do reach the network, they sometimes seem to have already made up their mind; they are upset by any fact or opinion that runs counter to their belief system. A position that differs with theirs is, by definition, biased. The anger, anxiety, and angst that are part of some talk radio has found its way to C-SPAN. A direct-redial phone and some patience are needed to get on the air to express it. Those who cannot get through use faxes or bombard the viewer services department, claiming the network has sold out to one side or the other. For some the deck is always stacked against them. A typical unlucky caller complained, "I identified myself as a Democrat and got cut off." Another believed that her call was rejected because she was "an African-American female living in a housing project." A third asserted that C-SPAN's lack of an 800 number was "an infringement on . . . democratic rights."[57] Discontent and suspicion bear no party label.

Asked about the quality of discourse engendered by its most strident callers, Brian Lamb pointed to Michael Kinsley's February 6, 1995, essay in the *New Yorker* magazine. Kinsley wrote, "Populism in its latest manifestation celebrates ignorant opinion and undifferentiated rage." He asserted, "As long as you are mad as hell and aren't going to take it anymore, no one will inquire very closely into what exactly 'it' is that makes you angry and whether you really ought to feel that way."[58] Reflecting on a class of callers, Lamb remarked: "What you find out is, time and time again, day after day, how little people know. How little they pay attention." He said: "Even our junkies miss it by a wide margin. They fax and call-in and say things that are so wrong that you really wonder, and I'm talking about the basics. You wonder who is out there that knows." When callers do get a point, he continued, sometimes they cannot let it go. "I love journalism and I love those questions," he said by way of example. But when callers play the same record over and over again about bias, he gets frustrated. The callers, he asserts, may be right, but he wonders why they cannot "accept it and go on from there." Having "discovered an important fact," they ought to learn from it, and factor it in when they read, but they don't. Lamb noted, "They still want to complain."[59] For some, it appears, the noun "journalism" does not exist without the adjective "liberal" placed in front of it. In some minds, to invite journalists on the air, even when they come from papers with long conservative traditions, is to take a liberal stance. Lamb is the lightning rod for such talk. For Lamb and others, the criticism is less frustrating than the lack of learning.

Despite the fact that some callers do not appear to be open to any ideas other than their own, that some whine, and that others seem not to know what to do with what they have learned, the call-in format plays an essential role in the C-SPAN mission. Over the past ten years approximately 150,000 Americans have been heard on C-SPAN. With the advent of e-mail and faxes, another 400 a month now have their opinions aired. Another 35,000 a year use the viewer services department to express their sentiments.[60] The number of calls, faxes, and letters ebbs and flows depending on the issues, but there are always people who want to be heard. Although C-SPAN has no record of the number of viewers who attempt to call but are unable to get through, the open-phone segment following the 1988 debate between Lloyd Bentsen and Dan Quayle is an indicator of the emotions that hot-button issues can illicit. AT&T had its circuits jammed by 26,000 callers trying to reach C-SPAN immediately after the exchange.[61] Callers, however, represent only 3 percent of the network viewership. Even though their opinions are an essential part of the programming mix, the network programs for the other 97 percent.

C-SPAN's call-in format is different, and it has made a difference. Unlike other hosts of such programs, C-SPAN hosts are not "agent provocateur[s]." They are not antagonists who "excite" callers "to greater extremes to elicit their deepest prejudices."[62] They do not invite public accolade, nor do they pass themselves off as experts or even entertainers. They are not in the business to make points; success is measured by the quality of the dialogue and by the amount of information provided. To understand the difference C-SPAN makes, imagine the *Washington Journal* hosted by Larry King, Rush Limbaugh, or Sam Donaldson. Think about what the program would look like in the hands of Ricki Lake, Geraldo Rivera, or Oprah Winfrey.

The *Washington Journal*

When C-SPAN began its call-in format in 1980, it was the only television network offering viewers a chance to talk directly to elected officials, policymakers, and media reporters. Although the format evolved over time, its pace was always leisurely by conventional media standards, its conversations were mostly civil, and its talk was issue-oriented. By the 1990s the network had abandoned both its 6:00 and its 9:00 P.M. call-ins because of scheduling problems. The lengthy House sessions frequently caused the cancellation of the early-evening version. The morning show, an hour and a half in length, provided the only consistent outlet for viewers to be heard on the air.

Once its sole domain, the call-in format no longer belonged to C-SPAN. The major broadcast networks, CNN, and newer cable outlets like CNBC, Court TV, Talk News Television, and others had all adopted the format. The age of "talk-show democracy" had arrived. Larry King introduced Ross Perot to America. Bill Clinton reached a different niche audience by appearing on MTV. Rush Limbaugh became a hero to "new wave" Republicans. Talk radio emerged as a major player in American politics. In a decade, news talk-radio stations grew from 250 in number to more than 1,000. Limbaugh commanded 20 million listeners on 660 radio stations. His television show appeared in 220 markets. His books made the best-seller lists.[63] "Sideshow" television, the daytime talk shows, occasionally entered the world of politics. Through the prism they provided, the conventional aphorism "all politics is local" became "all politics is personal."

By the spring of 1992 the Core Product Committee for C-SPAN 2000 recommended a change in the network's call-in format. It suggested expanding the morning program to three hours. The program should contain at least three segments, one emphasizing "Today on the Hill," another featuring a journalist, and a third focusing on a member of Congress.[64] As always, network officials moved cautiously on the recommendation. Finally on October 31, 1994, Susan Swain and Terry Murphy announced "the first substantive changes" in C-SPAN's call-in format in fourteen years. These changes would allow the network "to remain a competitive outlet for public affairs information." Two factors, they claimed, motivated the changes: the widespread usage of the format by other networks, and "the unproductive use of valuable resources" caused by preemption of the evening call-in by extended House sessions.[65] The climate was now ripe for change. The three January 1st groups concerned with Hill coverage were already in place. They would push for better "contexting," improved scheduling information, and a more creative use of video.

In early November 1994, a select group of staffers from the programming and programming operations units set out to design a format for what was hoped to be "a signature program for C-SPAN." Its two major components would be "timely, useful information about what's going on in Washington and the opportunity for the public to be heard."[66] Those components had been central to C-SPAN's mission from the beginning. Many of the elements composing what came to be called the *Washington Journal* were, indeed, newly formatted old ideas. Some of them, such as the use of daily newspapers to prompt discussion, had been around since the beginning. The old ideas would simply be given a new look.

The committee members had no solid evidence of viewer dissatisfaction with the old call-in format. What was available to them was a sense of what other networks were doing, a set of impressions gleaned from viewer call-ins and from data gathered by the viewer services department, and their own understanding of C-SPAN's enhanced technical capabilities. They had, after all, produced, participated in, and watched more of the call-in programs than any other Americans. Those seeking to develop "a new creative and informative approach" to the morning call-in would not have the opportunity to rely on focus groups to evaluate the suggested changes before they were implemented.[67] There would be no market testing in the conventional sense of the word. The *Washington Journal* would be launched with its creators knowing that whatever appeared on the air in January would look quite different in April.

Many of the suggestions made by the planners were in place by April 1995, when the program finally got its sea legs. Like the *Sunday Journal*, after which it was patterned, the *Washington Journal* looked and sounded more like a video magazine than a discussion of a single issue. The new program would begin an hour earlier, at 7:00 A.M. The first half-hour segment featured a daybook of events happening in Congress, at the White House, in the Supreme Court, and around Washington. It would also be used to advise viewers of the network's schedule for the day. The headlines of a half dozen or more East Coast newspapers would appear, as if the papers had been delivered at the viewers' doors. During the segment, viewers were invited to call in and describe what their own local press had chosen to highlight that day. The aim was to provide information and context. Viewers could see if what played as news in Washington also played in Peoria.

During the next segment, from 7:30 to 8:30 A.M., the focus shifted. The news, as papers reported it, would remain paramount, but the treatment would be different. No longer would experts on a specific topic be invited to discuss a single issue in depth. The core of the old call-in would be replaced by asking two "interesting" guests to share their opinions about the stories that had caught their eye that morning. The pace was faster by C-SPAN standards. More topics were covered. A variant of the old *Journalists' Roundtable*, a Friday staple in the old format, now often without journalists at the table, had replaced the single-issue structure that had dominated midweek before. Viewers could now compare the stories that had caught the attention of those in Washington with the stories in their own hometown presses. They could exchange views. An interview by phone with the author of one of the leading stories was often added to the mix. The old hour-and-a-half treatment of a single issue was condensed into a fifteen-minute segment of its own, a "Point-Counterpoint" exchange. What emerged was a tamer, more reasoned and reasonable variant of the productions of the broadcast networks, CNN, PBS, and CNBC. There were no shouting matches on C-SPAN.

The final segment of the new format would be the least defined and the most vulnerable to preemption each morning. If the assignment desk decided that a hearing, a presidential announcement, or a press conference was more interesting and informative, the *Journal* could end early. When uninterrupted, the final segment would be devoted to features and to open phones. Features usually consisted of interviews with opinion leaders who were in the nation's capital for a special event or of mini-versions of *Booknotes*. Open phones ensured that the program would reach a thirty-call minimum. Hosts were told that viewers' phone calls remained "a priority," and they were asked "to get as many" as they could "into each segment of the program."[68]

Overall supervision of the *Journal* would be given to Connie Brod. Brod had worked at the network from 1984 to 1990. She left to go overseas with her husband, a Foreign Service officer. Just about the time the decision was made to begin the *Washington Journal*, she returned to the network full-time after having freelanced for the better part of a year. During her earlier tour at C-SPAN she had produced the tenth anniversary special. She had been in charge when the network began its coverage of the courts, and she was the first producer assigned to international programming. "Whenever something's new," Terry Murphy told a group of recent employees, "we turn to Connie." Brod had always "pushed the envelope" when it came to the C-SPAN style.[69] It was time for her to do so again. New opportunities allowed for a change in style. New technological capabilities would enhance it. Before 1995, C-SPAN had, for the most part, practiced minimalist television. To fully exploit the possibilities that technology offered was, in the minds of some, to violate the mission. Better, more exciting pictures did not equal "accurately" conveying "the business of government." C-SPAN had begun its call-ins using a simple cloth background. It could not afford anything else. Over time, the notion arose that an attractive set could detract from content. The shift to a book-lined backdrop was innovation enough. Viewers watched everything and assigned meaning to it. Some callers from the South, for example, complained about the message sent by the Abraham Lincoln bust on the shelves. Typically, C-SPAN now rotates the busts used.[70]

By C-SPAN standards, the visual look of the *Washington Journal* represented a quantum leap. "Beauty shots" and "wallpaper" now took viewers off the set. During call-ins, cameras showed scenes of Washington residents coming to work. Between segments of the show, both live and canned shots were used to take viewers outside the studios to city landmarks. Remote interviews were conducted from the Hill. Clips from the C-SPAN bus took viewers across America. Because of a major equipment upgrade in the production facilities on the first floor of 400 North Capitol, the executive conference room on the sixth floor was temporarily used as a studio. Overlooking Capitol Hill, it provided a lighter and fresher backdrop for guests. By the spring of 1995 plans were under way to convert the area into a regular studio. The *Journal* looked different from the old call-in format.

The *Journal* sounded different too. More voices were heard and from more places. More topics were covered. More information was provided about what was ahead on the show and on C-SPAN that day. The appearance of readily identifiable segments made the program more viewer-friendly. Segments with a purpose in turn shaped the context of the calls received. Because the "Point-Counterpoint" segment was often taped in advance, viewers frequently heard two different C-SPAN hosts on the air in the same program.

In April 1995, C-SPAN conducted an in-house survey in an attempt to assess the *Washington Journal*. Part of its findings were contained in a memo from Connie Brod to all hosts. The memo was distributed to all editorial and technical staff involved in the show as part of a new team concept for production.[71] The *Journal*, Brod pointed out, had "gone through a variety of lives in its four months [of] existence" in an "attempt to find a format, style and rhythm that work." Central to the style was a reminder: "CALLS ARE OUR MOST IMPORTANT FEATURE." Several of the suggestions dealt with phone calls. Hosts were reminded that calls during the "Newspaper Headlines" section in the program's first half hour had to be treated differently than did those received during the "open phones." Hosts were encouraged "to engage the caller to find out a little more about the story" and to help viewers "put it in context." In practice during the first half hour, hosts attempted to steer callers away from general comments and toward their morning newspapers.

Balance between regions could be difficult to achieve, since viewers would have to be up at 4:00 A.M. to catch the first segment if they lived on the West Coast. At that time most daily papers had yet to arrive. Some on the East Coast quickly figured this out and used the West Coast access number to get on the air. As in the past, "THOSE PESKY REPEAT CALLERS AND FAXES" were a matter of concern. Hosts were asked to mention "the 30 day policy" during each show. Every fifteen minutes, a reminder of the policy appeared on-screen. Some callers still could not be accommodated by the new format. One who finally got on the air attacked the network for being "rude." C-SPAN had not picked up the phone when he had tried to call earlier. After it rang for a while, he was disconnected. Brod, the host that day, patiently explained that C-SPAN did not pick up the phone until the caller was about to go on the air. Doing so prevented the viewer from being charged while on hold. The telephone company, not C-SPAN, had cut him off. The caller didn't care.[72] Most of those who reached the network that day would not be so nasty.

Although calls were still vital to C-SPAN, the network was no longer carrying a conversation; it was producing a program. A rhythm had to be established to do this well. Hosts were asked to make the *Journal* more viewer-friendly by promoting guests and the issues to be covered in upcoming segments. They were requested to arrive in the studio at least fifteen minutes before airtime to brief "the newspapercam operator" about the headlines they would use and the order in which the headlines would appear. Brod claimed that without doing so, the program seemed "to lack a content and a flow." Guests for the "Newspaper Roundtable" were asked to arrive forty-five minutes before their segment began in order to select the stories they wanted to discuss. Knowledge of those stories and the order in which they would be introduced would help the host "context wise" and would make the job of the camera operator and the director easier.

The *Washington Journal* added "visually pleasing" to the C-SPAN taxonomy. "After all, we are TV," Brod told the hosts. "Beauty shots" would require a context for the program to run smoothly. The pictures had to have a purpose. Scenes outside the studio were used to foreshadow programming or to set a historical context, especially during open phones. Hosts were provided with examples of how to integrate the oral and the visual: "As we take our next call, a look at the west lawn of the capitol where officials are setting up for the NOW rally later today . . ." Lines—"We continue with our calls this morning and look at Ford's Theater, . . . the place where Abraham Lincoln was assassinated 130 years ago today"—would be used for historical context. The suggestions might sound simple, but the scenes and their utilization were new for C-SPAN.

Not quite six months into the show, Brod hoped to introduce additional innovations designed to make the program more interesting. She talked about trying to feature a newsmagazine writer who had an important article about to hit the newsstands that week. Viewers would be given an in-depth preview. In late May the network took delivery of a new microwave van. It would allow C-SPAN to go to some guests rather than have the guests come into the studio. Each interview offered the opportunity for a new look visually.[73] Most viewers applauded the changes. As always, some did not. On March 27, 1995, for example, flowers arrived for Brian Lamb, then hosting the *Journal*. The card read, "Unlike the scent of this bouquet, your new format stinks."[74]

Although it may not appear to be so in terms of broadcast network programming, the *Washington Journal* looked new and different. For C-SPAN, it was a major departure in terms of programming. The mission remained the same, but production values changed. Gone was a simple set. More cameras, and hence more scenes, came into play. Produced programming had always played second fiddle to "the viewer in the room" philosophy. The *Journal* enlarged that room and added a picture window looking out on Washington and the nation. The difference between the *Washington Journal* and the old call-in format was almost like the leap from black-and-white to color television. If the new format was designed to be viewer-friendly, it also brought more information in a more orderly fashion and managed to be visibly pleasing at the same time. The structure also proved more useful to the programming operations department. Prime-time and overnight scheduling would gain a bit of flexibility because the "Newspaper Roundtable" and the "Point-Counterpoint" segments could be re-aired separately.

Every change, of course, creates more change. When Swain and Murphy first announced the new format, they told C-SPAN workers: "This new program will necessitate some operational adjustments. It is our goal to identify those necessary changes by December 1 to allow the affected staff members ample time to prepare."[75] For some, the new format would mean live programming on Saturdays and all that this would entail. For others, the workday would begin an hour earlier. The first *Washington Journal* staffer to arrive came in to work at 2:00 A.M. It was Leona Jordan's job to make sure that the necessary videoclips were available and ready and to do a scan of the morning papers that might be used on the *Journal*. Producers Peter Slen and Kristie Schantz arrived between 4:30 and 5:00 A.M. Brent Betsill, who would direct the show and run master control, was at 400 North Capitol shortly thereafter. Everyone who worked with him to get the show on the air and keep it there had their clocks set forward. Field crews had to pick up their equipment and be on-site, ready for early-morning interviews. In the planning stages, the emphasis had been on more and better information; it would be up to master control and those behind the cameras to execute the new look. They had to turn ideas into good video and audio. From graphics to clips to phone interviews—all had to be integrated smoothly.

Perhaps the change in the C-SPAN style had the most impact on those responsible for making it work visually. The pace would be different. There would be less "sit there and watch it" and more action—more things to manage and more things that could go wrong. Betsill said, "I've got so many stimuli coming in all at once . . . it's a rush."[76] If the seven phone lines reserved for callers could light up all at once, so too could the director have three or four camera operators, several control rooms, and the producer all vying for his attention.

Like the initiatives to enhance coverage of the House and the Senate, structural changes in the call-in format represent a more sophisticated and complex way of accomplishing the C-SPAN mission. The *Journal* is indeed more viewer-friendly, more interesting, and more informative at the same time. The changes were made to help C-SPAN "remain competitive" in a media environment in which others had built the call-in format into their own programming. In fact, within the time slot scheduled for the *Journal*, little or no competition exists for the kind of talk that C-SPAN favors except on Sunday mornings. The new *Saturday Journal* did, however, offer an opportunity to attract new viewers to the format. It competed with endless cartoons. Through it, C-SPAN hoped to attract viewers in their early thirties and younger by featuring guests in that age bracket, the next generation of political leaders. Its first guests were Jesse Jackson Jr. and Michael Sununu. On a later show two mayors, one twenty-three years old and the other twenty-six, were featured. Whether the faster-paced new look will increase the average C-SPAN viewer's two-hours-a-week commitment to the network remains to be seen. Regardless of the innovations, the program still airs from 7:00 to 10:00 A.M. on the East Coast and from 4:00 to 7:00 A.M. on the West Coast, requiring alarm clocks and dedicated viewers in the latter time zone. Should C-SPAN take the *Washington Journal* to the 1996 conventions, the program will be time zones away from the capital. The clock that is chosen for the 7:00 A.M. start could well make a difference in audience demographics.

Remaining competitive in a multichannel environment might not be so much a struggle to gain viewers as it is a struggle to schedule guests who can both attract viewers and provide information not easily found elsewhere. As long as C-SPAN viewers vote the way they have in the past—78 percent of C-SPAN viewers voted in the 1994 congressional elections versus 38.8 percent nationally—politicians will welcome the opportunity to appear on the network.[77] Whether or not the major players will do so is another matter. The shorter time slots for guests, the enhanced capability to bring the cameras to them, and increased visibility for the *Washington Journal* will all help. It is one thing to be called "the network of record," but it is quite another to entice citizens to use that network. This is what the *Washington Journal* is all about.

CHAPTER FIVE

Extending the Mission

Presidential Campaigns and *Booknotes* as Signature Programming

C-SPAN's coverage of Congress and the network's development of the call-in format arose from its mission statement. Coverage of political campaigning is the largest extension of this mission. Although a portion of this chapter concentrates on presidential races, the efforts that C-SPAN devotes to their coverage are duplicated in congressional races and gubernatorial contests. Especially when they reflect national concerns, C-SPAN attempts to follow actions at the state level. This chapter will also treat a fourth strand of signature programming: the development of *Booknotes*. Of the four, it reflects the personality of the founder more clearly than the others.

The Third Strand: Covering Presidential Campaigns

For C-SPAN, a commitment to carry "other forums where public policy is discussed, debated and decided" would mean going on the campaign trail.[1] Sooner than some might think it would be able, the network did so. Fulfilling the mission was reason enough, but if C-SPAN was to be "a network of record," it had to be recognized as such. Publicly it professed not to be worried about ratings; after all, it did not even subscribe to the Nielsen service. Ratings are one thing, however, and recognition is another. As a nonprofit organization, C-SPAN did not have to attract advertisers, but it needed viewers to gain the status necessary to attract guests who were players in the electoral process. It needed mentions in the press. It would be helped by having something to show the cable industry that supported it. Even a low-budget soup kitchen that has no one there for lunch or dinner will soon lose its support, charity or not. C-SPAN could not and would not rest on the laurels its coverage of Congress had gained for it. For C-SPAN, this was not a case of a good reason, the mission itself, serving as a cover for the real reason, the search for recognition. It was simply a marriage between the ideal and the pragmatic. Fulfilling the mission would bring recognition.

The changing nature of political campaigns and of their coverage would prove a boon to the network. The 1980s brought well-crafted, thirty-second commercials, which ran three times longer than the average evening news sound bite. Campaign debates seemed to approximate the cola wars of that decade, with spin doctors waiting in the wings. Spin doctors were the political counterpart of the product-liability lawyer. Character had always been an issue in American politics, but Americans have a short historical memory. They seemed both fascinated and disgusted by the charges and countercharges that grew in number in the 1980s and 1990s. They had forgotten Thomas Jefferson's "dusky Sally." The strains of "Ma, Ma, where's my Pa? He's in the White House, Haw, Haw, Haw" were strange to them. Before Ronald Reagan and Lee Greenwood had the country singing "I'm proud to be an American," Indiana's Albert Beveridge had boosted William McKinley's campaign with "The March of the Flag." "The Bloody Shirt" was around long before Bill Clinton's draft status became an issue. Before there were "A Thousand Points of Light" there was a "Cross of Gold." Writing in the 1950s, the economist Kenneth Boulding put his finger on what politics would become in a mass-mediated age. He believed that the art of persuasion lay in the ability to perceive the weak spots in the images of others and in the skill of prying those images apart with well-constructed symbolic images of one's own.[2] In an era in which much of campaigning seems to consist of just such messages already tested by focus groups, there would be a place, and a need, for long-form coverage. C-SPAN would attempt to provide that context.

Journalism too had changed in the 1980s and 1990s. The media are, at the bottom line, a commercial business. They cannot survive without making a profit. Covering politics on the evening news was never a profit-maker except for the star anchors. Their reputation would carry the show. The higher the ratings, the more expensive advertising would be. In twenty-three minutes each evening, the networks met head to head in competition. In reality the anchor could do little more in that time than provide visual headlines and briefly assign meaning to them. Every world crisis brought the anchors to the scene as if their presence and a three-minute story made the difference. The networks' battle for dollars all but ended in-depth coverage. By the 1980s the documentary would be replaced by newsmagazines, sometimes looking as if they were run by the entertainment division. Gone too were the days in which the networks were each other's only competition. Gone also was their reputation for fairness in the eyes of many. All political drama needs a villain. Spiro Agnew's political career may have ended in disgrace, but his attacks on "the nattering nabobs of negativity" and their "querulous criticism" would provide a script for some talk-show hosts and candidates alike.[3]

If the networks faced stiffer competition, made fewer dollars from the news, and had their journalistic ethics questioned in the 1980s and 1990s, the print media had problems too. In the age of accelerated news, the race went to the swift, and print simply could not keep up. Newspapers had to learn to do something different and make that something interesting in the process. Like the networks, many would turn to survey data and focus groups in an attempt to bolster their audience and add advertising dollars. Like their cousins on television, they too would belong, in the minds of some, to "the liberal" media. In 1995 the *New York Times*, the *Washington Post*, and the *Los Angeles Times* all lost readership.[4]

From the outset C-SPAN attempted to provide another view of the political process. It featured whole events, not sound bites. It followed candidates where the commercial networks expected there would be no news, and it gave time to those who others knew would pull up lame in the sport of horse-race journalism. Philosophically C-SPAN avoided explanation; financially it could not afford to interpret events even if it had so desired. A lack of money, in fact, may have played an important part in imprinting a style of coverage that the network would carry through the 1990s. As far as the media go, C-SPAN is still a shoestring operation. Its 1996 budget is about one-quarter of the "close to $100 million" spent promoting the entry of the television comedy *Seinfeld* into syndicated reruns in the fall of 1995.[5]

FOR A MONTH IN OCTOBER 1980

C-SPAN had not yet been on the air a full year when Chair Bob Rosencrans suggested that he and his fellow board members make a "voluntary contribution" of $150,000 so that the network could cover the 1980 election. On December 13, 1979, the board approved the "special assessment."[6] The following September, Brian Lamb presented a plan calling for one hundred hours of programming that would include covering everything from campaign speeches to the American Legion–sponsored Boy's Nation to a meeting of the Black Caucus. He also promised "behind-the-scenes coverage of the campaign process." His budget was predicated on receiving $100,000 from the "special assessment." Concerned about the tightness of the budget, especially in terms of "promotional dollars," the board's Executive Committee set aside an additional $25,000 from the general fund to be used, if necessary, for publicity.[7] There were other concerns, of course. C-SPAN had only two cameras, plus the switching units and tape decks it could use as a result of its deal with the Close-Up Foundation. Forrest Boyd, of the International Media Service, provided the equipment necessary to put the National Press Club on television for "$200 an event." Lamb and five other full-time employees had been operating a single channel "running about 5 hours a day."[8]

When Congress adjourned on October 3, 1980, Lamb added eleven temporary employees, mostly technical staff. Even Senior Producer Gail Picker was initially hired for only one month. National Press Club speeches and call-in programming built around them would provide the backbone for the election coverage. Both President Jimmy Carter and third-party candidate John Anderson spoke at the club. So too did Moral Majority leader Jerry Falwell. Pollsters Lou Harris and George Gallup also appeared that October. The National Press Club booked the speakers, but C-SPAN had to book the call-in guests, and that was another matter. Both the Carter and the Reagan camps were generally uncooperative. The fledgling network simply could not get the kind of audience that would entice them. As might be expected, members of John Anderson's camp needed the exposure, but even then C-SPAN "really had to hustle" to book them. Seven of the eight days of call-ins in October were devoted to politics. The network attempted to balance the guests it could get. For beginners they did well enough. Among those who appeared were Paul Weyrich, Richard Vigerie, Morton Kondracke, Hodding Carter, Alfred Kahn, pollster Peter Hart, and a score of reporters from newspapers and magazines. Unfortunately for viewers, the shows were scheduled at different hours in the afternoon.[9]

The bulk of the month-long coverage was devoted to political speeches, carried live whenever possible. What it could not cover itself, C-SPAN got from independent producers or picked up from CNN, from Independent Television News Associates, or from Independent Network News, groups with which it had reciprocity agreements. Hired crews brought viewers a speech by Walter Mondale from San Diego and a talk by Ronald Reagan from Los Angeles. Lamb was able to buy a tape of a Jacksonville, Florida, address by Reagan from a local television station "for practically nothing."[10]

That October, C-SPAN also brought viewers two other events that would come to typify the network's approach to campaign coverage. On October 27, it visited the offices of the *Washington Star*. There it interviewed the executive editor, watched decisions being made, and talked to reporters as well. The cameras even peered over the shoulders of cartoonist Pat Oliphant as he prepared a drawing for the paper. Visits to newspapers would become a C-SPAN staple. On election night the network simulcast Larry King's radio talk show with featured guest political satirist Mark Russell. King appeared twice more in the C-SPAN schedule over the next several years. C-SPAN, not CNN, introduced him to television. Carrying radio talk shows would become part of the C-SPAN programming.

Compared with what the network would do in the future, October 1980 was barely a start, but Lamb did all this for $25,000 under budget. On December 11, 1980, he proudly told the board that the month's thirty hours of call-in programs had produced "800 calls from 350 cities in all fifty states."[11] The effort also won the network national recognition. Appearing on ABC's *Nightline* in the final week of October, the *Washington Post* television critic Tom Shales faulted the broadcast networks for turning "campaigns into Kentucky Derbys, candidates into cartoon characters, and elections into game shows." He claimed, "The best political coverage may have been on something called C-SPAN." The network, he said, "lets you see political speeches and rallies without the overly interpretative translations of the national news people." The board was heartened to hear Jeff Greenfield praise the cable industry for C-SPAN. Most viewers, Greenfield pointed out, "never have a chance to see and hear any political coverage other than the inherently truncated versions shown on newscasts." Cable television, he suggested, "may be the only way we'll find out what a campaign is all about."[12]

CAMPAIGN 1984

Despite the praise C-SPAN gained for its efforts in 1980, the task of becoming a recognizable source for political information would not be an easy one for a network that four years later still labored in "near total obscurity."[13] The strategy in 1984 would emphasize "applying C-SPAN's proven formulas to events taking place outside its traditional turf—Washington, D.C."[14] C-SPAN would not innovate as much as it would creatively package combinations of programs it had used before in order to cover events it had never telecast in places it had never been. On its list of objectives was "an attempt to build community demand" in cities where the network was not yet available. To do so, the network would have to go on the road. The cable industry would help. Through the efforts of the ATC cable company, C-SPAN was able to cover a live, ninety-minute exchange between Gary Hart and North Carolina high school students.[15] The arrival of a mobile satellite uplink would help even more.

C-SPAN began its election coverage with a weekly "two-hour political wrap-up show," called *Election '84*, in the same month it carried the Hart dialogue. It would be the coverage of the Iowa caucuses and the New Hampshire primary, however, that brought the network some of the recognition it sought. Those at C-SPAN marveled at "the raft of press clippings" they generated.[16] David Crook, of the *Los Angeles Times*, told readers that it was his job to watch television. In doing so, he found "an obscure little cable TV service called C-SPAN." There he and his wife, also a fellow reporter, watched the Iowa caucuses. What they saw told them "a lot more about the political process than a week of 'Nightline' programs or a month of Bill Moyers's Commentaries." For them, *Nightline* and Moyers were the best the broadcast networks could offer. Seeing C-SPAN's version of Iowa, he told his readers, assured him that Thomas Jefferson's "vision of an informed and enlightened electorate" was still alive. As "corny" as it sounded, he said, the programming made him "feel good about America." Here was "a lesson in democracy," one that "none of the networks with all their vote projections, fancy electronic maps and multiple satellite interviews even came close to presenting." He wrote, "The more I think about it, the less and less I believe that it matters one iota what Dan Rather, Tom Brokaw, Peter Jennings or their respective networks have to say about this year's presidential primaries."[17] Buoyed by such commentary, the network's *Weekly Report* claimed, "It is clear that the C-SPAN approach to covering elections has proven its worth."[18]

C-SPAN covered the events with the logistical support of the cable systems in Iowa, New Hampshire, and almost every other place it visited that year. Still, there would be more to prove and more help needed.

In 1984 C-SPAN would broadcast the national political conventions gavel-to-gavel. It would also embark on a fourteen-city tour called *Grassroots '84*. Both events would prove important in helping to establish its reputation, though for different reasons. The conventions came first. Politics and broadcast economics combined to ease C-SPAN's entry into convention coverage. The primary system, the broadcast networks asserted, meant that little news would be made either in San Francisco, where the Democrats would meet, or in Dallas, the site of the Republican gathering. Lower-than-hoped-for ratings at the 1980 gatherings were also a factor. ABC, CBS, and NBC all decided to forgo gavel-to-gavel coverage in 1984. Instead, they would carry the conventions only in prime time. CNN would provide gavel-to-gavel coverage patterned after the style used by the major networks at past conventions.

Despite the cutback in coverage, the broadcast networks expected to spend up to $20 to $25 million each on their programming. CNN estimated its costs at about $10 million dollars.[19] It would have a "$300,000 anchor booth" alongside those of ABC, CBS, and NBC and a "$2.5 million state-of-the-art production truck."[20] The Public Broadcasting Service (PBS) briefly considered competing gavel-to-gavel but dropped out of the media poker game when it figured its costs at $10 million dollars.[21] Estimates of the number of staffers the major networks would employ ran as high as twenty-one hundred. CNN brought a competitive workforce. Then there was C-SPAN. It budgeted $150,000 for San Francisco and slightly more for Dallas.[22] It would bring thirty-five staffers to the West Coast and add a few more at the Republican Convention. To cut costs in San Francisco, it would bypass the pool podium feed and use a camera of its own, saving $60,000. The network went without tally lights on its cameras. Using them would have let the operator know if his or her camera was the live one, but not using them saved stringing "miles of cables." C-SPAN would make do with headsets.[23]

Conventions presented a challenge. Given the age of the C-SPAN staff, it was doubtful if many had even seen a political convention before, let alone attended one. C-SPAN had five camera positions in the hall in San Francisco. Since the network was used to handling only two on its regular coverage, that worried Field Production Manager Terry Murphy. He now had five screens to watch, not two.[24] When a PBS staffer visited C-SPAN's small control room, he asked where the directors were. "There he is," said a crew member, pointing to Brian Lockman.[25] With nearly half the staff at the conventions, those at 400 North Capitol, coordinating the action and preparing for reruns, were shorthanded too. Connie Brod, who had been full-time for only two weeks, would serve as the bridge between the two sites. Both at the conventions and at home, some of the key players were new. Four producers had been added since February 1984.[26]

Fortunately C-SPAN had several things going for it at the conventions. Mike Michaelson's thirty years of service with the Radio-TV Gallery in the House meant that he not only knew the convention players but also knew the rules of the game. Michaelson seemed to know everyone who mattered. Without his help in negotiating with convention planners, in making the case for access on a par with the broadcast networks, and in securing the proper credentials, the situation would have been much more difficult for C-SPAN. With him, C-SPAN had a familiar face at both gatherings.

If Michaelson's role cannot be underestimated, neither can the support of the cable industry. In San Francisco, Viacom provided crews, equipment, and local contacts. Its president, John Goddard, was a C-SPAN board member. In Dallas, officials at Sammons did the same, plus provided space for the call-in program. Its president, Jim Whitson, would later chair C-SPAN's board. In San Francisco, the broadcast networks and CNN may have had three-story skyboxes filled with "state of the art television equipment and famous anchormen," but using local connections, C-SPAN operated out of the twenty-fourth floor of a building providing "a bird's-eye view" of the convention center on one side and a backdrop of the Bay Bridge on the other for call-in programs. The space was donated by the McKisson Corporation. Thanks to a relative of a C-SPAN staffer, the Metropolitan Parking Board allowed the network to station a satellite uplink on one of its properties "across the street" from the Moscone Center. Life in Dallas was made easier with the help of Sammons Vice-President Bill Strange. His "we're sold on C-SPAN, anything they want they can have" attitude was what any network would like to hear. The assistance in all facets by cable companies made the coverage possible.[27]

At both sites, C-SPAN covered the conventions gavel-to-gavel and more. One staffer's friend even joked that C-SPAN was the network where you could be sure to see the janitors sweep up each evening.[28] "More" would mean something in addition to simply replacing the broadcast networks' end-to-end coverage. While at both conventions, C-SPAN carried platform committee meetings as well as increased the number of daily call-in shows from three to as many as eight. In doing so, the network gave viewers both more information about the gatherings and a feel for the politics of the cities they visited. C-SPAN did more than book big-ticket guests. In San Francisco this meant that viewers could discuss the role of Chinese-Americans in politics with a local official and talk to Eldridge Cleaver about his candidacy for a House seat. In Dallas, Ben Sargent, the cartoonist for the *Austin American Statesman*, drove four hours just to appear on the network. There too, those watching could meet Elena Hagii, a national board member of ACORN, a community activist organization. In each city, C-SPAN took viewers to the editorial boardrooms of local papers to hear how they planned to cover the conventions. In both towns, viewers could sit in on local talk radio shows. At each site, they would get visual postcards of the community. Like delegates, viewers could take time to tour the convention cities.

The conventions would test C-SPAN's philosophy regarding bias. During the network's coverage of an event, "the camera generally just stays on the speaker," Executive Producer Brian Lockman said. That type of shot had already become a signature for C-SPAN, a sign of the network's "principle of unbiased coverage." The conventions, however, differed from a committee hearing or a National Press Club speech. There might be as many as three stories to be told at once. There were speakers at the podium, delegates in funny hats, and "media types . . . running around looking for stories." Conventions convey many messages; choosing which to develop makes a difference. C-SPAN "tried to stay on the podium as much as possible," but it also selected other shots to capture the convention mood. In a six-and-a-half-hour period in San Francisco, Lockman estimated that he used "over 1,800 cutaways" from the dais in order to catch the spirit of the gathering.[29]

Good video, of course, does not always mean accurate coverage. This was a constant concern for C-SPAN as it attempted to give viewers a seat in the convention hall. Where one sits makes a difference, hence negotiations with convention planners about camera placement. Making full use of camera angles too would be relatively new for C-SPAN. The network shot Mario Cuomo's speech from below "to heighten the drama and the intensity," according to Lockman. The camera looked down on Gary Hart to emphasize his "sway over his diehard followers." One of C-SPAN's best efforts, Terry Murphy recalled, showed a woman crying during Jesse Jackson's speech. All of this was tame stuff for the commercial networks, which constantly sought the dramatic, but C-SPAN struggled to try to match the mood to the content without misrepresenting the action. By the time of the convention in Dallas, crews already had some experience with shot selection and with camera angles. There the problem of balanced coverage was a little different. To supplement its own cameras, C-SPAN decided to use a pool feed provided by the Republican Party free of charge. Not surprisingly, the GOP had the best camera positions in the hall. Also not surprisingly, it had a point of view. Lockman, who produced the coverage there too, would pick and choose with caution. The experiment went well enough in Dallas, but at a later convention pool feeds would be a problem.[30]

Despite its low budget and small, inexperienced staff, the 1984 conventions worked for C-SPAN. "We got more direct feedback—letters, phone calls—out of the two conventions than anything else we've done in the past five years," Lamb told a reporter from the *New York Times*. "Our time," he said, "came in San Francisco—a lot of people who hadn't been paying attention to us until then suddenly discovered who we are."[31] Some realized C-SPAN's worth at the conventions, others found out by reading the press, and still others watched and liked what they saw. At the conclaves, print journalists from publications like the *Wall Street Journal*, the *Boston Globe*, and *Time* magazine all had the televisions in their convention offices turned to C-SPAN. In San Francisco, Viacom cablevision was able to pipe C-SPAN's signal into sets that Moscone Center officials had placed in the halls there. In both cities, cable officials worked to see that C-SPAN's signal could be received in the major hotels housing the delegates and the press.

The print media spread the word about C-SPAN's efforts. After the network's first day in San Francisco, David Bianculli, of the *Philadelphia Inquirer*, wrote: "In one evening, C-SPAN cemented its place as a valuable, necessary part of the political equation at these conventions. Regardless of what the commercial networks were televising—whether it was commentary, floor reporting, or *TV's Bloopers & Practical Jokes*—viewers curious about what the speaker of the moment was saying could always turn to C-SPAN."[32] The comments of Michael Dougan, of the *San Francisco Examiner*, must have pleased the network. "C-SPAN provided a clean video *verité* experience that made viewers feel that they were there," he wrote. "It was, in many respects, the best show on the dial."[33] His words almost matched Brian Lamb's description of C-SPAN's goal. "We're just going to put the viewer in the delegate's seat and let him see the convention as it happens. We promise no analysis whatsoever," Lamb told another reporter.[34]

Viewers indeed liked what they saw. "You have put C-SPAN on my map," wrote the chair of the Republican Party of Texas. "I have seldom watched your channel, but that will all change now because of the great work you did at our national convention in Dallas." A viewer from Arlington, Virginia, called C-SPAN "a national treasure" and added, "C-SPAN will be with me as long as I have the strength to turn on a TV set." A Brown Summit, Kentucky, resident wrote, "It was wonderful to watch what was actually going on rather than having to wade through the slanted comments and opinions of the all-knowing, seemingly omnipotent network anchor people." She asked, "Just who do they think they are, anyway?"[35] House Historian Raymond W. Smock told Mike Michaelson that C-SPAN's coverage "was the best thing to happen to political conventions since the advent of television." It freed him from "commercials," the "professional pundits," and the "media stars." He could "hear all of the speeches, both the major and the minor ones, uninterrupted," and for that Smock was thankful.[36]

If there was a disappointment in 1984 for C-SPAN, it was the failure of many PBS affiliates to take advantage of Lamb's offer to show C-SPAN's coverage, gavel-to-gavel, on their stations. He made the feed available at $40,000 per convention to PBS. It would then be made available to PBS affiliates for no charge. Most were simply not interested. Although as many as 30 eventually took the feed, according to C-SPAN sources, shortly before the Democratic Convention Lamb told a *Baltimore Sun* reporter that only 7 of the 295 affiliates were committed to the coverage. "All seven," the *Sun* said, were "the second smaller public TV stations in their markets."[37] In the final count, affiliates in seven of the top-ten markets—Los Angeles, Chicago, Philadelphia, Detroit, Washington, Cleveland, and Dallas–Fort Worth—all passed up the opportunity. A spokesman for the PBS station in Philadelphia termed coverage "not very interesting." His counterpart in Cleveland labeled it "dull and boring."[38] Sardonically, Tom Shales, of the *Washington Post*, claimed PBS simply "didn't want to disrupt those important plant-potting shows and Boston Pops reruns."[39] Bob Brewin, of the *Village Voice*, explained that it was another case of "N.I.H. (not invented here)."[40] Lamb had encountered some opposition from C-SPAN's board about the offer in the first place. The board had wondered why the network should give coverage to competing channels. The decision to do so, however, offered the network an opportunity to reach homes not yet subscribing to its service. In 1984, C-SPAN was carried by only one-fifth of the cable systems in the country and in half of the cable-ready homes in the nation.[41] Increased visibility might well increase demand.

Buoyed by its convention successes and with a mobile uplink now available, C-SPAN took to the road. From September through just before election day, it traveled to fourteen cities across the country. Called *Grassroots '84*, "the video journey across America" was designed to "offer viewers a perspective on American political opinion" that went "beyond the headlines and political polls." Whereas other media approached the campaigns "from the candidates' point of view," C-SPAN would "attempt to examine the concerns of the people who go to the polls." The sites to be visited—beginning with Mission Viejo in California and ending with Cleveland, Ohio—"were selected to present a broad section of the nation," according to C-SPAN's fact sheet.[42] In actuality the trip would mix marketing with a different look at the campaign, a look suited to the resources available to C-SPAN. Each site visit would be sponsored by a cable company and underwritten to the tune of $15,000 per city.[43] The cable companies would also supply local support in terms of providing facilities and booking guests for the call-in segments. Towns would be visited for three days. Seven producers took charge of two sites each. Two interns handled preliminary logistics while Brian Lockman ran the conventions.

The trip would be significant for several reasons. Although the board had a hands-off policy when it came to programming, the extra dollars could make possible special ventures if the board so wanted. Certainly in terms of local publicity and goodwill, *Grassroots '84* would be beneficial to the companies supporting it. The chamber of commerce in each community must have been pleased too. The journey would also represent one of a number of efforts C-SPAN made to secure outside help during the mid-1980s. It did so by bartering on-air mentions with those who helped make the trip possible. The Pace Arrow motor home that transported the crew from place to place was "on loan from the manufacturer," Fleetwood Corporation, and had to be returned in "tip top condition." Budget Rent-A-Car supplied cars for the crew changeovers. C-SPAN also sought to trade recognition for hotel rooms. Even though not all of its efforts to make such deals proved successful, the in-house *Weekly Report* asserted that "enough" were a success "to justify the effort."[44]

Grassroots '84 would follow the same formula that C-SPAN used elsewhere: interviews with local officials and citizens; call-in shows and rallies; town meetings; and debates if they could be scheduled while the network was in town. A fair portion of the programming would be devoted to "local color." Sometimes C-SPAN looked like Charles Kuralt's *On the Road*. In Mission Viejo, California, the crew visited the Olympic Swimming and Diving Headquarters. In Monterey it took viewers on a walking tour of a winery and let them watch a fishing crew at work. In Seattle the crew boarded a ferryboat to interview citizens on their way to work. C-SPAN talked to Chef Paul

Prudhomme in New Orleans. Viewers participated in a call-in show in Tulsa with the chiefs of the Creek and Cherokee Nations. They toured Notre Dame University in South Bend, Indiana, and strolled down Chocolate Avenue in Hershey, Pennsylvania. They walked through a steel mill in Harrisburg, Pennsylvania, and saw a city of contrasts—New Haven, Connecticut, home to Yale University and the seventh-poorest city in the country. They listened to comedian Pat Paulson on the campaign trail in Traverse City, Michigan. Certainly viewers gained the flavor of the communities that C-SPAN visited.

Seventy-eight of C-SPAN's eighty-two employees in 1984 were under thirty years old. They would confront all kinds of situations on the road. The motor home broke down in Tennessee and was reported stolen at another stop when the driver took it to a service station without telling anyone. At several stops crew members were asked if *Grassroots '84* was a band. At a small gas station in rural Texas, a man walked up to the mobile satellite and asked, "What does that thing represent?" In Little Rock, Arkansas, another man came up to the crew and said, "Say, you guys are the Ghost Busters, right?" If the technology was new in some parts of the country, so was C-SPAN. At a Seattle bank, a crew member waited patiently for money wired from Washington, D.C., to support the trip. The teller could not find it under "Sea Span." A similar delay occurred while the crew waited for a rental car in California. The network's name had been changed to "C-SPAM."[45]

It is hard to imagine that a motor home towing a mobile satellite represented "a network of record," especially when the trailer looked like "some little flatbed that you hauled firewood on." Brett Betsill, who did two legs of the trip, has only "fond memories" of a group of people "in their mid to lower twenties tooling around the country in this camper." He noted, "We had a great time."[46] Not everyone's memories are quite as pleasant. Garney Gary, the lone African-American on one leg of the trip, recalled being asked at one stop if he was along to keep the motor home clean.[47]

Half of the producers for *Grassroots '84* were women. C-SPAN was more than a decade ahead of the broadcast networks on that count. One of the woman producers, Carrie Collins, captured a sentiment shared by many who went on the road. After the trip, in response to a caller on a program, she said: "I was impressed by how proud people were of their areas and how pleased they were to have us there. Partisan politics aside, the bottom line was that it was their area and they were going to make it better." Brian Lamb captured the other half of the equation. "I learned something very sad on this trip," he said. "Time and time again [I saw] that people in this country feel that they're being discriminated against no matter what walk of life they come from: blacks, Catholics, Jews, Italians, Hispanics, white-Anglo-Saxon Protestants—everyone at one time or another seemed to feel that." Perhaps *Grassroots '84* had indeed captured the spirit of the country, as embodied in Collins's and Lamb's recollections.[48]

At C-SPAN too, two moods existed after a year on the road for the first time. In an address before the Washington Metropolitan Cable Club, Board Chair Jack Frazee recounted the devotion that the network always seemed to find in its chair. Frazee had been in Manchester, New Hampshire, where he had scouted out sites for "man on the street" interviews. He had helped set up a makeshift set in Chicago, overlooking the newsroom. He had been at both conventions. He had answered phones for a call-in show in Denver during *Grassroots '84*. Working alongside the young C-SPAN staff, he said, "You can't help but share in their enthusiasm of their successes." In Frazee's mind it had been a successful year, providing "more than 4000 hours of first run public affairs programming." With re-airs, C-SPAN offered its affiliates 8,760 hours of "quality cable programming . . . at a cost per television hour of just $622." By comparison he pointed out, "In 1983 ABC offered its affiliates less than half that much programming (4,342) at a cost per hour of $598,800."[49]

C-SPAN certainly had done well in 1984. It would go on the road again the next year for *States of the Nation*. But there was also a darker side to the successes in programming. "We were too frugal in 84–85," Bruce Collins told Jim Lardner as he prepared a piece on C-SPAN for the *New Yorker* a decade later. "We tried to get too much work out of too few people; we just brought people to the breaking point. . . . I guess we lost perspective for a little bit."[50] Success on the road would lead to a movement to unionize at home.

CAMPAIGN 1988

C-SPAN was in a better position to cover politics in 1988 than it had been four years earlier. It now had two channels available, a larger staff, a 40 percent increase in its campaign budget, more and better equipment, some name recognition, and experience. Gone, for example, were the days of towing a mobile satellite uplink around on a flatbed. The network's Ku-band satellite trucks now reduced preparation time for telecasting from four hours to thirty minutes.[51] C-SPAN would be truly mobile this time—no "plopping down" at a site for three days at a time unless it wanted to do so. This time its staff was not confused with a folk-rock group. The crew members had been to

Iowa, New Hampshire, and the conventions before. The mission would not change; it would simply be easier to execute.

In 1988, C-SPAN no longer acted like the new kid on the block, but it was still the poorest. Out of its $10.5 million budget for that year, $1.4 million would be dedicated to campaign coverage. To put that figure in perspective, consider that CBS still planned to spend $348 million for news in 1988 despite cuts and layoffs. CNN's yearly earnings in the same time period reached $89 million.[52] In short, C-SPAN would have to be penny-wise; it did not have the pounds with which to be foolish. The way in which the network dealt with the costs of broadcast networks' pool feeds is a case in point. The feeds had been a bigger-ticket item in the past than C-SPAN liked, but the nature of certain major events forced it to accept them or to have no video and audio at all. In 1984, the network had protested the costs and turnaround time for receiving videotapes of the presidential debates. After calculating their own costs, the broadcast networks providing the pool feed charged C-SPAN 7 percent of that figure for the right to carry the signal. There was no limit, C-SPAN officials believed, to what broadcasters would spend "to produce their products." Executive Vice-President Mike Michaelson told his fellow employees, "As we know, the presidential debates in Louisville and Kansas City carried a price tag of around $300,000 for an hour and half production, the inaugural $400,000 for an hour and 15 minutes, the conventions, over $1,500,000." Costs were bound to have risen. "At a 7% figure," he asserted, "today's C-SPAN fees would be extravagant." Michaelson was also wary of the broadcast networks' manner of costing an event. The network providing the pool feed for the 1985 inaugural attempted to charge C-SPAN $3,600 for the audio portion of the coverage. C-SPAN later learned that the other network had gotten the audio for free from government agencies. C-SPAN would negotiate for a better pricing system for 1988.[53]

In the late 1980s, the broadcast networks too were worried about costs, especially in their news divisions. C-SPAN had heard that the "big three" were considering ending pool feeds altogether for the 1988 conventions unless doing so could be made cost-efficient. If they ended the feeds there, they might do so elsewhere as well. Efficiency for the major networks might well translate into higher fees. Michaelson predicted the major networks' actions in a bargaining process: "[They will] probably try to link [C-SPAN] more directly with M.S.O.s [multiple systems operators] suggesting that they foot the bill for us, are rich and powerful and should provide the funding necessary to allow us to play in the so-called 'big leagues.' " In Michaelson's opinion, that notion simply did not square with "the facts of life in the cable industry." He told his colleagues, "We should be prepared to challenge this concept and lay our budget on the table if necessary."[54] In 1988, C-SPAN would cover as much as it could with its own cameras, and it pressed for fairer rates when the pool feed was the only source available, as in the case of presidential debates. It was a matter both of cost and of the type of camera shots the network preferred. Higher pool costs could threaten C-SPAN's own fee structure, and pictures that emphasized drama at the expense of substance ran counter to C-SPAN's mission.

If C-SPAN's financial concerns could be measured in thousands of dollars, the broadcast networks' problems made that seem like pocket change. The big three all underwent major shake-ups during the period between the 1984 and 1988 elections. Most agreed that little news would come out of the conventions and that less and less would be discovered by following every scripted speech delivered on the campaign trail. ABC, CBS, and NBC cut back on blanket coverage and concentrated on what they believed was news. Gary Hart's *Monkey Business* was news. The plagiarism by Senator Joseph Biden (D-DE) of a speech by Neil Kinnock and Biden's hyperbolic account of his academic career were both worth reporting. The Willy Horton advertisement was news. Iran-Contra was news. Even the news process itself was news, as the case of the now-famous Dan Rather interview with George Bush would prove. News was a search for "defining moments" or even their creation. The Bush-Rather exchange "failed as news" in the conventional sense of the word, media critics Neil Postman and Jay Rosen observed, but "it succeeded as drama." They noted, "It failed as discourse [but] it succeeded as entertainment."[55] Ironically, the interview raised CBS's ratings and advanced Bush's candidacy.

Disgust with conventional coverage would lead many disgruntled viewers to turn to C-SPAN's no-commentary, no-analysis format for political events and to its call-in programs to chastise journalists and campaign operatives for their behavior. C-SPAN would again be an alternative network in 1988. The network was about information, not analysis. It left the analysis to the experts invited to participate in call-in programs, and it let viewers talk with and to them. C-SPAN's political producer, Carl Rutan, claimed that its goal in covering the campaign was "to go out and to do more sooner than anybody else."[56] The "more," of course, would be done in C-SPAN style, long form. There were regularly scheduled call-in shows dealing with the election, and there was coverage of conferences and forums on issues, on image making, on policy, and on process; the network even telecast a seminar on how to run a campaign. C-SPAN showed political commercials and provided an opportunity for voters to discuss them with journalists and campaign operatives. The network visited newspaper editorial rooms and sat in on talk radio programs, as it had done in the past. The struggle for publicity in a crowded field led most of the major candidates to participate in C-SPAN's call-in format. Even fringe candidates had their moments. C-SPAN agreed to provide five minutes to anyone who met the Federal Election Commission qualifications for candidacy. Thus America got to hear from Billy Jo Clegg of Shawnee, Oklahoma, from Isabell Masters of Topeka, Kansas, and from Anthony Trigona of Middletown, Connecticut, among others. Some—like Buffalo, New York, native Michael Levinson—certainly added flavor to the campaign. Levinson wanted Muhammad Ali as his running mate and Richard Nixon as his "Secretary of State without portfolio." To top it off, Levinson promised "a giant four hour inaugural address" with "every line a delicate, sensible rhyme."

Road to the White House

Given the nature of C-SPAN's coverage of events, it would be difficult to provide programming continuity for the campaign. The network did not plan or sponsor candidate appearances, it only aired them. Their schedule, not C-SPAN's, dictated much of what could be shown. Viewers had difficulty finding out what would air and when, except in the case of major events, which the broadcast networks were most likely to cover too. Of course call-ins were regularly scheduled, including one specifically dedicated to the campaign, but often guests would not be booked in time to provide adequate publicity. Viewers knew that the "Event of the Day" had a set evening time slot, but they could not find out what that event would be. Extended House sessions took priority over anything else, sometimes filling the time allotted for the day's highlight.

Very few average viewers had the luxury to watch every campaign event C-SPAN covered. Political insiders, campaign staffs, and journalists could and did watch with some frequency. After Bruce Babbitt's "shoddy performance" at a Houston debate in July 1987, "the campaign installed C-SPAN in his home." His assistant press secretary, Vada Manager, explained: "That was part of our campaign for recovering. We wanted to show him how members of Congress and other candidates use television." Watching an early book-signing event for Pat Robertson, campaign staffers for Bob Dole got a feel for the Virginian's appeal. "What impressed me," a staffer said, "was not just that there was a crowd, but that they had all brought the books. There's a message there." George Bush's deputy campaign manager, Rich Bond, pointed out, "The electronic presence in politics is becoming more pervasive, and C-SPAN is paving that road for better or worse." Undoubtedly, this presence helped journalists assigned to cover politics. It was cheaper to watch a candidate at a small rally in Iowa than to fly there. Most such talks would be low-yield news events anyhow. If, on rare occasions, news was made, purchasing the tapes from C-SPAN would still be cheaper. Even though others chose not to follow C-SPAN on the campaign trail, the ability to watch enhanced the network's reputation. As Andrew Rosenthal, of the *New York Times*, noted, "The Washington based Cable Satellite Public Affairs Network, once known as 'the network that dares to be boring' . . . found new prominence and respect in the 1988 presidential election season."[57]

Neither praise from the press nor use by political operatives could help viewers find programs, however. Although C-SPAN, using two channels, often re-aired events as many as four times, this was still not enough. The viewer had to learn when the event would be re-aired. To be more viewer-friendly, C-SPAN set aside a regularly scheduled time on Sundays to replay any presidential debates that had aired during the week.

It would be a new format called *Road to the White House* that came as close as anything else C-SPAN produced to being a weekly summary of campaign events. As a signature program, the ninety-minute weekly broadcast represented only a small portion of the network's treatment of the 1988 election. Importantly, it had both a regularly scheduled airing and a regularly scheduled re-airing time. Viewers would know how to find it. Carl Rutan was assigned responsibility for the program. At thirty-two, he had been with C-SPAN for five years. Since 1986 he had been the network's political editor and had coordinated its coverage of the senatorial and gubernatorial contests that year. When the series began, in January 1987, it added several new features to C-SPAN's traditional campaign coverage. The first, made possible by enhanced technical capabilities, was a "new emphasis on events leading up to the primaries and caucuses." Rutan went to Iowa and New Hampshire fifteen times in the first ten months of the program's existence. Second, by miking candidates along the way, the network allowed viewers to hear more than speeches; they could listen to exchanges they would ordinarily hear only if they traveled with the office seeker. Finally, the show recommitted C-SPAN to listening to voters along the way. In addition to producing *Road to the White House*, Rutan was also in charge of the coverage of the Iowa Caucuses.[58]

Starting in Iowa and New Hampshire, "the general idea was to get out in these remote areas and show how many candidates were running early in the campaign."[59] Indeed, 1988 was made for C-SPAN, given the large number of hopefuls in the field. Getting in the field early meant telecasting minor skirmishes while the broadcast networks sat on the sidelines waiting for the decisive battles. In the beginning C-SPAN was virtually "the only source of information" for voters outside of Iowa and New Hampshire. Still Rutan had some difficulty attracting the attention of campaign staffs, especially their Washington, D.C., headquarters. "C-SPAN doesn't get the respect that a *New York Times* reporter or the broadcast networks do," he pointed out. "I have to fight a bit harder to get what I want." Traveling with the candidates would prove to be "the great equalizer." By early November 1987, C-SPAN had brought "more than 35 hours" of television from Iowa. A sampling would appear on *Road to the White House*.[60]

To bolster its position with the political campaigns, the press, its viewers, and the cable industry, the network staged "The Great C-SPAN Fly Around" near the end of 1987. In three days, C-SPAN would blanket Iowa. Thanks to the cooperation of Heritage Cable and other cable operators, C-SPAN officials visited eight of the largest television markets in the state. Jim Cownie, president of Heritage, chaired the National Cable Television Association. David Oman, the company's vice-president, was a member of the C-SPAN board. The company's private plane ferried the C-SPAN staff across the state. At each spot, cable system operators helped set up interviews with the press, politicians, and local citizens. In a way it was like *Grassroots '84*, only at a much faster pace. The cable industry helped C-SPAN with the coverage of four caucuses in February 1988. With two channels now available, C-SPAN carried two rural and two urban caucuses, a set of Democratic and Republican gatherings each. Whereas the major networks were more interested in the statewide outcome, C-SPAN wanted to help viewers better understand the process. "We don't cover the caucuses [because] it's a gimmick," Brian Lamb said. "We do it because every political reporter in the country writes and talks [about them] but no one knows what they are."[61]

While Carl Rutan had emphasized the fact that C-SPAN wanted to cover remote rural areas in places like Iowa, the network's press liaison, Nan Gibson, put a different spin on the coverage. Iowans, she said, wanted to dispel the media characterizations of their state as nothing but "pig farms and corn fields." She believed C-SPAN helped to do that. C-SPAN, after all, was not in Iowa to tell people what it thought. "We want[ed] the people of Iowa to tell us what the state is all about."[62]

But it was Rutan's message, not Gibson's, that made it through to some reporters. Steve Daley, of the *Chicago Tribune*, took the network to task based on what he had heard and seen of it. In Iowa, a C-SPAN representative, most likely Rutan, told him: "We've got a great idea. . . . Every day we're going to [put] our truck in the most rural campaign setting in the state we can find, right up until the 8th." Daley found a C-SPAN crew covering a speech by Dick Gephardt in a barnyard near Mount Vernon, Iowa. There the candidate talked about farm issues to a crowd of 125 sitting on bales of hay inside a large machine shed. According to Daley, "the man from C-SPAN" thought it was "good TV." Daley thought otherwise. "The cable service had the pictures it wanted, and those pictures delivered images and impressions to a potential audience of 33 million cable TV households around the country," the columnist reported. The problem, he pointed out, was that "C-SPAN's quest for the unvarnished rural setting" misrepresented the state. In 1984 only 7 percent of the approximately 200,000 caucus voters were "working farmers." The real action took place in the cities, which looked like cities everywhere. Iowa was not simply a one-issue state. Citizens there were also concerned "with the Persian Gulf and the arms race and corruption in Washington." In looking for "the perfect family farmer" in Iowa or "the bucolic Yankee spouting front porch wisdom" in New Hampshire, C-SPAN got it wrong, Daley thought. "One of the most pervasive elements in national political reporting, particularly the TV side," Daley contended, "is the search for the specific example that constitutes the rule." To his mind the real story was about "the Iowa bounce" received by those who did better than expected in the caucuses. "That's what makes it so important, and what makes getting it wrong with the C-SPAN satellite truck in the barnyard so significant," he concluded.[63]

In a sense Daley was both right and wrong about C-SPAN's performance. Clearly Rutan had developed a rural strategy, for Iowa at least, that would help distinguish C-SPAN coverage, but the network could not pick the places candidates chose to speak; it could only follow them there. There would be more than bales of hay in C-SPAN's Iowa coverage, but one would have to watch more than a program or two to understand the network's programming philosophy. C-SPAN was not like the broadcast networks. It seldom sought a representative anecdote; instead it provided viewers with a selection and then let them do the analysis for themselves. C-SPAN was not interested in providing interpretation. It simply wanted citizens to have the raw materials that would allow them to judge.

Iowa and New Hampshire are special stops along the road to the White House. The symbolic impact of performance in those two states far outweighs their actual worth in terms of delegates. They represent the first pole in the political horse race, the first real opportunity to count the votes. Their results were news. Neither vote was like a Florida Republican straw poll, although C-SPAN would cover that too. C-SPAN was not interested in calling a horse race. It was after something different, and that difference would be revealed in its programming. "What I'm trying to do is to give people the flavor, give them the whole stump speech," Rutan said. He wanted to portray what a campaign looked like, to "let people know that a candidate tells the same joke a hundred and fifty times, and let people be there for the big events—the debates and the policy speeches."[64]

There were debates galore to cover during the early primary season, and C-SPAN telecast most of them. There would be variants of set campaign speeches too. C-SPAN even caught Bruce Babbitt and Al Gore—speaking fifteen minutes apart at a Southern Leadership Conference gathering—telling a similar joke. Campaign flavor would come from "man on the street" interviews in places like Montezuma, Iowa, and at the Made-Rite Sandwich Shop in Des Moines. Everywhere it had the opportunity, C-SPAN would let the voters speak. While attempting to provide coverage of Iowa and New Hampshire, it would send twenty-seven-year-old Susan Bundock, a five-year veteran at C-SPAN, to Michigan to bring viewers both the regular and the rump conventions of the state Republican Party. Through C-SPAN, viewers could see the result of the attempt of a fragile alignment of Pat Robertson and Jack Kemp backers to take over the party from Bush allies. Having supervised the coverage, Bundock returned to New Hampshire, where she was placed in charge of C-SPAN's operations for the primary. Within its limited resources, the network tried to remind its viewers that Iowa and New Hampshire were not the political universe even if its early coverage strategy seemed to make them appear so.

Through its early coverage and the *Road to the White House* series, C-SPAN put average citizens on the campaign trail. Sometimes the journey looked as if it were sponsored by the food industry. Much of the contest for Iowa and New Hampshire beyond formal debates was staged at picnics and in restaurants. Viewers could see Jesse Jackson and Dick Gephardt at an Iowa corn boil. They would be there when Neil Bush stood in for his father and Elizabeth Dole for her husband at a GOP picnic in East Andover, New Hampshire. Candidates Al Haig and Jack Kemp would be there too. Viewers could attend a Lion's Club pancake breakfast in Boone, Iowa, with Vice-President Bush and have a seat at Lilliemae's Restaurant in Marshalltown, Iowa, with Pierre Du Pont. They could follow Bruce Babbitt into the Tin Palace in Durham, New Hampshire. The Dunkin' Donut Shop in Nashua was a favorite stop long before Bill Clinton recommended it to Newt Gingrich in 1995. On *Road to the White House*, there were farms to see in Iowa and fishing docks in New Hampshire, town meetings to attend and coffee socials to visit. Every stop offered talk of the formal kind and candid conversations too. Viewers would hear most of the talk, since many of the hopefuls agreed to wear a wireless microphone at C-SPAN's request.

Miking candidates allowed viewers to hear Pierre Du Pont discuss drug testing with a group of high school football players. Viewers listened to Paul Simon as he was being interviewed by Japanese journalists, and they picked up the small talk as Michael Dukakis sampled cooking-contest entries at the Iowa State Fair. Sometimes wary at first of the small microphone, candidates soon seemed to forget it was there. At least Senator Joe Biden did. Biden's appearance at a coffee in his honor in Claremont, New Hampshire, on April 7, 1987, was the third stop of the day for Carl Rutan and the C-SPAN crew. They covered a morning event in the northern part of the state, drove south for about one hundred miles to be at a luncheon, and then made the return trip to be with Biden. As that gathering appeared to be about to end, someone Biden knew on a first-name basis asked a string of questions. He wanted to know what law school the senator had attended and where he had placed in his class. Before he could get a third question out, Biden cut him off. "I think I probably have a much higher IQ than you do," Biden barked. He then launched into an explanation of his academic career.[65]

Although Biden's explanation simply did not match the facts, the matter rested there. It was, after all, a small event early in the campaign, one that broadcast news did not even cover. But on August 23, 1987, again before C-SPAN cameras, Biden debated six other hopefuls at the Iowa State Fair. There he Americanized the British Labour Party leader Neil Kinnock's family story and told it as his own. It took the press almost three weeks to suggest plagiarism. Both the *Des Moines Register* and the *New York Times* placed excerpts of a speech by Kinnock and Biden's talk side by side. Soon a number of articles pointed out Biden's unattributed use of the works of others in earlier speeches. It was only a matter of time before someone recalled the

New Hampshire exchange. The networks had already obtained a tape of the Iowa State Fair talk from C-SPAN. Now Eleanor Clift got the New Hampshire segment from *Road to the White House* and revealed its contents to *Newsweek* readers. "The final blow" that led to Biden's departure from the race, columnists Jack Germond and Jules Witcover wrote, "was a C-SPAN video tape of an incident in New Hampshire the previous April."[66] This may be the only segment from the *Road* series anyone recalls from 1988. C-SPAN was not out to make news, only to give voters information. "All we're doing," said Brian Lamb, "is taking these cameras and putting them in front of events that we have no control over, and what they do is up to them."[67]

Iowa and New Hampshire made C-SPAN's coverage distinctive. Of its efforts in his state, Democratic Party Chair Joe Grandmaison said, "This year, 'local events' in New Hampshire truly became 'national.' C-SPAN's coverage not only assisted the nation's voters in becoming better informed about the candidate, but permitted a better understanding of our state's unique role in the nominating process, and the seriousness in which we go about attempting to fulfill our responsibility."[68] From the Granite State, C-SPAN would move on to Super Tuesday.

On to the Convention and Beyond

If C-SPAN followed the back roads for the first two delegate races, Super Tuesday would put the network in the political version of the Indianapolis 500. The network's coverage was supervised by Barry Katz; thirty years old, Katz was a C-SPAN old-timer, having been at the network since 1980. Again C-SPAN sought its niche. Rather that attempt to cover the election, as the networks were doing, Katz spread crews throughout the South before the event. The week before Super Tuesday, C-SPAN gave viewers five hours of coverage a night using a combination of live and taped video from twenty-six different southern sites. For C-SPAN the highlight occurred the first night, when it was able to cover the presidential debate in Williamsburg, Virginia, with its own cameras, thanks to the Democratic Leadership Council, which sponsored the exchange. Katz had to keep in touch with party leadership in the primary states, schedule satellite time, shift his crews, hire others, and oversee editing the tape obtained. By primary night, when the broadcast networks all had special programs designed to divine the results, Katz's work was finished. Earlier that day, C-SPAN had devoted its morning call-in program to the contest, and that evening it provided open phones for callers to talk about the results. Post-election analysis was left to the callers on the program the next morning.

There would be other primaries to cover and speeches to broadcast as the field narrowed in the coming months. As always, C-SPAN was on the road: expressing delight that it had been able to retrieve the rented wireless microphone it had hooked to Michael Dukakis in State College, Pennsylvania; being turned down by Jesse Jackson's aides in Harrisburg when it tried to place a microphone on him too; and showing George Bush, before a group of children in East Harlem, wearing "Action Jackson" buttons.[69] As the primary results began to make it obvious to Americans who the presidential nominees would be, *Road to the White House* showed them a series of withdrawal and endorsement speeches. It presented interviews with national, regional, and state campaign directors. One segment featured former presidential candidate Walter Mondale explaining the American political process to foreign visitors at the Carter Presidential Library. Just as Dole, Gephardt, and Gore departed the field, so too would Billy Jo Clegg. Third-party candidates remained in the race, and C-SPAN granted them a small amount of time by its own standards, but coverage would be gavel-to-gavel. Thus viewers saw veteran Eugene McCarthy announce his candidacy on the Consumer Party ticket. Twice before the election both Ron Paul, the Libertarian Party standard-bearer, and Lenora Fulani, the head of a fusion party, gained airtime. Both the candidates and the audiences they addressed added a different dimension to what was at issue in the campaign. Paul appeared before a rally at the University of Washington and before the National Organization for Reform of Marijuana Laws. Fulani was shown raising funds in Shaker Heights, Ohio, and at a press conference sponsored by the Nation of Islam.

The conventions in Atlanta and New Orleans proved less a challenge for C-SPAN than those held four years earlier. The network's crew had been there before and had some idea about what would work and what would not. Planners also had two in-house documents that Mike Michaelson had drawn up, based on the 1984 experience. The documents gave advice on everything from selecting drapes and partitions for work space in the hall to purchasing equipment, from the cost of pulling cables in the convention halls to the process of securing credentials, from how to conduct preconvention on-site inspections to ideas for a close-of-the-convention party for the staff. No detail would be missing, including recommendations for obtaining floor-seat passes for "visiting VIP Board members, staff and guests."[70]

Three issues in Michaelson's planning proposals are significant if one wants to understand the C-SPAN mission and how it is achieved. The first deals with using pool feeds, the second with choosing camera positions, and the third with locating working space. "C-SPAN should provide its own floor coverage in 1988, *rigorously rejecting* taking a second position to commercial networks," Michaelson argued. He also rejected the notion of "*taking* the entire feed" from in-house systems controlled by the political parties. "*For obvious reasons, C-SPAN should* do its own thing." Put simply, the network wanted to be treated as an equal in negotiations, and it wanted to maintain its independence.

Michaelson then turned his attention to negotiating for camera positions and for a spotter position on the podium. Before 1984, a first-tier position to provide head-on shots of the speakers "was the sole prerogative of the [broadcast] networks." In San Francisco C-SPAN had broken that hold by gaining a spot "next to the network pool cameras." At the GOP gathering in Dallas it was "forced to take a feed from an 'in house' Republican camera at a fee of $250.00." For the usually frugal C-SPAN, this was not an issue of dollars but a concern for controlling the pictures available to it. Although there had been few problems in Dallas in 1984 because the podium "was rather a 'quiet' place," Michaelson pointed out that "a wide open convention in 1988 would make this shot extremely important." He concluded, "Having it controlled by a *political party*, might be very undesirable." C-SPAN also wanted to place its cameras on the podium. It had done so at the Democratic Convention but not in Dallas, where the space was allocated to the broadcast networks and to the party cameras. "If either or both conventions resort to their own cameras," Michaelson asserted, C-SPAN should "take a strong position in opposition to the usurpation of the podium position by a political party and the commercial networks." C-SPAN's "proven track record, growth and purpose," he believed, "would serve it well in any position negotiations." Still the network should "*insist on equal treatment for positions and locations with the commercial networks.*" For the most part, C-SPAN secured the camera positions it wanted at both conventions. Again it decided against building a skybox. "Although costs were a predominant reason" for not doing so in 1984, Michaelson reminded planners, "CEO Brian Lamb has usually taken the position that duplication of the commercial networks and others does not serve C-SPAN's best interest and had recommended studio space outside the convention halls parameters."

When it came to securing such space, Michaelson believed that the 1988 conventions could be a problem. "Finding locations within walking distance of the convention complex is one problem, finding a location 'free' is almost out of the question, unless, we get lucky as we did in 1984," he noted. Once again the cable industry came to the rescue, especially in Atlanta. Cox Cable owned systems in both convention cities. Its parent company, Cox Enterprises, also owned the *Atlanta Journal* and the *Atlanta Constitution*. Both had their offices three blocks from the convention hall. When David Andersen, the vice-president for public affairs at Cox Cable, suggested that the newspapers provide space for C-SPAN's call-in shows and for its master control facilities, Jay Smith, the publisher for both papers, could only envision "inconvenience and disruption." In the end, corporate cooperation won out. C-SPAN moved into sixth-floor space that had originally housed the sports department. The area provided a view of the newsroom from one side and of the state capitol from the other. It offered a perfect backdrop for call-in programming. The shows in turn brought guests to the building. In effect, some of the people whom the *Journal* and the *Constitution* reporters wanted to interview would come to them. The press connections undoubtedly helped C-SPAN secure local officials for call-ins. While at the newspaper offices, C-SPAN telecast an editorial meeting, giving viewers an opportunity to see how the two papers would cover the convention. It interviewed key staffers, including Jay Smith. The reaction he received as a result of C-SPAN's presence would be satisfying to both parties. "I've gotten a lot of favorable responses from friends who have seen our reporters on the air and who have seen scenes from our newsroom," he told a C-SPAN representative.[71] By the end of the convention, he wrote to thank Andersen for his "gentle persistence." He said, "It paid big dividends."[72] In New Orleans, Cox representatives would perform a different service by seeing that convention hotels could receive C-SPAN.

Both conventions provided a different media mix in 1988. The commercial networks went again with smaller staffs and abbreviated coverage ("commentary" might be a better word). As Tom Shales, of the *Washington Post*, put it, "Now that the television networks have begun their convention in Atlanta, the question is whether the Democrats will get many words in edgewise."[73] In New Orleans Jeff Gralnick, ABC's news producer, denied the charge that the networks cut away from the speakers too often. "You have to look at what's happening on the podium and see where it figures into the editorial line that you as an organization have developed," he told Shales. "We have to see where it fits into the theme of our coverage," he said, "not the theme of their convention."[74] C-SPAN's cameras also roamed from the podium, but the speaker's voice was always heard. Some of the shots that C-SPAN audiences saw were of anchors chatting in their skyboxes and of reporters seeking interviews on the floor.

As they did in 1984, the broadcast networks took a beating from the print media for their performances. Richard Harwood, the *Washington Post* ombudsman, rejected ABC News President Roone Arledge's suggestion that the conventions were "anachronistic." He rejected too another network official's contention that the conventions themselves made politics "distasteful to the American people." The networks' "real concern," Harwood argued, "has nothing to do with the health of the American political system. It has to do with money." Coverage cost money, especially in terms of lost advertising dollars. What nearly two-thirds of the viewing public did not care to see the commercial network owners, managers, and affiliates did not want to cover.[75] Criticism for the big three did not automatically translate into praise for C-SPAN or even a mention of the network. Viewers who wanted an alternative would have to look for it. The year 1988 would bring Americans more suppliers of information. With one of the conventions in its hometown, CNN would be back, with a budget comparable to the money expended in 1984. Its convention staff would approach three hundred in number.[76] Local television stations would add one thousand more reporters and crews to the three thousand present in 1984.[77]

Everyone's anchor, it seemed, had to be on the scene. Both parties would supply gavel-to-gavel coverage to those stations that decided against traveling to the convention sites. They would arrange satellite interviews for them as well. Mark Goode, the producer for the GOP-originated feed, told Tom Shales that "many, many network affiliates" used the services his party provided. "That means," Shales concluded, "a network affiliate could pre-empt Rather or Brokaw or Jennings or some panting correspondent to pick up a podium speech by someone from their own region."[78] The parties themselves made the gatherings as media-friendly as possible. They hired experts to ensure that the settings were as visually pleasing as possible and took the experts' advice on how to dramatize the proceedings. Most talks would be screened before delivery by a "central speech unit." "We just want to make sure all the notes are sounded in this symphony," one party official told Tom Rosenstiel, of the *Los Angeles Times*. So great was the attention paid to the media that one unhappy delegate speculated, "Next time you think they'll just rent a sound stage in Hollywood, hire a couple thousand people to pose as delegates, a couple hundred to act as demonstrators and film the thing."[79]

C-SPAN budgeted $233,000 to cover the Democratic Convention and just under $300,000 for the GOP gathering.[80] It took nearly eighty people to both. In a move uncharacteristic of the network, it spent advertising dollars outside of trade publications to highlight its brand of coverage. It depicted itself as a choice, not *the* choice, as other news outlets might. "Admittedly," the advertisement read, "C-SPAN's convention coverage is not for everyone. But for those who want to watch the event for themselves, there is nothing like it. No interruptions. No commentary. Just pure political process in action." Pure process would be supplemented by over thirty call-in programs per convention, platform hearings, a look at the host city, and an examination of "how millions of TV viewers and newspaper readers get *their* stories." C-SPAN promised not only to air convention proceedings live but also to re-air them entire each night. The nine-point program described in the advertisement introduced two new features in the 1988 format. Using the Sky Channel, C-SPAN would bring the conventions to Europe. It would also offer the telecast in Latin America on Pan American Satellite 1. Someone living in Peru could get gavel-to-gavel coverage. More germane to the audience that the advertisement supposedly targeted, C-SPAN would also run "a collage of conventions past." Thanks to an agreement with NBC News, C-SPAN had access to its archives, and thus viewers were able to see a series of highlights, including entire speeches, from both the Democratic and the Republican Conventions starting in 1948.[81]

C-SPAN delivered what it promised at both conventions, giving viewers a sense of what taking part in the action might be like. It took them on a tour of CNN headquarters in Atlanta and featured Ted Turner on a call-in show one morning. By following C-SPAN's cameras, viewers could also attend delegation receptions and parties. With C-SPAN, there would be no thirty-second peek into the room, with an anchor explaining the action. At both conventions, those watching would get a chance to meet delegates to find out who they were, what they did, and how they got involved. There would be no attempt to locate a Jackson delegate unhappy with Dukakis's choice of Lloyd Bentsen as a running mate or a Bush supporter displeased with the selection of Dan Quayle. C-SPAN interviewed delegates to help viewers find out more about them, not to use them as pawns to heighten action or to flesh out a story line.

During the period between the conventions and the elections, C-SPAN planned and executed a series of interviews with old and close friends of both the candidates. Providing that context, producers believed, would help voters better understand the George Bush and the Michael Dukakis not seen on political commercials. "Logistically and financially," it proved worth the expenditure to purchase a "new C-Band portable satellite dish" and to locate it at the Continental Cable studios in Cambridge, Massachusetts. The move would free up C-SPAN's "Ku truck" and save the cost of shipping crews and equipment to and from Boston. It would also facilitate covering events sponsored by the Kennedy School of Government at Harvard, now standard fare on C-SPAN.[82] The move was yet another case of cable industry cooperation. Continental CEO Amos Hostetter was a member of the C-SPAN board.

C-SPAN aired the presidential and vice-presidential debates live using commercial network feeds. Rather than predict the outcome and pick winners and losers, C-SPAN showed facilities being set up, covered pre- and postdebate rallies to let viewers see "the spin doctors" at work, and even brought newspaper reporters onto call-in shows to share their experiences on a debate panel. After every contest it would open its phone lines for viewers' comments. Following Lloyd Bentsen's "I knew Jack Kennedy" response to Dan Quayle during the vice-presidential debates, C-SPAN's open phone lines lit up as usual. The number of calls taken on the air gave no evidence of the real interest the exchange provoked. C-SPAN knew how many callers got through; the next morning the network would be surprised to learn how many tried and did not get through. The day after the debate, a representative of AT&T called C-SPAN to ask for advance warning of any programming like that of Wednesday night. "The number of calls 'blew away' the AT&T network," he said.[83]

All of this was pretty heady stuff for a network still uncertain about its viewership. Throughout the campaign C-SPAN had attempted to provide viewers with something more than what the commercial networks brought them and with something different. It gave them a full library of information, everything from the events themselves to the opportunity to talk to those who made the events happen. Throughout, it had attempted to balance its coverage, and it kept detailed records to make sure that it did. C-SPAN coverage, as intended, would buck journalistic conventions. In an era of sound bites, when political staffs met each morning to decide the day's theme and then waited until the evening news to see if the networks delivered their message to the voters, C-SPAN stuck with the long-form coverage. Returning from a trip to Minnesota and the Dakotas, the *Washington Post* columnist Michael Barone marveled at how "little political information" was available outside the beltway. "The amount of national political news" he saw on his trip "was less than appears on page 4 of *USA Today*."[84] Citizens obviously could see much more if they had cable and knew about C-SPAN.

Despite its efforts, in 1988 many still saw C-SPAN only as the channel carrying Congress. It was neither fully understood nor reckoned with by a large number of those in the world of politics and journalism. Perhaps the most fitting commentary on its status was provided by the *Road to the White House* program on November 6, 1988. On the Sunday before the campaign ended, it featured a panel, sponsored by the American Association of Political Consultants, on media and the political campaign. Participants included Mary Alice Williams, who had coanchored much of CNN's coverage, NBC President Lawrence Grossman, ABC political director Hal Bruno, CBS executive producer Warren Mitofsky, and political consultants Bob Squier and Charlie Black. C-SPAN was represented by its cameras. The network recorded everything and said nothing.

CAMPAIGN '92

In 1992, C-SPAN would once again set out to establish itself as "the network of record." As C-SPAN Vice-President Susan Swain noted, a presidential contest was the network's version of the Olympics. According to one trade publication, "virtually all of the elements" of C-SPAN's 1992 coverage were in place in 1988.[85] There would be adaptations, of course, as we shall see shortly. Not the least of them would be a new person in charge of campaign coverage, Steve Scully. Carl Rutan, his predecessor, had never been completely comfortable with the C-SPAN approach to journalism.[86] Scully quickly adapted to both the network's mission and its style. Like Lamb himself, Scully even pictured his parents when he thought of his target audience.[87]

As always at C-SPAN, the best new faces were seldom in the foreground. The staff changes at C-SPAN—and there were others—were inconsequential compared with what happened to the *news* in 1992. Acceleration of the news cycle, increased competition, and a growing skepticism over "the relevance of network TV news reporting" all played a role in how Americans saw the election.[88] An accelerated news cycle meant that voters could, and often did, see events before newspapers could report them or before the evening news network anchors could attempt to explain them to the public. The response in the print media was to turn to more analysis and to follow the stories the tabloids had made the news in order to sell papers. Competition would be a key word. On the cable front, MTV, Comedy Central, and Lifetime all offered "unique takes on election year events." CNBC, BET, and the Discovery Channel also offered perspectives on the campaign.[89]

More media outlets meant more news or what passed as news. Television and print media are commercial operations, after all. Lurking behind every decision would be the dollar. That is not to say that only stories that attracted audiences were worthwhile printing or airing; it is to say that no business can survive without making a profit. There were still dollars to be made in the news business. For network television, they would come from selling advertising. The network evening news with the highest Nielsen ratings could, and did, earn $15,000 more per thirty-second spot than its competitors in 1992. That figure translates into close to $30 million more a year in revenue.[90] Still, the fight for those dollars would become more difficult in 1992. The network monopoly on the news was a thing of the past. By the time the evening news program went on the air, local stations had already reported the world and national news that the big three were about to bring, often using the same pictures. In major cities, before viewers saw Rather, Jennings, or Brokaw they could watch up to three hours of news from a local affiliate. Throughout the 1980s the executive producers of network news had used the anchor position as a competitive weapon against their own affiliates and against other news outlets. Former NBC executive Reuven Frank called the anchor the producers' "million-dollar baby" in a "five and ten-cents store."[91] In 1992 Peter Jennings negotiated a new contract. The five-year deal totaled $35 million, double C-SPAN's entire budget for 1992. His network, however, would still cut back on money for the news department, $7 million in 1991 and over three times that amount more in the following year. The downsizing trend was evident at both CBS and NBC as well.[92]

If 1992 brought increased competition in the news business, it also redefined the news. Of course the shift was not a sudden sea change; it had been coming for a long time. Now the tide had gone out, due to a number of factors. Distrust of the old media in particular, and of journalism in general, grew almost as rapidly as did alternative sources for the news. In the process, the distinction between the news and infotainment seemed to disappear. At various times during the campaign, all three major candidates seemed to run against journalists and the practice of journalism as well as their opponents. On occasion, the tabloids and the local television stations set the news agenda for the networks and the print media. In his biting critique of the press, Howard Kurtz wrote: "The plain truth is that there are no rules anymore, no corner of human behavior into which prying reporters won't poke. All of the media, from the prestige press to the sensationalist rags, had been infected by a tabloid culture that celebrates sleaze."[93] *Hard Copy* had become hard news. Downsizing, and sometimes the disappearance of whole newspapers, marked the 1990s as owners and editors struggled to redefine their role. Some may have laughed at the "MacPaper," *USA Today*, in the beginning, but adopting its look and style became a tool for survival for some dailies.

If both the commercial networks and the newspapers had trouble being seen as creditable sources in 1992, politicians aided and abetted the problem. In 1992, candidates began to use both talk radio and its television variants as whistle-stops on the campaign trail. Doing so supposedly allowed them to get close to the voters without having to confront "gotcha" journalism. Ross Perot was open in his disdain for conventional media and launched his campaign on the *Larry King Show*. To introduce "the real Bill Clinton" to America, Clinton's advisers "aggressively" sought out popular talk shows. Johnny Carson, Barbara Walters, Oprah Winfrey, Phil Donahue, Sally Jesse Raphael, Larry King, David Letterman, Arsenio Hall, and even Don Imus and Rush Limbaugh were all under consideration.[94] The candidate played his saxophone and spread his message on some of their shows and fielded questions from an MTV audience. Sunday-morning public policy venues were not the place to be.

If the media were in trouble in 1992, so too were professional politicians, and the candidates knew it. Ross Perot ran against the entire Washington establishment. Party made no difference; anger with "politics as usual" did. This was a constant theme for Perot. Bush adviser Fred Steeper picked up the message in focus groups. "The voters are disgusted," he wrote. "They are disgusted with politicians, partisan politics and government gridlock." He reported, "They sound like their tolerance for politics-as-usual has worn out."[95] It was clear to the Clinton group that voters wanted to see "the person . . . beyond the 30 second spots," the consultants, and "his people."[96] Alternative media and public debates provided that opportunity. Against the backdrop of anger and frustration with the conventional news sources and politicians in general, C-SPAN began its coverage of the 1992 campaign.

The Luxury of Lingering

As had always been the case, C-SPAN would operate on a campaign budget far below that of other news outlets. Its commitment of $2 million dollars to the campaign was $1.5 million less than the grant CNN had received from the Markle Foundation to supplement its coverage. For the early part of the campaign, Steve Scully was C-SPAN's political unit. He and Terry Murphy would decide what to air. From Scully's perspective, at least, that would have its advantages. He would not have to go through several layers of administration to get a decision. The turnaround time from idea to execution would be minimal.

Although the *Road to the White House* series would not begin until May 3, 1991, C-SPAN was already covering prospective candidates on its *Politics '91* series. Paul Tsongas (D-MA), for example, had been on the hustings early. *Politics '91* featured a one-on-one interview with him, as well as covering his appearance before the staff of the *Los Angeles Times* Washington bureau. The program also served as a vehicle for introducing the issues likely to play a part in the campaign. *Politics '91* brought highlights of the meetings of both the Republican and the Democratic National Committees and carried interviews of participants attending both parties' House Caucus meetings. Other programming added context to the upcoming campaign. For example, C-SPAN telecast an academic forum: "New Hampshire, One Year and Counting." The network featured Governor Douglas Wilder (D-VA) on *An American Profile*. *Booknotes* presented the author of *Grassroots: A Year in the Life of the New Hampshire Primary*. In a sense, C-SPAN was on the road before the road officially opened.

In the early part of 1991, however, *Politics '91* was overshadowed by C-SPAN's gavel-to-gavel coverage of the Keating Five hearings and by its treatment of the Gulf War. It would send a crew to Saudi Arabia on January 7, 1991. The day after the attack to regain Kuwait, the network went to a nearly round-the-clock call-in format. Later it would follow, literally, a military police unit to the Gulf. Viewers were on the plane with the unit, thanks to the C-SPAN cameras. The war, in the early months of 1991, seemed to promise a dull *Road to the White House*.

Road to the White House would again be a signature program for C-SPAN, serving as the network's version of a sound bite from its *Election '92* coverage. Although it often presented entire speeches, some were edited C-SPAN style. "If we have two forty-five minute speeches we want to use," Scully told an interviewer, "we'll condense one into twenty minutes and another into twenty-five minutes." He had had experience at a local station finding "the minute or minute and a half" bite that would be extracted from a forty-five-minute news conference for the evening news. Twenty minutes, he believed, gave the viewer "a pretty good sense" of what a speaker had said.[97] The entire speech would appear somewhere on C-SPAN's schedule anyhow. Still the preference was for gavel-to-gavel coverage, live when possible. For important events that appeared under the *Road* logo, one could expect a call-in segment. A regular Monday evening call-in was eventually dedicated to the campaign.

Road to the White House was regularly scheduled twice weekly during the campaign. Both programs would be re-aired. Blocking out the time was initially a concern. "We didn't like the idea of setting aside ninety minutes a week in two different time slots," Lamb commented. "But we learned if you do define a time, more people will tune in. In a sense, I guess it's an important breakthrough for us."[98] What *Road to the White House* could not cover, *Election '92* would. Still, the *Road* series was as close as C-SPAN came to packaging its campaign coverage. Especially in the early primaries, the show gave the fullest possible views of the election process.

Just how full this view was can be seen by looking at what was shown on the five days before the Iowa caucuses and the week before the New Hampshire primary. The Friday before Iowans met in caucuses, the *Road* series featured Mike Glover, an Iowa political writer, giving an overview of the state of politics there. Viewers heard Pat Buchanan discussing his campaign, and they listened as Bob Woodward talked to students at Drake University about the role of press coverage in a campaign. The Sunday program brought half-hour segments with supporters of Governor Bill Clinton and Senator Bob Kerrey (D-NE) and a rally for and speech by Paul Tsongas. A Monday special edition presented live coverage of the caucuses for the third straight presidential primary year. The following day the chair of the Bush-Quayle '92 Iowa Steering Committee appeared to discuss GOP activity in the state and to look ahead to future contests. Then the show went back to New Hampshire, where each evening before the primary, C-SPAN simulcast the political portion of ABC affiliate WMUR-TV's evening news. The *Road* show took a comparative look at campaign advertisements, held a live call-in segment with Pat Buchanan, and followed Jerry Brown at a meet-and-greet in Nashua. It carried live the Conservative Victory Fund Dinner as well as a Democratic fund-raiser, The Hundred Club Dinner.

C-SPAN's cameras were there, gavel-to-gavel, when George Bush formally announced his candidacy before the New Hampshire legislature. The network re-aired the CNN-sponsored Democratic candidate debate. From February 11 to primary day on February 18, an hour and half each evening was devoted to call-in programs with candidates, party leaders, journalists, and New Hampshire residents. On election night C-SPAN returned to

WMUR-TV to simulcast primary returns and followed that with two hours of open phones. Somehow Brian Lamb also managed to moderate an hour-long teleconference on issues and process. The show brought together high school students from four schools around the state.

Buried in the frenzied final week in both states were two of the elements that distinguished C-SPAN's coverage in 1992. First was the extensive use of simulcasts with local network television news programs. C-SPAN would use, in Terry Murphy's words, "the best, strongest news operation in the local market" to provide a unique frame of reference for its programming.[99] Steve Scully would do all of the bookings. The ability to use network affiliates in long form is a telling commentary on what had happened to the power of the national news organizations. The idea was not a new one for C-SPAN, however. The network had first applied the strategy during the 1990 congressional elections, linking up with stations in California, Florida, Massachusetts, and Texas. It alone had brought David Duke's concession speech in the Louisiana gubernatorial race in 1991 as part of its simulcast of the coverage of election results by NBC affiliate WDSO-TV (New Orleans). What simulcasting allowed viewers to see is best illustrated by WMUR's programming on primary election night. Before Brokaw, Jennings, and Rather began their special coverage of the returns, C-SPAN viewers had been to Bush, Buchanan, Clinton, Tsongas, and Brown headquarters. ABC's Jeff Greenfield, Cokie Roberts, and Jim Wooten had already been on WMUR "nine or ten times" before their own national coverage began.[100] C-SPAN continued to use this format throughout the campaign with commercial network affiliates in Baltimore, Atlanta, Houston, Tampa, and Boston.

Although *Road to the White House* aired only one meet-and-greet session in the final days in Iowa and New Hampshire, following candidates over a period of hours would be a C-SPAN trademark in 1992, as would the use of the wireless microphone. Susan Swain explained the process to a reporter for *Broadcasting* magazine: "We might walk up to a campaign press manager or the candidate himself and ask if we can hook up a wireless microphone while they're walking around working the crowd. Most of the candidates in the field are very used to C-SPAN. They've been around them for a long time. They make the decision on the spot. A lot of it's spontaneous, especially in the primaries."[101] Scully recalled, "Every time we put a wireless mike on the candidates in 1992, they started out realizing that we're there and we're shooting." He added, "But we're not there for just five minutes like other camera crews; we spend a whole day with these folks so by the second or third stop they become almost oblivious to the fact that we're there and you really see these folks as who they are." The technique, he believed, helps voters to learn what makes the candidates "tick." Viewers can see if "they are warm or cold" and learn if they "listen to voters" or simply "shrug them off."[102]

Both Scully and the candidates realized "there's definitely a danger wearing a wireless mike." The one conversation overheard in 1992 that best reveals that danger is one that C-SPAN chose not to air. At a gathering in Bedford, New Hampshire, on November 15, 1991, C-SPAN caught Bob Kerrey telling Bill Clinton an off-color joke about Jerry Brown and two lesbians. C-SPAN had a verbal agreement with Kerrey to broadcast him talking to voters. In C-SPAN's mind, the joke was part of a private conversation; it fell outside of what they had "agreed upon," according to Scully.[103] Speaking for C-SPAN, Bruce Collins, its legal counsel, tersely stated: "It is C-SPAN's policy that any private conversation recorded by its crews in the course of the network public affairs coverage will remain private. The conversation recorded in Bedford, N.H., involving Senator Robert Kerrey was private." Despite Collins's statement, the network was "inundated with requests for copies or transcripts of the tape." Other news agencies approached C-SPAN staffers, hoping to get at least "informal or off the record confirmation of Kerrey's remarks."[104] Kerrey remembered the incident but still allowed himself to be overheard. After he had finished reading *Curious George* to a group of five-year-olds, one little girl pointed to C-SPAN's boom microphone and asked him, "What's that?" "That," he replied, "is something that can get you into trouble."[105]

Scully tells the story of the Kerrey incident, if asked, but his favorite vignettes were more quixotic. He readily recounted the hours spent in the cold when other crews had quit videotaping Tom Harkin as he stood deep in a ditch helping to excavate a sewer line and cursing like a trooper. Scully smiled as he recalled Bill Clinton finding a way to refuse a sea urchin at a New Hampshire fish market. He remembered Paul Tsongas swimming laps in a pool while C-SPAN cameras watched. His image of George Bush was far different from the *Saturday Night Live* spoof of the C-SPAN meet-and-greet style that featured Bush responding "Persian Gulf, Persian Gulf" to every question. The video verité style, in Scully's mind, was not "rocket science." He noted, "It's just a chance to show these folks as they truly are." Scully kept meticulous records of the time allotted to candidates on the *Road* series to ensure as much balance in coverage as he could.[106] He could not, of course, command candidates to make appearances in order to equalize coverage.

The wireless microphone would work well for C-SPAN, especially in the early days in Iowa and New Hampshire, the time for what Scully calls "retail" politics. As the race became more defined, "wholesale" politics began, and it became more difficult to employ the strategy. The coverage of "retail" politics earned C-SPAN plaudits. Tim Russert, NBC News senior vice-president and a longtime network fan, called the coverage "the video version" of Theodore White's famous *The Making of the President* series.[107] Bob Beckel, who managed Walter Mondale's 1984 run for the White House, labeled the *Road* program "the best single coverage of any presidential campaign" he had ever seen.[108] For Marvin Kalb, who headed Harvard's Center on Press, Politics, and Public Policy, the program offered viewers "the little things you would never get on the regular news." He told Susan Baer, of the *Baltimore Sun*, "They're not significant news, but they're so insightful." The early *Road* series was significant to one candidate. "When a lot of other folks didn't believe," Tsongas press secretary Peggy Connolly said to the same reporter, "C-SPAN was willing to fill the air waves with Paul Tsongas."[109] Tsongas himself told Scully that "if it had not been for C-SPAN," his campaign "would not have gotten the message out at a critical stage" of his effort.[110] From April 1991 to December 1991, Tsongas received close to nine hours of airtime. Why did he receive that much time? He was in the race early. "He was an announced candidate. He had a good campaign organization. He had money. He was moving up in the polls and he had a message." That was enough for Scully. "If I had my druthers," he said, "I'd cover them all." Availability, resources, and balance guided his choices. Candidates needed "to reach a threshold of, say, to 20 percent name recognition in the polls," they "needed to qualify for matching funds," and they needed to have "a campaign organization," Scully said.[111] Susan Swain added one other consideration: they needed to be on the ballot in a fair number of states.[112]

Looking back on 1992, Swain pointed to the coverage of third parties as a network "hallmark." "We really did cover all of them," she told a group at an Annenberg Washington forum in 1993.[113] Those who failed to meet the criteria mentioned above still got their voice heard more than once. Whether they were fringe candidates for major party nominations, represented a third party, or stood as independents did not matter. Where else could one see Larry Agran, Tom Laughlin, and Charles Woods? Who else would give Gene Smith and J. R. "Zachary" Taylor airtime? Who else would bring the nation "highlights" of the Libertarian Party convention, highlights that included four hours and ten minutes of platform debate and just under that time for the party's nomination process? Who else, while airing the series of debates that took place among Democratic Party candidates early in the campaign season, would also cover the Independent Party of Massachusetts debate or the Rainbow Coalition "Debate on Democracy" featuring Larry Agran, Leona Fulani of the New Alliance Party, and Tom Laughlin? Who else would cover a third-party representative debate using the questions asked in mainstream presidential contests? And then there was Ross Perot. C-SPAN aired all of his major speeches and carried his infomercial. Between April 1991 and July 29, 1992, he was on C-SPAN twenty-nine times for a total of 1,043.67 minutes.[114] His appearances counted with his voters. His interview on *An American Profile*, which aired on March 18, 1992, tops the list for transcript requests at C-SPAN. The next three spots also belong to him.[115] Coverage of candidates running on other than the Democratic and Republican tickets would be yet another way of "contexting" the race. The effort in 1992 was but an expansion of what C-SPAN had begun four years earlier.

Two new elements would be introduced in 1992 coverage. Albeit small ones, they too would represent attempts to broaden viewers' understanding of the process. C-SPAN not only would offer viewers a chance to compare commercials used in the 1992 election but also would air commercials from previous campaigns. It showed past presidential debates as well. Even though C-SPAN might not appeal to an MTV generation, it went into high school classrooms after each presidential debate to listen to what those about to cast their first vote had to say.

Scully admitted that covering "raw form politics unedited" can be "sloppy as far as the way the process works." But it was also C-SPAN at its best, better than in 1988. For Thomas J. Meyer, writing in the *New York Times Magazine*, C-SPAN was "television for the post modern age, with a formula that's revolutionary in its simplicity: Turn on the camera and let it run. In the era of the sound bite, the network offers an outlandish alternative: the sound gulp, the sound meal, the sound *diet*." Meyer also praised CNN but added, "It is C-SPAN that had made the most radical breakthrough in broadcasting: showing life as it happens, without commentary, spin or editing."[116]

The Conventions and Beyond

By 1992 the major commercial networks had given up all pretenses of carrying the national conventions. They had brought viewers only ten to twelve hours from each site in 1988 and would bring approximately the same amount in 1992. CBS actually brought less, breaking away to offer its viewers the

revenue-producing all-star baseball game the second night of the Democratic Convention. NBC telecast three hours a night, but two hours were
carried only on PBS, with whom the network had engineered a cost-cutting
arrangement for coverage. The commercial networks provided commentary,
not coverage. To put their efforts in perspective, consider that the Comedy
Channel gave over two hours each night to its own brand of convention coverage. From the networks' standpoint, the issue was a matter of money and
news. It was estimated, for example, that CBS lost "$1 million in ad rev-
enues" when it pushed back its previously announced Wednesday schedule
to broadcast Mario Cuomo's nomination of Bill Clinton.[117] That figure becomes more meaningful when one realizes that this dollar amount was about
three times what C-SPAN would budget to cover the whole convention.[118]
As for the news, commercial network news executives and their anchors believed that the conventions were unlikely to produce any. In New York, Tom
Brokaw, for example, sardonically told Robert MacNeil and James Lehrer,
"There's a danger that some news may break out at the convention."[119] For
him and others, the conventions would be one part party advertisement and
one part festival and circus. If news was to be made, the press would have to
make it. The gavel-to-gavel coverage promised by CNN and C-SPAN would
relieve the other networks of the problem of sticking with floor action.

There would be a difference once again between CNN's and C-SPAN's
coverage. CNN brought "a small army of about 300 people" to New York,
armed with just over twenty-five cameras. C-SPAN came with "60 workers and 13 cameras."[120] CNN would be like the big three had been in earlier
years, controlling the agenda as much as covering what happened on the floor.
CNN would produce a program, whereas C-SPAN would turn the cameras
on and let them run. CNN was a new ABC, CBS, and NBC. "The broadcast
networks," Brian Lamb told a reporter, "have to worry about 'producing'
the conventions and using their on-air talent and commentators to make the
conventions interesting to a mass audience, even though [with the current
methods of selecting presidential nominees] there is very little suspense to
conventions today." C-SPAN's "self appointed mission," Lamb explained,
"is to cover political events from start to finish, to give viewers the sense of
being there themselves and making up their own minds about what's going
on." He added, "We never have to decide what we're going to cover based
on how interesting the event is going to be."[121]

As in 1988, both parties provided satellite news services for stations unable to be in New York or Houston. The new kids on the block in 1992 would

be Comedy Central and MTV. Indeed MTV had made an effort to follow the campaign all year long as part of its "Choose or Lose" programming. Its major coup that year was its ninety-minute forum with Bill Clinton. Its chief correspondent at the convention was twenty-four-year-old Tabitha Soren. She was assisted by guitarist Dave Musteene and rapper M C Lyte. Even Larry "Bud" Melman was on the floor in New York, commenting for *Late Night with David Letterman*. The year 1992 would be a different news year.

As it had each convention year, C-SPAN made some adjustments in its programming in 1992. The trademark features of its coverage remained, of course: gavel-to-gavel coverage; nearly forty hours of call-in shows from each city; and an expansion of its in-the-room look to an in-the-convention-town one. At the Democratic Convention in New York, however, costs forced the network to run preconvention-week programming from Washington, D.C., instead of from the convention city and to accept a pool feed of the floor action, a feed that it would supplement with two cameras of its own. The use of the pool camera feed led C-SPAN's camera-angle-bias watchdog, Professor John Splaine, to urge that "only C-SPAN cameras and crews be used for coverage" of the Republican Convention. Bluntly, he wrote to Brian Lamb and to the network's two senior vice-presidents, "C-SPAN's field crews are the public policy professionals, no one does it better than they do." The purpose of the network pool coverage differed from that of C-SPAN's coverage. His suggestion would be heeded.[122]

To do one of the things it did best, context the floor coverage, C-SPAN brought five handheld camcorders to the convention. At $2,200 each, they cost less than one-tenth the cost of standard news cameras. Although the quality of the picture they produced was not quite as good, it would be good enough for video verité. C-SPAN used them to produce convention vignettes on the streets outside the convention centers, in the halls, and around the city. Viewers could see what it was like to actually be there. They could even learn about the production of political buttons and see Jerry Brown as he prepared to spend a night in a homeless shelter.[123] The network used both channels to air the vignettes. In fact, it used its second channel better at both gatherings than it had in 1988, running archival footage, re-airing speeches, and taking phone calls.

As always, the materials produced from working around the fringes of gavel-to-gavel coverage brought viewers a glimpse of the process, a glimpse that, before, they would never have been able to see without participating in the process—and maybe not even then. Two examples will suffice. Who else would place a cameraperson on the bus with Jennifer Carrey, an eighteen-year-old delegate from New Hampshire, to trace her trip to the Democratic Convention? C-SPAN viewers even saw her checking into the hotel.[124] In Houston, Brian Lamb spent time with three female delegates, ordinary folks, not party moguls. Where else would one learn that it cost one of them "$1000 air fare and hotel"? But, she said, "We're sharing rooms and having fun." Lamb asked another delegate one of his favorite questions, "Who inspired you to be a delegate?" She responded, "I'll start to cry if I talk about it, but it was my dad, who's a Democrat, who told me to always stand up for what I believe."[125] Neither account may have meant much to anchor desks or to academics who study the process. Those at the desks were looking for "a story." Perhaps Jerry Brown would lead a rump group in New York and break away from the party. Perhaps George Bush would dump Dan Quayle from the ticket. Now that would be news! Some scholars of the process might have little interest in the young woman from New Hampshire or the three delegates in Houston. Did they affect the polls? Did they constitute a representative sample? These four women were outside the political big leagues, but for viewers, they humanized the process. Anyone could participate. Real people, people like those whom viewers may have known, were important too. Their coverage would be part of C-SPAN's grand national civics lesson. Politics was about more than polls, consultants, and spin doctors. In a democracy, it was about average citizens too.

On the surface, C-SPAN's coverage of the conventions went off without a hitch. The network was no longer an amateur when it came to carrying conventions. Its crews were experienced. For several of the crew, even if they were only twenty-nine years old, as was Jay Berk, New York and Houston would be the third time around. Below the surface, however, things did not go smoothly. In Houston, things large and small were a problem. Perhaps inadvertently, convention organizers had left C-SPAN representatives out of several important planning sessions. The work space given to the network was poor. Master control was assigned to an area that suffered power failures. To add insult to injury, Brian Lamb was among a number of journalists who exited the convention hall the first night only to find his car towed "from a lot the police had instructed them to park in." To make matters worse, the officers could not tell Lamb to which of three possible locations his car had been taken. He finally found it in a lot ten miles away. Cab fares and fines hit the three-figure mark. It should have been a good week for Lamb. There was "a whole page" in *Time* magazine devoted to him as well as "high marks" from the *Los Angeles Times*. Lamb, however, dislikes personal attention, especially when he believes it comes at the expense of credit he thinks is due the network. The final blow came on a call-in program he hosted when a female viewer accused the network of favoring the Democrats. Bias is the only charge viewers can make that will let them know the usually complacent network head has a pulse. Lamb had had enough; he left Houston early.[126] Republican stalwart Peggy Noonan did not notice either his departure or his bias. Writing for the *Washington Post*, she said: "The authentic sound of the Republicans in August was the authentic sound of the Democrats in July: 'Richmond, Virginia, you're up next.' The man and woman to whom all eyes turned in trust and for inspiration were C-SPAN's Brian Lamb and Susan Swain."[127]

Lamb of course would survive the incident. If one were to ask him, he would say he really did not matter anyhow. C-SPAN would go on to cover the presidential debates in its characteristic fashion. Again the difference between C-SPAN and the commercial networks would be the contexting. C-SPAN alone provided retrospectives of earlier contests, even inviting Walter Mondale to its set to discuss his encounter with Ronald Reagan in 1984. It showed the debates being set up, focused on reporters preparing for coverage, carried the moderator's instructions to the audience, listened to professional spin doctors from both camps, and then, after the debates, listened to those whom Terry Murphy, C-SPAN vice-president for programming, called "the ultimate spin doctors—the audience," the C-SPAN callers.[128]

On election night, C-SPAN again relied on simulcasts with network affiliates in California, New York, and Illinois in an effort to emphasize Senate contests. It used feeds from each presidential candidate's headquarters, put its cameras in the newsrooms of *USA Today*, and tried to stay out of the way. Above all it allowed Americans to talk. "This is a place for reflection more than for reporting statistics," Susan Swain told a reporter. Brian Lamb put it simply, "People want to talk about what they see." He concluded, "In many ways our job is more important tomorrow than it is tonight."[129]

Tomorrow would come soon enough, but in 1992, amid a changing media environment and in a period of shifting political styles, C-SPAN had remained faithful to the spirit of its mission. It had become more sophisticated at the same time. It relied less on cable affiliates for production purposes but used them well in 1992.[130] In one sense it was ahead on the political communication curve. Since it had first aired Larry King's radio program in the early 1980s, it had brought viewers a steady diet of radio talk-show simulcasts. It had a sense of their growing influence. Its use of network affiliate television programming beyond sound-bite length was both unique and cost-effective. Insiders at C-SPAN would point to the extensive use of the wireless microphone as a distinguishing factor in its 1992 electoral coverage. For the third campaign in a row, the press had announced that the network had come of age. This time, officials at C-SPAN had reason to believe the statements. In its own gavel-to-gavel style, it had done its part to slow down the acceleration of the news cycle, giving viewers both text and context and plenty of both. It probably gave viewers more than they could handle, with nearly twelve hundred hours of election coverage and with the constant re-airing of pivotal events. Lamb had set out to break the stranglehold of the commercial network news when he had first conceived of C-SPAN. The network had indeed played its part once again in doing so. In an ironic way, C-SPAN had started out to supplement what Americans could see and read elsewhere, but for some viewers, it became the source, not the supplement. What they saw led them to like C-SPAN more and to distrust the mainstream media in the process. Election coverage in 1992 cost just over $2.3 million.[131] C-SPAN was still a low-budget operation for three reasons. First, it had no network star to pay. Second, long-form television required much less money for editing. Third, unlike the commercial networks, C-SPAN needed no fancy bells and whistles to attract attention.

Still, many viewers thought C-SPAN's gavel-to-gavel, no-commentary format was potentially harmful both to journalism and to the citizenry in general. In his *Strange Bedfellows: How Television and the Presidential Candidates Changed American Politics, 1992*, Tom Rosenstiel barely mentioned the network. When he did, it was to illustrate the costs of a programming format that does not "filter out by any journalistic norms of editing what it airs." As a case in point he noted that since C-SPAN had begun, the use of Special Orders speeches in the House had doubled. Using that vehicle, Congressman Bob Dornan (R-CA), "without a shred of evidence," had made Bill Clinton's visit to Moscow while an Oxford student an issue in the campaign. C-SPAN, Rosenstiel pointed out, reached a potential sixty million homes, and that "changed everything." He asserted, "It did not take long for Dornan's daily diatribes to become the buzz of Washington." Merely letting the cameras roll allowed rumor and innuendo to become campaign news. C-SPAN provided the outlet. Rosenstiel also faulted the network for letting its cameras run while a local reporter on a radio talk show in Boston released exit-poll results on election night before the polls had closed. The major commercial networks had agreed not to do so because such information might affect those who had not yet voted. When officials at ABC heard of the leak, Peter Jennings supposedly said, "There goes C-SPAN's virginity." To which Roone Arledge, president of the ABC news division, replied, "It's about time."[132]

At C-SPAN there would be a word of caution too. Reviewing election-night coverage, John Splaine thought the production lacked focus. "It was difficult to know what was happening," he wrote to Terry Murphy, or "where C-SPAN was going with its programming." He asserted, "During the major national events, C-SPAN needs to have its mission clearly in mind." It should "stay with its mission and keep focused." Splaine cautioned, "Don't try to do too much!" Despite the faults he saw that evening, Splaine concluded, "C-SPAN's coverage throughout the election was exemplary."[133] Many Americans agreed.

AN EARLY STARTING GUN: THE 1996 RACE BEGINS

Road to the White House for 1996 began broadcasting in April 1993. "We had no intention of starting *Road to the White House* this early," Steve Scully told Elizabeth Kolbert, of the *New York Times*. "The reason we started was because events started."[134] The earliest episodes of the *Road* series ran ninety minutes monthly, taking viewers to receptions and speeches and following some individuals, like former Defense Secretary Dick Cheney and former Housing Secretary Jack Kemp, as they tested the waters and then decided not to get in. By September 1993, Senate leader Bob Dole (R-KS), his colleague Phil Gramm (R-TX), and Bob Dornan had all been on *Road to the White House*. So too had Lamar Alexander (R-TN), who showed signs of attempting to imitate Paul Tsongas by being in Iowa and New Hampshire early and often, as the senator had done in 1988. As the paths to the primary states became better traveled, the *Road* program settled into a weekly Sunday night slot in February 1995.

The *Road* show would again be event-driven; the campaign cycle would dictate what it would cover. In the season for announcements, for example, announcements would compose the show. Based on his experiences in 1992, Scully had a number of ideas for the program. He explained what he would "like to do in the weeks in which there is a lot going on." He said, "I'd like the first ten minutes or so to take a look back on some of the main things that have happened and the last fifty or forty-five minutes to be a focus on a particular candidate."[135] Again C-SPAN would try to be "almost invisible" on *Road to the White House*: airing entire speeches when possible, following candidates with wireless microphones as they met voters, and listening as they appeared on talk shows. Scully was especially interested in showing candidates before groups who did not belong to their regular constituency, to see how ideas get adapted to people and people to ideas. He wanted to let the show evolve as the campaign evolved—from the testing of the waters "to the campaign announcement season, to the meet-and-greet, to the let's see what their message is portion, to campaign advertising, to the fall when the speeches become much sharper, to the debate season." Above all he was concerned about balance and fairness. If the New Hampshire Republican Party gave equal time to radio talk-show host Alan Keyes and to Bob Dole, so would he. Of course, given the length of the *Road* show, there would be editing. For C-SPAN, that usually meant cutting out question-and-answer sessions or lengthy introductions of the head table before editing out any part of a speech. Almost without exception, any event so edited had already appeared in full as part of C-SPAN's regular programming.

Campaign dynamics played a large role in who got covered, for how long, and when. When the race heats up, Scully pointed out, he hopes to "really concentrate on the main players." But he added that C-SPAN would not neglect others "because the polls have proven time and time again that they're wrong." Still, who to cover, and when, would be in the hands of the candidates and the voters. Scully planned to watch his budget, approximately $2.5 million for 1996, closely. Planning might well prove difficult. "You can plan one thing and something completely different can happen," he pointed out. He explained that if someone had asked people in February 1991 who Ross Perot was, they probably could not have answered. "Most would have never heard of him." Based on "a hunch" that something like the Perot campaign might happen again, Scully spent less funds early in the 1996 election season. By the spring of 1995, representatives of the "Draft Colin Powell" movement had appeared on *Road to the White House* several times, and C-SPAN had covered a nearly hour-long speech by Ross Perot in Des Moines, Iowa, as well as given backers of the Patriot Party a chance to be seen. Scully also planned to cover the Libertarian Party. "That doesn't mean that we will be giving them the same amount of attention as the Democratic or Republican Party," he said, "but our feeling is twofold: we want to have as many voices as we can and go back to what our mission is: let the voters make up their mind as to what they think about a particular candidate, ideology, or political party." He added: "We are not out to try to help or hurt any party or movement. We are just out to cover them and let the decision be made by the voters."

Whether the single-weekly-program, sixty-minute format would remain in place was a matter of conjecture. It had been cut from ninety minutes to sixty early in the season to make a better fit in the Sunday evening schedule. The schedule, in turn, had been regularized as part of C-SPAN's effort to make the network more viewer-friendly. Viewer friendliness also would be the reason for a more "produced" program in 1996. It is well to remember that for C-SPAN, productions are not a flash-and-powder thing. Production simply meant "more explanation of what's going on." Scully pointed out, "We need to show events, but we also need to help viewers understand their significance." He asserted: "Nothing will change. You'll still have the speech by the candidate or the meet-and-greet sessions or the editorial board gathering." In addition, viewers will see "a week's events as they happened and interviews with journalists" and players in the game. One thing was certain, however. "We're not going to make our own views known." C-SPAN staff would simply try to explain events in the manner they had adopted in the beginning of 1995 for coverage of Congress and on the *Washington Journal*. That would mean "contexting" in the broadest sense: more archival tape. Now sixteen years old, C-SPAN had some footage of its own. "I'm a firm believer," Scully said, "in explaining why things are important, and why things happen the way they do." In C-SPAN style, however, Scully would not do the explaining; instead, invited guests would.

It is well to remember that the *Road* series represents only the tip of C-SPAN's campaign programming, even if it is a signature program. In 1996, C-SPAN planned to do more live coverage of events on its weeknight schedule. With the luxury of two twenty-four-hour channels, it could not only telecast the events but telecast them again and again. As an example of what that means, the candidacy announcement by Dick Lugar (R-IN) aired seven times using both channels. In late June 1995, Scully learned that he had two new tools available to him. The board approved funds for a second C-SPAN bus and authorized the establishment of bureaus in several places outside of Washington. In the first case, Scully had already planned to use the original bus to enhance programming. Its availability had allowed him to conduct interviews in Iowa. Among the ideas considered was using it for "voter on the street" interviews and using it at both conventions. The second bus opened up the possibility of covering both coasts. Funding for the bureaus included money for camera crews as well as producers, again enhancing what C-SPAN could do. In addition C-SPAN, like many of the candidates, was on-line. It too could use the Internet to provide context.

Once again C-SPAN geared up to provide text and context for viewers. Once again it prepared to do its job with less. In April 1995, "two people, plus the editing staff and editorial input from others in the company," represented the network's campaign commitment of resources at the editorial level. Scully said that CBS, at the height of its efforts, had "a budget of almost $20 million for 1996," compared with a budget of about $2.5 million allocated at C-SPAN. CBS also had, Scully remarked, "almost three hundred people covering just presidential politics" in its news division. C-SPAN had Scully and others like him. He would plan for 1996, go to New Hampshire and Iowa, make preliminary visits to both convention cities, host the *Sunday Journal* as well as some of the "Point-Counterpoint" segments of the *Washington Journal*, and handle C-SPAN's political desk. As late as July 1995, that was his job. "It's a small company," he said, "a small operation, and we'd like to have more people, but we can only do so much." What it could do would depend on how the candidates chose to use the media. Talk shows, satellite uplinks, faxes, and the Internet were all available to them.

The Fourth Strand: *Booknotes*

As a signature program, *Booknotes* is by far the least complex of those we have noted thus far but, in its simplicity, may be the most revealing about the C-SPAN style. It is on the air less time than any of the others. It is the only program identified with a single host. It is not event-driven. The selection criteria are simple: books must be in hardback, must be public policy related, and must be available to readers. Timeliness is not necessarily a factor. No author can appear more than once.[136] *Booknotes* is the perfect niche in niche television. Perhaps more important, the way Brian Lamb approaches the program and his manner in asking questions shed light both on the personality of the network and on the personality of the man who founded it. Over time, the consistency in questions says something about the man who asks them. Long before *Booknotes* appeared, Lamb told a reporter, "I think I'm the most inquisitive person I've ever met." He revealed: "I'm not comfortable talking. I'm more comfortable asking because I have this philosophy that you don't learn anything with your mouth open."[137] Of course, asking is also a way of telling.

The beginning of *Booknotes* may actually have been a sign that C-SPAN had arrived. Lamb's penchant has always been to try to do what one does better, rather than to strike out in new directions. The new program signaled that there was enough comfort and confidence to move beyond the meat-and-potatoes, core-mission programming. Lamb's rationale for beginning *Booknotes* sounds like much of his talk about the founding of C-SPAN. Others were not doing a good job getting beyond sound bites when discussing books on television. Viewers needed more information to make better decisions, in this case in selecting books to read. "There are a lot of authors interviewed elsewhere on TV," he told Frazier Moore, of the Associated Press, "but after the typical six-minute interviews, you'll never know if the book is any good or not. You won't even know if the author can write. You won't even necessarily know if the author can speak. After watching an interview like that, I find I'm left wanting more."[138] More information is a passion for a man who admits how little he actually learned in college. He just was not ready then. Time after time, whether it is before a group of college interns or before a group of university professors, he makes that admission. One senses that *Booknotes* is one way Lamb has found to go back to school. In the selection of book titles and in the questions he asks, one can even chart a series of courses Lamb might take if he had the luxury to do so—classes in foreign policy, history, journalism, and the craft of writing, with a dash of the visual arts tossed in for good measure.

As is always the case with new programming, the seeds for the idea of *Booknotes* appeared long before the actual program did. One could catch glimpses of the idea in some of the discussions that occurred on call-in shows. Today a mini-version of *Booknotes* has become an often-used element on the *Washington Journal*. The first full-length attempt to feature a book on a C-SPAN network program occurred in the fall of 1988. On September 14 of that year, producer Eileen Quinn taped a two-and-a-half-hour conversation between Brian Lamb and Neil Sheehan, whose *Bright Shining Lie*, a book about Vietnam, had been sixteen years in the making. The subject had always interested Lamb. "After a lifetime of research and writing," he said, "I felt that fifteen minutes on the air wouldn't be enough time to begin a discussion" of Sheehan's work. C-SPAN ended up presenting the interview in three half-hour segments. The idea of *Booknotes* emerged from that experience.[139] In-house producers first thought of making the program ninety minutes in length and considered, in C-SPAN style, occasionally using "open phones to allow viewers to comment on the book."[140] The first program in the series aired on April 2, 1989. Lamb interviewed Zbigniew Brzezinski; the book was *The Grand Failure: The Birth and Death of Communism in the Twentieth Century*. The set was a throwback to C-SPAN's earliest days: "a black curtain, two cameras, and two humans sitting there, [talking]."[141] The program would run sixty minutes in length and would be regularly scheduled each Sunday at 8:00 P.M.

<h2 style="text-align:center">SELECTING THE BOOKS</h2>

In the media backgrounder distributed for the fifth anniversary of *Booknotes*, Lamb asserted, "Books are selected not on the basis of whether they are good or bad, but whether they have some impact on the political system."[142] From the beginning, "impact" would be defined loosely by the producers who helped to select titles and by Lamb himself, leading to the selection of books such as Nathaniel Branden's *Judgment Day: My Years with Ayn Rand*, Charles Fecher's edited edition of *The Diary of H. L. Mencken*, Robert Remini's *Henry Clay: Statesman for the Union*, Peter Robinson's *Snapshots from Hell: The Making of an MBA*, and Howell Raines's *Fly Fishing through the Midlife Crisis*. Mostly, however, the choices would be mainstream politics and policy, with media-related and historical titles tossed in. Clifford Stoll's *The Cuckoo's Egg: Tracking a Spy through the Maze of Computer Espionage* remained an in-house favorite. Stoll did yo-yo tricks and distributed cookie recipes on the air.

Often neither the title nor the topic would put a book on *Booknotes*. One could not bypass a book by Margaret Thatcher, Helmut Schmidt, Richard Nixon, or Robert McNamara. The selection process was difficult, however. By the mid-1990s, at least thirty-five to fifty books, or inquiries about submission, came in to the network weekly. There were some easy cuts: paperbacks were automatically out, and the rare novel that made its way to C-SPAN could be put aside. It was up to the producer to winnow down the number for Lamb to consider. Sometimes other employees made suggestions. Sarah Trahern, the program's producer for much of the 1990s, often sneaked away, armed with a load of books, to a favorite restaurant for a long lunch. She would read as much as she could, enough at least to see if the text added something to one's understanding of the political system as she saw it. The first task for any writer would be to engage Trahern, to make her want to know more. A Georgetown graduate with an interest in American Studies, it was she who was drawn to Douglas Brinkley's *The Majic Bus: An American Odyssey*, an account of a bus turned into a classroom to take students across the country to visit literary and historical landmarks. In the beginning, Lamb wasn't much interested in the book; in the end, he was glad that Trahern had made a case for it. At frequent intervals, Trahern passed on a small number of books to Lamb, he sometimes brought some of his own, and the negotiation process began. Lamb never relied on reviews to help make the choice; he did not even read them. In the long run, titles from more-established presses had the best chance of making it, simply because these presses were better at marketing their product to the network. It did not hurt an author to have an aggressive agent. Still, Lamb and the producers who worked with him were never afraid of selecting a lesser-known writer whose work came from a smaller publishing house, and they often did.[143]

THE INTERVIEW PROCESS

Whether on *Booknotes* or on the *Washington Journal*, Lamb displays one unique trait at the network. He tries to avoid a meet-and-greet with the guest if possible. Introductions most often come just before *Booknotes* begins. It is the producer's job to instruct the guest, set the author at ease, and see that everything is ready to go. Lamb comes on to the set, carrying the book, which is usually heavily underlined and often marked with marginal notes. The title page, and the blank space before it, are filled with questions, references to page numbers, and other jottings. The show opens with a question from Lamb. He can ask close to one hundred in an hour-long program. He worries about the queries he poses, just as he worries about everything else at C-SPAN. During the taping of Jimmy Carter's appearance on the program, for example, John Splaine sat by, doing what he called a "question trail," tracking the seventy-four Lamb asked that day. Splaine and Lamb then talked "a good forty-five minutes to an hour" about the kinds of questions he had asked. "Even the most experienced hosts get caught characterizing things or defining things," Splaine explained.[144] Lamb wanted to be sure he had avoided doing so. As a result, one usually hears just the basics from Lamb: who, what, why, where, when, and how. A position on an issue followed by a question mark just is not his style. Listening to his queries, one quickly learns to whom the hour belongs. It is the author's, not Lamb's. Those who have watched, like Howard Rosenberg, of the *Los Angeles Times*, quickly pick up on Lamb's style. "Lamb may be the purest journalist on U.S. airwaves," he wrote. "Aware that *Booknotes'* viewers are much less curious about his views than those of the authors, he maintains a refreshing BBC style dispassion."

As an interviewer, he "rarely reacts to the interviewee." Rosenberg added, "Once Lamb asks a question, he's invisible until the author stops speaking and it's time for another question."[145]

Lamb's *Booknotes* Notations

Lamb obviously reads the books in preparation for *Booknotes*. Here are just two just pages of the heavily marked-up copy of *The Lincoln-Douglas Debates*, edited by Harold Holtzer, featured on the August 22, 1993, program.

From the questions he asks it is clear that Lamb likes books. He told one reporter: "If you've figured out how to master a subject and then pass it on in a book . . . well, that's a tremendous accomplishment. I think it's one of the finest things you can do."[146] He likes books, and he reads them thoroughly. It is wise for an author to reread what he or she has written before appearing on *Booknotes*. Lamb is likely to preface a question, as he did one to David Halberstam, with "You write on page 949 . . ." or , as he said on another occasion, "In note six in the back of the book . . ." This happens frequently. Viewers are not surprised to hear him say, "Let me jump to the middle of the book, page 272 . . ." Most authors are ready for that kind of questioning; a few are not. In the papers of someone reviewing *Booknotes* for another project is a notation beside the name of the author of one of the featured books: "Did she even write this book? Couldn't remember it."[147]

Lamb's questioning style is disarmingly simple. He asks the kinds of things that many in the viewing audience, if present, might be embarrassed to raise because they might display their ignorance. Then too, the queries might be so direct that some would consider them out of line. The latter never appear so, however, when they come from Lamb. From him they sound earnestly innocent. Questions of the first type are spread throughout the *Booknotes* series. To Lewis Lapham, author of *The Wish for Kings: Democracy at Bay*, he asked: "You write about the World Bank and first class and business travel, What's the World Bank anyway?" The title of Neil Postman's *Technopoly: The Surrender of Culture to Technology* provoked this query: "Can I ask you about technocracy? You had the tool-using and then we're a technopoly? What's a technocracy?" It never bothers Lamb to ask for a definition he does not know or thinks his audience might not know. It did not phase him a bit to ask Nathan McCall, author of *Makes Me Wanna Holler*, "What's Jonin?" He was not thrown off-stride when McCall corrected his pronunciation of slang terms. Every such question is premised on the notion that someone else might not know. If they do not understand terms, they might not know people either. An amazed reporter told his readers that Lamb had once asked an author who had mentioned Lenin's name, "Who is Lenin?" The reporter wrote, "You have the feeling that if someone mentioned Abe Lincoln, Lamb would ask who he was, too."[148] Not quite, but in preparation for his interview with Harold Holtzer, whose edited version of the Lincoln-Douglas debates sparked their reenactment, Lamb listed a whole series of terms he hoped to get defined and individuals he hoped to get identified. Next to the title page he scrawled, "Slavery, Egypt, Mexican War, Kansas-Nebraska Act, Dred Scott Decision, Popular Sovereignty, Canvass, Black Republicans, Hibernians, Henry Clay, James Buchanan," and "Who's Deacon Bross?"

Some questions might seem offensive coming from someone other than Lamb. They can be jarringly direct. William Bennett's *Book of Virtues* was a national best-seller, containing stories every American should read. He had just told Lamb the book was about "self-discipline and courage, compassion, faith, friendship . . . about the hard realities that constitute the virtues." He said: "We need to teach the virtues in school. They are very important things for young people to learn." After two questions, Lamb turned to a picture of Bennett's family in the back of the book. Bennett identified his young sons, nine and four years old, and Lamb asked, "Have you tried out any of these stories on these kids?" Many others wanted to know the answer, but few could actually ask it. When Richard Nixon appeared for a special two-hour *Booknotes*, Lamb, who had never met him before, asked bluntly, "Did you write this book?" When Shelby Foote, author of *Stars in Their Courses: The Gettysburg Campaign, June–July 1863*, described his working library of approximately 350 books, Lamb asked, "Have you read all those books?" He asked Hugh Pearson, who wrote *The Shadow of the Panther: Huey Newton and the Price of Black Power in America*: "You say that some people wanted money before they would talk to you for the book. . . . Did you pay any of them?" However blunt, the questions are characteristic of *Booknotes* interviews. They go to process, not to policy. It is up to the viewers, not Lamb, to accept or reject an author's point of view.

One can generally count on Lamb to pursue two lines of inquiry on every edition of *Booknotes*. He is interested in the process of writing and in the ways that an author's background may reveal what Lamb thinks is important for the viewers. Lamb seems fascinated with how books get written. He constantly asks how long it took the author to finish the book, how he or she went about writing it, when it was written, and what problems were encountered along the way. Note the string of questions to David Halberstam, author of *The Fifties*: "Where did you get the idea for this book?" "Did you select the way the book was laid out and the way the chapters followed?" "Were you working on any other book in the middle of trying to do this?" Several questions later he returned to the topic: "Where do you write?" "What do you write on?" "Are you writing another book?" Typically, when Haynes Johnson appeared to discuss his *Sleepwalking through History: America in the Reagan Years*, the same line of inquiry was evident. "You say you started the book four years ago," Lamb said. "This is a general question, but how much work did you do on this book?" Again he would ask where it was written and under what circumstances. Finally he asked: "What did you leave out that you wanted to get in? And by the way, what controls—I don't know how many pages. Let's see, it's something like 524 pages." Lamb opened his interview with Larry Sabato, who wrote *Feeding Frenzy: How Attack Journalism Has Transformed American Politics*, with a statement: "You dedicate your book 'For my piranhas, who remind me why I chose to teach.' " Having read the acknowledgments, Lamb knew who they were, but his audience did not. Finding out they were "mainly undergraduates, some graduate students," who had worked with Sabato during "the three years" the author spent "researching the book" might help viewers understand the task of putting together such a volume. From Johnson they found out that even with two researchers and help transcribing tapes of oral interviews, the author spent "every spare moment of [his] life—every Saturday, every Sunday, every morning, every night on this book." Halberstam explained how using a separate apartment as a place to write preserved domestic tranquility in his home. Both he and Johnson stated that using computers was essential to their tasks.

So great is Lamb's penchant for exploring the writing process that two comedians produced their own version of *Booknotes* for a local cable outlet. C-SPAN aired it as part of its fifth-anniversary special on the program. "Do you write on a computer?" asked the character playing Lamb. "No, longhand," the pseudo-author replied. "Pen or pencil?" "Pen." "What kind of pen—Flair, ballpoint?" The satire wasn't far from wrong. When Lamb asked Shelby Foote where he wrote—"in a house or in an office?"—Foote told him he worked in a room where he could sleep close to his desk. Writing, Foote went on to explain, was "a very deliberate thing." Five or six hundred words a day was "a good day" for him. "I write with a dip pen," he said. "What's a dip pen?" Lamb asked.

To those in a society that always seems to seek the real reasons, less virtuous than good ones, behind every action, Lamb's interest in the process of writing might serve as a clue that he is gathering information on how to write a book himself or is looking for material to expose the world of publishing. Certainly there is a book somewhere in Brian Lamb, should he ever decide to break his self-imposed vow of silence about the world of politics and politicians. Surely he has learned enough about the world of writers, agents, and publishers to tell something revealing. There are simpler, and more realistic, reasons for the line of questioning, however. Lamb is interested in the process of writing. He feels that questions about it help viewers to understand, help them to choose what to read. More important, the questions he raises often reveal the person behind the prose.

There is an understated, perhaps never stated, theme that runs through the C-SPAN concept. It has to do with the American character. One sees it especially in the series *An American Profile*. It also is revealed by a line of questioning on *Booknotes*. Lamb always asks about parents, about growing up, about education, about family, about people who influenced the author. There are constant probes into why those to whom a book was dedicated were chosen. One can hear these sorts of questions regardless of who the author is. Taken together, they might make an interesting monograph. They are a staple of *Booknotes*, a kind of C-SPAN variant of Bill Bennett's *Book of Virtues*. The questions Lamb asks likewise reveal something else. If a viewer pays attention to them, he or she can get a glimpse of what kind of man Brian Lamb is. What comes through most clearly is a thirst for information, a willingness to listen and not to talk, and a belief that once gained, such information will make a difference. There is a sense too that the way to comprehend ideas is to get to know the people who espouse them.

Booknotes and Its Influences

For a network that does not subscribe to Nielsen ratings, it is not surprising that records are neither kept nor even sought on the impact of *Booknotes* on the public, on the authors, or on book sales. The little data that does exist in-house at C-SPAN consists of several quotations gathered for the fifth-anniversary special, an anecdote or two, and a suggestion to check publishers for data. Although the number of viewers who participated in selecting the "Best of *Booknotes*" to be run over the Christmas season of 1995 are not that high, they too are impressive. Brian Lamb, when pressed, recounted four recent incidents that he believes are indicators of the program's effect. The first two seem to surprise him. He tells the story he heard from Steve Scully. Scully, it seems, was at the White House on Saint Patrick's Day in 1995. Paul Newman walked up to him that evening and was "goin' off about *Booknotes*," but it was not just "goin' off," since "he knew it inside out." Then there was the Christmas card that David McCullough sent Lamb. McCullough had been on *Booknotes* discussing his book *Truman*. Along with George Will and David Halberstam, both *Booknotes* veterans, he had been invited back for a C-SPAN conversation. The card read: "Dear Brian: I've had more comment—all *favorable* on the conversation you did with David H., George Will and me than anything I've done on television. It has been amazing and of course a huge pleasure. So many thanks again."[149] Lamb said, "Remember what he's done. Now if that isn't the strangest thing you've ever heard." If Newman and McCullough surprised Lamb, two other incidents pleased him. He learned that his board's first chair, Bob Rosencrans, had donated videotapes and transcripts of *Booknotes* to Columbia University

Library, where he was a trustee. But the story that seemed to please Lamb the most was about a seventeen-year-old recent visitor from Pennsylvania who "never missed *Booknotes* on Sunday night." Lamb said: "There's no question that that kid's life will be changed because of it and it won't be because of me. It will be because of that author he was turned on . . . like a teacher turns people on to a subject."[150] For Brian Lamb and the C-SPAN mission, "that kid" was worth every bit as much as a Paul Newman or a David McCullough.

The Strands Together

In pulling the four programmatic strands together, one is struck by a number of things. First, it is apparent that C-SPAN arrived and grew in a rare moment of time when conditions in Congress, changes in technology, and a shift in public attitudes toward the media all made growth possible. How the network grew is instructive. So much had been done by so few with so little. That frugality may indeed have ensured their adherence to the mission. A lack of available funds helped solidify a no-stars philosophy. Long-form, viewer-in-the-room television meant fewer costs calculated for editing and less dollars expended for fancy graphics.

In a commercial world, money may be everything. C-SPAN is not a profit-making organization. Still it would not have survived without the dedication, commitment, and loyalty that its staff brought to programming, as is evidenced in the discussion of each of the formats examined. Clearly, to work at C-SPAN and to enjoy it, one must be a team player and, at least in the early days, a utility infielder.

One is stuck too by looking at how C-SPAN covered Congress, pioneered the call-in format, went on the road at election time, and developed *Booknotes* and by how well its conservative, incremental applications of innovative ideas actually worked. One is taken also by the network's willingness to let ideas develop, to experiment with them without fanfare or publicity.

Most impressive is the long-term commitment to the mission the network set for itself. At every turn, actions seemed to pivot on the notion that the network was obligated to give citizens more information and to context events in such a manner that viewers were better able to judge for themselves. It is clear that C-SPAN has been able to become more viewer-friendly in the process, as is evidenced both in the recent innovations in congressional coverage and in the *Washington Journal*. It took the network over a decade and a half to begin to explain the legislative process as the House and the Senate debated issues. Part of the reason was caution and part concerned money. It was better to go slow than to be accused of bias by innovating too quickly. Even when innovating, however, C-SPAN never abandoned its commitment to long-form programming. The standard was set early: no sound bites, no interpretation. It remains so today.

C-SPAN programming not only reveals something about the network itself but also indicates its attitude toward its audience. Despite the occasional frustration with callers, C-SPAN treats its viewers with respect. Its programming posits a faith in their ability to make choices when given the information necessary to do so. At the very least, the network suggests that whatever those choices may be, they will be better ones than those made when only the commercial networks brought the news. No network exists for itself. Its programming is developed for audiences. In the very beginning, before the first caller reached a C-SPAN call-in show, those at the network wondered if anyone was listening. Just who was listening, and is listening today, we will learn in the next chapter.

CHAPTER SIX

The C-SPAN Audience

The prosecutor's question struck at the heart of the matter: "Just how did you become a junkie?" The nervous defendant explained: "It all started a couple of years ago. I guess boredom was to blame. The old ways of getting a handle on the world just were not working. I was surfing one day and came upon this strange, yet somehow familiar scene. I had seen the place before, but it had been filled with crowds of people. Now one person seemed to be trying to persuade me to believe in a new way of doing things. I moved on but was drawn back. Pretty soon I was coming back on a regular basis. Today I can't imagine not getting my daily fix. It has crowded other things out of my life."

The prosecutor shifted to an understanding tone: "You are not alone. I have heard variants of your story time and time again. Few people explicitly plan to get into your situation. You may just have been more susceptible. An increasing number of people have begun to surf through their television channels, finding C-SPAN by chance. They recognize the House or Senate chamber and wonder, Why am I being allowed in there when it's not a State of the Union message or a historic debate? You have simply joined the growing contingent of C-SPAN's loyal viewers. Sure it's habit-forming, it may even be addictive, but it has not proven to be harmful."

As one journalist put it: "Deadly as it may seem, C-SPAN has an intoxication all of its own. It has created, across the country hundreds of thousands of C-SPAN junkies–hard-core loyalists who savor a good congressional floor fight the way true baseball fans love a pitchers' duel. Like that other Creator, Lamb made C-SPAN in his own image–committed to the idea that TV news should not be delivered in small, glib nuggets."[1] Another journalist explained: "C-SPAN is addictive. You can pretend it's the good old days and you're watching the Watergate hearings. You can pretend you're watching a ball game and check your scoring against the box scores of the morning paper. Sometimes when your hearing is repeated, you watch it again, to see your favorite parts."[2]

C-SPAN's audience represents a mix of viewers. The self-described "junkies" are regular viewers. C-SPAN is "their" channel. It is the backup channel they turn to when they have no particular viewing plan. Other viewers are more selective, relying on C-SPAN for coverage of particular events. Most television viewers continue to "surf" on by, often oblivious to C-SPAN's existence and stopping only when they happen to see a familiar face. Virtually all of C-SPAN's audience initially discovered the network by chance. Our discussion of the C-SPAN audience will consider its changing composition, the impact of the audience on C-SPAN, and the impact of C-SPAN on its audience. We will consider both empirical and impressionistic data.

When C-SPAN talks about its audience, it is most likely to think in terms of impressionistic vignettes about individual viewers. After hearing hundreds of stories about loyal viewers, C-SPAN decided to celebrate its tenth anniversary by chronicling their stories. *America's Town Hall*, written by Brian Lamb and his staff in 1988, profiles 104 regular viewers. The vignettes cover the gamut, from the well-known (entertainer Frank Zappa, Representative Newt Gingrich, and President Ronald Reagan) to teachers, businesspersons, and retirees. Some of the stories are dramatic. Shirley Rossi, of Pueblo, Colorado, deserves the title "queen of C-SPAN." She watches twelve hours a day and is an instant celebrity when she visits the C-SPAN studios. For those who feel that one person cannot make a difference, the story of Lawrence

White should be an inspiration. After seeing C-SPAN at a local fair in 1982, White, of Culver, Indiana, became so excited that he bought a satellite dish so that he could watch the network. Although this guaranteed access for himself, White began a protracted battle to persuade his local cable system to cover C-SPAN so that his friends and neighbors could get access as well. When the local cable system refused to include C-SPAN because of limited channel capacity, White began recording C-SPAN programming and running it on a local-access channel. White's efforts generated considerable local publicity, and public support began to build. Five years after his first introduction to C-SPAN, First Pic Cable added C-SPAN to its offerings. In the typical idealistic outlook of many C-SPAN regulars, White commented, "I didn't do it to become a celebrity, I did it because C-SPAN is a cause I believe in and want others to believe in it, too."[3] It is hard to imagine any other television network generating such an evangelical fervor.

For every Shirley Rossi and Lawrence White, there are thousands of other, less-committed C-SPAN viewers. Their stories are best told using public-opinion polls and empirical audience analyses. The available empirical data varies significantly in terms of content and comparability. Polls commissioned by C-SPAN are most often used to prove the value of the network to the cable industry. A yearly analysis by Statistical Research Incorporated (SRI) is called *C-SPAN Pricing Study*. Aside from outlining audience demographics, this study asks questions such as, "How much would you be willing to pay for C-SPAN?" Brian Lamb and other C-SPAN staff react almost viscerally to the idea of ratings: "We exist in a world that too often says, 'Show me your numbers or you don't matter.' We are the single place, including public television, that never makes a decision based on numbers. Never . . . If we knew of something that might interest even a couple of thousand people, it's probably worth covering. Most people in television would say that's crazy, no one's going to watch. But when your master isn't the Nielsen ratings, 'no one' is a relative term. . . . If you go down the list of what everybody else in TV does, we don't do any of it."[4]

C-SPAN's hesitancy about ratings probably stemmed from fear that ratings would be so low that no one—including themselves—would be impressed with the network's significance. The fact that C-SPAN allows no commercial advertising undermines the need for determining ratings of specific programs in order to set advertising rates. Over the years a virtue has been made out of these necessities. The corporate programming philosophy reflects the advantage of not having to lower standards by "chasing eyeballs"—a rather inelegant slam against the network's commercial opponents. In recent years, as channel capacity limitations have made adding channels a zero-sum game in which channels are dropped to add new ones, C-SPAN's interest in its general audience size has increased. C-SPAN recognizes that attracting too small an audience will make it a likely target for cable systems hunting for a spot to add a new channel.

Over the years several academic studies have attempted to determine the kinds of people who watch C-SPAN as a channel of choice, again without rating specific programs. A number of these studies received C-SPAN funding, indicating C-SPAN's curiosity about its general audience. Maura Clancey, who originally worked on surveys conducted by the University of Maryland, was employed for a period as C-SPAN's director of research. She later went to work for SRI and now oversees the yearly C-SPAN survey. Despite eschewing ratings, C-SPAN makes available the most positive findings, which indicate a growing audience of politically aware viewers. In front-page stories and color graphics, C-SPAN press releases and internal publications have long highlighted poll results.

As C-SPAN became an established entity, polling initiatives—such as the Times-Mirror Center for the People and the Press—included C-SPAN in their analysis of the contemporary media mix in America. Over the years consulting firms, such as Fleishman-Hillard, added to the understanding of the C-SPAN audience by polling congressional staff and monitoring the relative position of C-SPAN as a Capitol Hill information source. The following analysis will utilize the rich but varied data collected on the C-SPAN audience.

Finding C-SPAN

The unpredictability of much of C-SPAN's programming makes it difficult to build an audience based on program loyalty. C-SPAN cannot control the length of House and Senate sessions and must work its schedule around them. Detailed C-SPAN programming information is not published in many newspaper television listings. The philosophical goal of being timely (see chapter 4) leads to programming choices being made only a few hours before broadcast. Viewers constantly talk about how they "happened on C-SPAN" while they were looking for something else or how they simply discovered the network while they were surfing through the channels. C-SPAN is a network for those who have plenty of time. People often tell Brian Lamb that they discovered C-SPAN when they were sick, so often in fact that he is tempted to call it "The Sick Network."[5] Other viewers outline intricate patterns of acquaintances and how they learned about C-SPAN from friends and friends of friends. Some programs, such as *Booknotes*, with its regular schedule, have developed a small but loyal following via word of mouth.[6]

A portion of C-SPAN's growth, and perhaps even of its demographics (see later discussion), has been facilitated by contemporary methods of television viewing. With a limited number of channels and virtually inviolate schedules, the television era of the pre-1980s was dominated by "intentional" audiences that expressed loyalty to particular networks or particular programs through their viewing habits. Conversations were sprinkled with comments along these lines: "It is Tuesday night, it must be time for *Milton Berle*"; "We can't go out Sunday night or we will miss *Ed Sullivan*"; and "Don't call me at eight o'clock when *Dallas* comes on." Once a program was chosen, the viewer tended to stick with it until the end. Advertisers recognized this behavior pattern and "sandwiched" their most crucial messages *between* segments of programs, when the audience was least likely to switch channels.

The expansion of television offerings and the almost simultaneous arrival of the remote control set the stage for "channel surfing" or "grazing."[7] In 1983, only 21 percent of viewers reported using a remote control. By 1988, that figure had grown to 43 percent and by 1991 to over 91 percent.[8] The multitude of offerings—many of which were not listed in the newspapers' television guides—made planning one's viewing schedule a frustration. The remote control allowed easy channel switching. Now the remote is to television what page-flipping is to magazines.[9] Channel surfing varies significantly by age and gender (see graph 5). To some degree, channel surfing is a "male thing," especially among older viewers. As comedian Jerry Seinfeld succinctly put it, "Women nest and men hunt."[10] Better-educated viewers are also more likely to surf.[11] Whereas channel surfing is most often used for *avoiding* commercials, as opposed to seeking particular programming, switching channels improves the chances that a viewer will find C-SPAN. Although there is no hard evidence that C-SPAN viewers channel surf more than other niche audiences, an avalanche of anecdotes clearly indicates that a high percentage of C-SPAN viewers are addicted "zappers."

At the same time that many viewers are finding C-SPAN by channel surfing, its long-form approach with limited action feeds into another developing viewing pattern. Whereas television once required almost complete attention because of its audio *and* visual components, C-SPAN's slow-moving

Graph 5

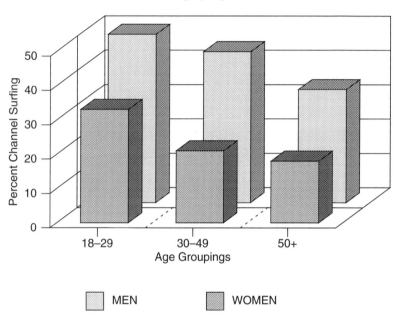

Channel Surfing by Age and Gender

Surveys indicate that channel surfing is a "guy thing" and is more common among younger rather than older viewers. *Source:* Times Mirror Center for the People and the Press, "The Role of Technology in American Life," May 1994 (press release, Washington, D.C.).

aspect allows it to be used as background stimuli. As media critic Walter Kirn put it, C-SPAN is "the video equivalent of New Age music . . . Environmental television . . . a free floating soup of info you can tune into or out of at will."[12] Viewers can read, work on the computer, or do housework while C-SPAN is on without feeling that they should be doing something else or that they will miss a key point.

Audience Size: Counting Eyeballs

C-SPAN executives have long been hesitant to enter the game of "counting eyeballs." When asked questions about the size of the network's audience, they tend to demure and to describe audience composition while dodging estimates of actual viewership. Cynics might conclude that they are worried about the small segment of the audience C-SPAN captures at any one moment. On the other hand, idealists are more willing to accept the official argument: "If we constantly worried about the size of our audience, we would stray away from our original purpose of providing politically balanced, long-form, public affairs programming."[13] The realists would be quick to point out that no one really knows C-SPAN's audience size. As a noncommercial network, C-SPAN does not need ratings figures to establish advertising rates, and ratings services generally do not measure the C-SPAN audience. Even if they attempted to do so, they would have difficulty applying their typical program-by-program approach because the C-SPAN schedule varies significantly day by day. The problem of actual measurement is exacerbated by the purported tendency of C-SPAN viewers to be channel surfers. Rather than turning on particular programs, they surf to C-SPAN with their remote and stop for a few minutes to determine their level of interest. If they choose to watch, it is seldom for an entire program.

Politicians use their own audience estimates for political purposes. Representative Robert Dornan (R-CA) enjoys sprinkling his speeches with comments about the "C-SPAN audience that at this time of day is probably at a peak of 1 million to 1 1/2 million people."[14] To add force to his assertion, Representative Dornan once included a visual image: "Over 1 million Americans are following this debate on C-SPAN. That is the Rose Bowl filled 100 times."[15] C-SPAN staff are at a loss to determine where his estimates come from. As Brian Lamb said: "He never asked us. . . . Bob Dornan's figure of 1 million is just not true. It does not make sense when Larry King only draws 700,000–800,000."[16] Representative Paul Kanjorski (D-PA) upped the ante by referring to the "7 or 8 or 10 million Americans who watch C-SPAN," presumably assuming that if anyone ever watches C-SPAN, he or she watches it all the time.[17] In the kind of exaggeration that would put the C-SPAN public relations staff to shame, Richard Ray (D-GA) addressed his remarks to his "colleagues and 50 million or 60 million people . . . looking at C-SPAN today."[18]

C-SPAN itself emphasizes its "potential" audience by carefully charting the number of affiliates carrying its signals and the number of households this represents (see graph 6). C-SPAN carefully tracks these figures, since they are the basis on which it receives revenue from the cable companies. Among cable channels, in 1995 C-SPAN was sixth in terms of market penetration (behind CNN, ESPN, TBS, USA, and the Discovery Channel), with over sixty-two million households. C-SPANII was twenty-second, with thirty-nine million.[19]

C-SPAN officials are quick to point out that only a small portion of the potential viewers are tuned in at any one time. In one of the rare times that Brian Lamb estimated C-SPAN viewership, he approximated the figure at "50,000 to 100,000" but indicated that for significant events, it could "go up to 3 million."[20] In reality, there is only limited data regularly collected on the viewership of specific C-SPAN programming. We know that the commercial networks' share of the viewing audience has declined (see the prologue), and it has been argued that niche networks like C-SPAN, which "began with tiny market shares," soon "boomed, doubling and tripling their modest shares . . . finally reaching millions of households and holding their viewers in front of the set for long stretches of time."[21]

Graph 6

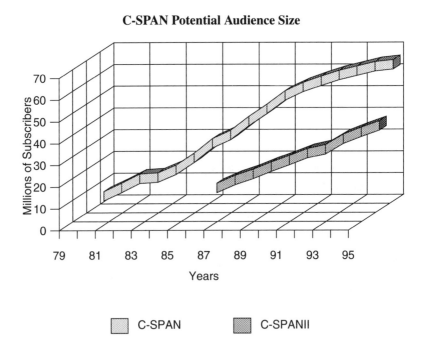

C-SPAN Potential Audience Size

Over 60 million households now have access to C-SPAN, and close to 40 million receive C-SPANII. *Source:* C-SPAN Marketing Division.

The available hard data comes from a series of often noncomparable public-opinion polls assessing C-SPAN familiarity and usage over time, with emphasis more on the demographics of the audience than on its size.[22] Initially the only people interested in evaluating the C-SPAN audience were academics and C-SPAN employees. Many of the early polls applied only to limited geographic areas or to cable subscribers. The first national poll to assess C-SPAN viewership was not conducted until 1987, eight years after the network began broadcasting. A significant sign of C-SPAN's "arrival" as a recognized force in American politics is the fact that national polls are now beginning to include questions about C-SPAN usage in their analysis of the contemporary American media mix.

AWARENESS OF C-SPAN

Brian Lamb developed C-SPAN with a kind of "Field of Dreams" philosophy: "If we build it, they will come." For "them" to come, however, the audience needs to know that C-SPAN exists. The early studies of cable households indicated that it took almost five years for over one-half of those with access to the C-SPAN signal to know it existed.[23] By the time of C-SPAN's tenth anniversary on the air, only about one-quarter of the national adult population knew of its existence.[24] Awareness expanded dramatically during the next few years. By 1992, almost two-thirds of adult Americans were familiar with C-SPAN,[25] and by 1995 awareness jumped to 73 percent.[26] The term "C-SPAN" is now part of the political lexicon for all but the most uninformed segments of the American public. The expansion of cable television penetration, the existence of two twenty-four-hour-a-day channels, the clear usage of C-SPAN by public officials, and C-SPAN's "arrival" as a reference point for popular humor on television and in cartoons have all contributed to its visibility (see chapter 2).

C-SPAN VIEWING

Along with increased visibility came increased C-SPAN viewing. The percentage of the population reportedly having viewed C-SPAN "sometime during the previous year" increased from under 10 percent in 1987 to over 40 percent in 1994 (see graph 7). Estimates of the number of households represented by such figures are impressive. Using a more limited definition of viewing, a large national survey, conducted independently of C-SPAN in 1994, indicated that 8.6 percent of the U.S. population had watched C-SPAN "during the last week"—a figure that has been growing on a regular basis. Table 1 compares the public use of C-SPAN with the public use of other media sources.

Graph 7

C-SPAN Viewership Growth

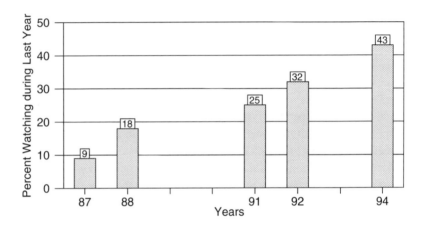

The percentage of television viewers who report having watched C-SPAN has grown each year. *Note:* Data were unavailable for the years 1989, 1990, and 1993. *Sources:* National surveys conducted by the University of Maryland Survey Research Center (1987, 1988), Statistical Research Incorporated (1991, 1992), and Luntz Research Associates (1994).

VIEWER ATTENTIVENESS

It is one thing to infrequently surf by C-SPAN and watch for a moment or two, but it is quite another to use the network regularly as an information source. Not only has the number of viewers increased dramatically over time, but viewer attention to C-SPAN has also increased. Among C-SPAN viewers, average viewing per month has increased from sixteen hours in 1987 to over twenty-one hours in 1992 (see graph 8). This figure would be impressive if the pool of viewers had remained constant, but the increased viewing time occurred during a period when viewership tripled. Either new viewers are coming with a much higher degree of interest or established viewers have significantly increased their viewing.

Table 1
C-SPAN Viewership Compared with That of Other Media Sources

	1992	1995
% Of the adult population viewing within the last week		
C-SPAN	6.1%	8.6%
CNN	36.7%	36.9%
ANY CABLE TV	59.5%	62.6%
% Of the adult population viewing yesterday		
EARLY NETWORK NEWS	14.1%	33.6%
LATE NETWORK NEWS*	4.8%	5.1%
% Of the adult population reading a newspaper each day		
DAILY	60.2%	57.0%
SUNDAY	66.5%	66.0%

*ABC's *Nightline*
Sources: Multimedia Audience Report (New York: Mediamark Research, 1992), pp. 2–57; Mediamark Research Inc., personal communication with the authors.

Brian Lamb has little misconception that C-SPAN is for everyone. Without recrimination, he realistically pointed out: "There are enough people out there to keep us in business . . . but I think it is fragile. There are enormous numbers of people who just don't want to pay attention at all. They want to make money and live. Bowling, golf, whatever—our stuff just doesn't matter to them."[27]

PROGRAM CHOICE

Although C-SPAN violently eschews ratings, there is an interest and practical necessity in knowing who is watching which type of programming. Among C-SPAN viewers, House and Senate proceedings are most popular, with over 40 percent reporting "often" or "occasional" viewing. Hearings are

Graph 8

C-SPAN Audience Attentiveness

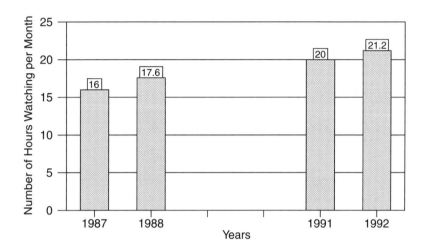

Not only are more viewers watching C-SPAN, but they are spending more time each month viewing its programming. *Note:* Data were unavailable for the years 1989 and 1990. *Source:* National surveys conducted by the University of Maryland Survey Research Center (1987), Statistical Research Incorporated (1991, 1992), and Luntz Research Associates (1994).

the next-most-popular category. About one-quarter of C-SPAN viewers report regularly viewing call-in programs, with specialized programs such as *Booknotes* regularly capturing the interest of about 10 percent of the C-SPAN viewers.[28]

Audience Demographics: Just Who Is Watching?

A number of predetermined factors related to opportunity and motivation constrain the representativeness of the C-SPAN audience. On the opportunity side, viewers must have access to cable television. Although the country is increasingly being cabled, the pattern of cable access presently represents a bias toward wealthier and urban and suburban communities. Initially, cable companies developed in areas with poor broadcast television reception or small media markets. As cable began to offer its own exclusive programming, the focus shifted to areas dominated by households having the discretionary income to afford cable even if they could receive broadcast signals. Given the cost and difficulty of laying cable, consideration was given to physical proximity and population density. Thus the second wave of cabling in the United States focused on the upscale suburbs, with urban areas being the last frontier of cable.

On the individual level, early cable subscribers tended to be upscale in terms of income and education. Even though the expansion of cable access has led to a more representative set of subscribers, they still do not mirror the population as a whole. In Todd Gitlin's view, the proliferation of cable choices "costs money. To those who have options shall more options be given. The supermarket is far greater on the better side of the tracks." C-SPAN, the Discovery Channel, and HBO penetrate into some 60 percent of the households, according to Gitlin, but he adds that this is not everyone. "The bifurcation of culture between a glutted middle road and a rutted low road holds true."[29]

In terms of motivation, C-SPAN viewers not only desire cable television but also must be interested enough in public affairs to seek out C-SPAN programs or to interrupt their channel surfing long enough to take in part of a program. On many cable systems, C-SPAN and especially C-SPANII are relegated to the high end of the channel selections, reducing the likelihood that they are even an option before the channel surfer lands on a channel for the duration of a program.

ASSESSING THE AUDIENCE MIX

Analyzing C-SPAN's actual audience is a tricky process. Initially, one must determine what it means to be a "viewer." There may well be a large difference between those who glance at C-SPAN once in a while as they surf by and those who intentionally view on a regular basis. The data seems to indicate that regular viewers represent an extreme demographic version of casual viewers.[30] Although casual viewers are somewhat better educated than the population as a whole, regular C-SPAN viewers are considerably better educated.

Two questions stand out. First, we want to know the population subgroups that tend to view C-SPAN. Second, it is important to determine the overall composition of the C-SPAN audience. Since population subgroups vary in size, the two answers may imply very different things. For example, polls indicate that 52 percent of college graduates have viewed C-SPAN at some time, as compared with only 31 percent of noncollege graduates. But since college graduates compose about 28 percent of the population, only about 30 percent of the C-SPAN audience is from this subgroup.[31]

Although the C-SPAN audience has changed a bit over the years by becoming more representative of the population, our emphasis will be on the most current data in the hope of understanding where C-SPAN is right now and perhaps where it is going. In terms of propensity to view, watching C-SPAN is positively correlated with education and income (see table 2). Age also plays a role, since middle-aged viewers are more likely to watch C-SPAN than are those younger or older. Men are considerably more likely to be viewers than women. Whites show a greater propensity to view than blacks. The two surveys from the 1990s reported in table 2 indicate a consistent demographic pattern and significant increases in viewing across groups.

The Aggregate Social Mix

Although differences in viewing tendencies between demographic groups are interesting, the more politically important question seems to be the overall mix of the C-SPAN audience. Those who hope to use C-SPAN for political purposes (see chapter 7) want to know with whom they are communicating.

Survey data from a variety of polling sources reveals a consistent profile of C-SPAN viewers. Viewers are considerably better educated and wealthier than the population as a whole. C-SPAN viewers are also less likely to be members of a minority group. Despite the fact that women outnumber men in the population, men make up a larger percentage of the C-SPAN audience. The C-SPAN audience is largely middle-aged, with viewers tending to be older than the population as a whole (see table 3).

The uniqueness of C-SPAN viewers as upscale—well-educated, affluent, and extremely interested in politics—has declined to some degree as the audience base has expanded. Surveys that distinguish between long-term viewers and newer viewers indicate that newer viewers are much more like the population as a whole. In an analysis of four national surveys in the 1980s, Maura Clancey concluded: "The C-SPAN audience is beginning to lose some of its distinctiveness. Several of the demographic characteristics

Table 2

Population Subgroup Tendencies to Watch C-SPAN, 1992–1995

Percent indicating watching C-SPAN "within the last seven days"

	1992	1995
TOTAL	6.8	8.6
EDUCATION		
College Graduate	9.0	12.5
Attended College	8.1	10.1
High School Graduate	5.7	7.2
Non H.S. Graduate	2.6	4.9
HOUSEHOLD INCOME		
Over $75,000	9.2	12.7
$50–$75,000	7.5	10.1
$30–$50,000	5.1	9.1
$10–$30,000	3.0	6.9
Under $10,000	2.3	4.0
RACE		
White	7.6	8.9
Black	6.3	6.7
Hispanic	5.9	8.7
GENDER		
Male	7.5	10.7
Female	4.9	6.6
AGE		
18–24	4.2	5.8
25–34	5.0	6.8
35–44	7.2	8.7
45–54	7.8	10.2
55–64	6.7	11.9
65+	6.1	9.3

Sources: Multimedia Audience Report (New York: Mediamark Research, 1992), pp. 2–57; *Television Audiences Report* (New York: Mediamark Research, 1992), pp. i, 3–177; 1995 data, Mediamark Research Inc., Spring 1995 survey, computer printout.

Table 3
The Social Demographics of the Aggregate C-SPAN Audience

	C-SPAN VIEWERS (% of each group classified as "viewers")			NATIONAL POPULATION (% of population in each group)
	1987[a]	1992[b]	1995[b]	1992[b]
EDUCATION				
College graduate	37	29	29	19
Attended College	30	23	31	19
No College	33	48	40	62
HOUSEHOLD INCOME[c]				
Over $75,000	9	22	21	12
$50–$75,000	15	26	22	17
$30–$50,000	35	28	27	27
$10–$30,000	31	19	24	32
Under $10,000	4	5	5	11
RACE				
White	na[d]	83	89	82
Black	na	11	9	11
Other	na	6	2	7
GENDER				
Male	54	58	60	48
Female	46	42	40	52
AGE				
18–24	8	11	9	14
25–44	50	41	40	44
45–64	31	31	33	25
65 and over	10	16	18	16

[a]Figures for 1987 come from a national survey done by the University of Maryland. Individuals classified as C-SPAN "viewers" indicated they had watched C-SPAN within the last year.

[b]Figures for 1992 come from national surveys reported in *Multimedia Audience Report* (New York: Mediamark Research, 1992), pp. 2–57; *Television Audiences Report* (New York: Mediamark Research, 1992), pp. i, 3–177. This data is also reported in Harold Stanley and Richard Niemi, *Vital Statistics on American Politics*, 4th ed. (Washington, D.C.: Congressional Quarterly Press, 1994), p. 55. Figures for 1995 come from Mediamark Research Inc., Spring 1995 survey, computer printout. "Viewers" are defined as those having watched C-SPAN within the last seven days.

[c]The family income figures were not corrected for inflation.

[d]na = Not available.

that set C-SPAN viewers apart—very high levels of education and income, and a preponderance of older, white male viewers—are disappearing."[32]

THE POLITICAL MIX

C-SPAN viewing is not an isolated political activity. Members of the C-SPAN audience tend to be heavy users of all public affairs media. Viewers are more likely to read a daily newspaper and watch television news. The messages seem to sink in, since C-SPAN viewers are almost twice as informed about political facts related to Congress than are nonviewers. C-SPAN viewers also use more informal methods of exchanging political views and information. More than twice as many C-SPAN viewers (24 percent) as nonviewers (9 percent) discuss politics with their friends and family.[33] Significantly, the characterization of C-SPAN viewers as being well-informed about Congress remains clear independent of the viewers' unique demographic characteristics and levels of media use. C-SPAN either draws or creates sophisticated Congress watchers.[34]

In terms of partisan self-identification, the overall C-SPAN audience is relatively balanced. Republicans are more likely to watch C-SPAN, but until recently their lower percentage of the population and the lower tendency of C-SPAN viewers to identify themselves as partisan independents meant that the *aggregate* C-SPAN audience was consistently more Democratic than the population as a whole. By 1992, the national tendency toward independent self-identification was even more dramatic among the C-SPAN audience, with over 40 percent of the general viewers of C-SPAN being unwilling to identify with one of the major parties. In choosing a party, C-SPAN viewers in 1992 were somewhat more likely to call themselves Republicans than was the population as a whole.

Voting patterns of C-SPAN viewers indicate significant autonomy in behavior despite partisan loyalties. In 1984, the C-SPAN audience favored Ronald Reagan over Walter Mondale more than the population as a whole and more than the overall cable audience. In 1992, C-SPAN viewers also opted for the winner—Bill Clinton—at a percentage somewhat higher than the general population.

In 1992, self-identified "regular" C-SPAN viewers were significantly more Republican than all C-SPAN viewers and the population as a whole. During the summer of 1992, these regular viewers also represented a hotbed of support for Ross Perot. As we will discuss later, the 1992 vote by all types of C-SPAN viewers may well have been a reflection of their growing dissatisfaction with the status quo.

When it came to the final vote in 1992, C-SPAN viewers in general acted much like the national population. C-SPAN viewers were only slightly less willing to support Republican George Bush or Independent Ross Perot. This result is particularly interesting because the general C-SPAN viewers were significantly more likely to call themselves "independent" than they were to identify themselves as an affiliate of one of the major parties.

In the 1994 elections, polls by Frank Luntz found that frequent C-SPAN viewers tended to be much more likely than the population as a whole to support Republican congressional candidates.[35]

C-SPAN VIEWERS: POLITICAL ACTIVITY

If one thing distinguishes C-SPAN viewers, it is their political activity. The C-SPAN audience is clearly not composed of political "couch potatoes." No matter what the measure, C-SPAN viewers act on their convictions. Viewers are almost twice as likely as nonviewers to discuss politics, contribute to political campaigns, and write their members of Congress.[36] Perhaps most important in a democratic system, cable viewers in general and C-SPAN viewers in particular tend to vote.[37] As graph 9 reveals, C-SPAN voters consistently report voting at rates 25 to 35 percentage points above the national averages in both presidential and off-year elections. Like levels of knowledge, the unique participatory character of the C-SPAN audience remains independent of audience demographics and attentiveness to other media.[38] In Brian Lamb's words, "You don't waste your time watching C-SPAN if you don't vote."[39]

Statistics tend to take on a life of their own. The 1988 University of Maryland survey indicated that 84 percent of C-SPAN viewers were registered to vote and that 92 percent of those registered reported voting. It soon became the conventional wisdom among C-SPAN staff, politicians, and the media that 92 percent *of all C-SPAN viewers* voted in 1988. The figure was reported so many times that everyone believed it. The actual 78 percent of all viewers reporting they had voted was impressive in itself, but in a 1992 article the *Los Angeles Times* increased the estimate to a totally unrealistic 98 percent.[40] In an era marked by concern over declining voting participation, observers seem to grasp at any straw indicating deviation. The bottom line is that C-SPAN viewers do vote at significantly higher levels than the general population.

Graph 9

Reported Voting Participation by C-SPAN Viewers and the General Population

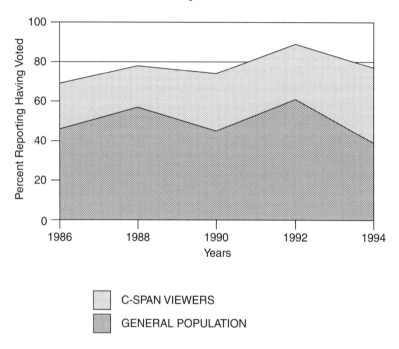

C-SPAN viewers are significantly more likely to vote than is the general population. *Note:* Self-reported voting supersedes actual voting by about 10 percent. It is reasonable to assume that C-SPAN viewers do not inflate their participation any more than does the general public and that comparisons across groups are legitimate. *Source:* Viewer data is based on national surveys conducted by the University of Maryland Survey Research Center and by Statistical Research Incorporated. General population figures are based on national surveys reported in Harold W. Stanley and Richard G. Niemi, *Vital Statistics on American Politics*, 4th ed. (Washington, D.C.: Congressional Quarterly Press, 1994), p. 80.

C-SPAN JUNKIES: A LOOK AT THE REGULAR VIEWERS

Defining "viewers" as simply those who casually drop in on C-SPAN programming may well miss that small component of the national public reporting that they regularly watch C-SPAN. Of American adults, 6 to 8 percent identify themselves as regular viewers. To some degree this C-SPAN "junkie" contingent is simply an extreme caricature of the more casual viewers (see table 3). They are similar to regular viewers in terms of gender distribution, income, and partisanship. However, some characteristics of casual viewers are not valid representations of regular viewers. Regular viewers come equally from all racial groups. Undoubtedly reflecting available time for viewing, the regular viewers tend to be much older. Forty-four percent are over sixty years old, more than twice the figure for the national population. Given the more limited opportunities for higher education a half century ago, it is no surprise that these older regular viewers are much more evenly distributed in terms of education, with 35 percent not having graduated from high school. Significantly, the polling figures are not simply tapping the more general audience for public affairs programming. Regular C-SPAN viewers are older and somewhat less educated than the viewing audiences of competing public affairs programming such as CNN, National Public Radio, and the *MacNeil-Lehrer Report*.[41]

Information Level of C-SPAN Junkies

On factual questions about the political process, C-SPAN junkies are better informed about politics than are members of the general population—a clear vindication of C-SPAN's primary goal.[42] They feel that they "have a real understanding of what is happening in Washington, and regard those depending on network TV reportage almost patronizingly."[43] The self-described C-SPAN "addicts" perceive C-SPAN's success as an example of how the media have underestimated the public. As one television writer and regular C-SPAN viewer put it: "C-SPAN is proving that there is a huge and serious audience of people interested in serious topics. All the evidence is there that the public is smarter than editors and TV producers believe."[44]

Some interest groups use C-SPAN as part of their alternative media mix to inform their membership about politics. Ralph Reed, of the Christian Coalition, explains: "Many of our members have satellite downlinks. They can be reached immediately either directly by [our] leaders, or through faxes and phone banks, and told . . . about an upcoming C-SPAN broadcast."[45] Although the number of such focused and tuned-in viewers is small, their level of information and commitment increases their potential influence.

In one sense, C-SPAN is a testimony to Brian Lamb's unbridled optimism about the public's capacity to "get it." From the earliest days he argued: "Let's just turn the cameras on and let the people figure it out. They are smart enough to figure it out." After fifteen years, Lamb is a bit less confident. He is willing to admit: "What you find out is, time and time again, day after day, how little people know."[46]

Activating the C-SPAN Junkies

Regular C-SPAN viewers are akin to fans of a sports team. They watch "their" network with fierce loyalty. As one editorial put it, "C-SPAN . . . is to politics what ESPN is to sports."[47] Like "Monday-morning quarterbacks," a significant number of regular viewers participate in the call-in programs, and many have strong opinions about what C-SPAN is doing right and wrong.

There is a basic principle in politics—and in life—that people are more threatened by the loss of something they have become accustomed to than they are motivated to take action to secure something they do not have. Although more politically active than the rest of the population, the bulk of C-SPAN junkies do not spring into action until "their" network is challenged. It is the C-SPAN junkies who have gone to bat when their access is threatened. When C-SPAN was forced to change its satellite transponder in 1982, four hundred to five hundred cable systems dropped coverage. Retired insurance executive William "Bud" Harris, of Cherry Hill, New Jersey, refused to take the interruption sitting down. Contacting C-SPAN to get names of other disappointed viewers, he founded "Friends of C-SPAN," the first viewer-membership group for any cable channel. Harris became president, and Shirley Rossi, of Pueblo, Colorado, became the national vice-president. The goal of the organization was to get the network restored on those systems that had dropped it and to expand access. The project generated hundreds of letters and thousands of phone calls. Bud Harris's multi-page missiles to offending cable operators cajoled, threatened, and attempted to embarrass them into reinstating the distribution of C-SPAN. Local "Friends" chapters emerged in several cities. Such viewer support is credited with pushing many cable systems to reinstate C-SPAN. When the threat subsided, it became difficult to maintain a national organization.

When the "must carry" decision in 1992 again diminished access, a number of local groups again organized to put pressure on their cable systems. They mounted telephone initiatives and barrages of letters to newspaper editors in support of carrying C-SPAN. Slowly, cable system operators began to realize that the C-SPAN audience was not a group to trifle with (see chapter 2).

THE ELITE AUDIENCE

Although C-SPAN's primary purpose is informing the public, it has become a vehicle for elite communication as well. Presidents and prime ministers find C-SPAN useful for monitoring the political process. A broad range of political "movers and shakers" also have discovered C-SPAN as a vehicle for acquiring important information.

The First Viewer

Perhaps no one was more surprised than the C-SPAN staff when President Ronald Reagan, shortly after participating in a session with Close Up Foundation students at the White House, contacted a 1983 call-in program on C-SPAN to correct his comments. He began his call by saying: "I just came upstairs to the study here and turned on the set and there you were. The thing I called about was the exclusionary rule. Evidently there I did not finish the answer, you're so conscious of the clock going by."[48] The president stayed on the line for eight minutes, answering additional questions from several students.

This was not a one-shot occurrence for Reagan. Another time he called to challenge the comments of a reporter on a C-SPAN program.[49] Reagan called himself "a fan of C-SPAN's national call-in shows" and reportedly watched congressional debates and hearings while he shaved in the morning.[50] In dealing with governors and members of Congress, he often prefaced his comments with the admission, "I saw your speech on C-SPAN."[51] David Gergen, one of Reagan's key press aides, called C-SPAN "one of the most interesting developments" in providing information to the president. He noted that Reagan found C-SPAN to be a "wonderfully helpful tool and source of information."[52] In a speech to the National Cable Television Association convention in 1984, Reagan commented: "I'm a fan of C-SPAN's national call-in shows. I am continually struck by the sophistication and intelligence of the questions that are asked. . . . I think it shows once again that . . . the people out there are way ahead of us."[53]

The ability to monitor other political players and receive "heads-up" information reduces the isolation of the presidency and gives the president a powerful political tool—information. In describing a meeting with state governors, journalist Howell Raines pointed out, "Mr. Reagan was ready for the governors' criticisms today largely because of C-SPAN."[54]

Since Reagan was the first president of the C-SPAN era, his viewing habits received considerable attention. Both Presidents Bush and Clinton and their staffs watched C-SPAN regularly. Clinton seldom watched C-SPAN during the day, but he reviewed congressional debates late into the evening.[55] In 1993, Clinton used the C-SPAN replay to critique his State of the Union Address. "Well after midnight," the president and Mrs. Clinton "headed off to watch a rerun on C-SPAN of the hours-old hit everyone was talking about."[56]

Presidential "wanna-be" Ross Perot is also a regular viewer. During a talk-radio appearance he commented: "I'm a longtime avid C-SPAN watcher. I watch C-SPAN to get good information."[57]

The Political Elite in Congress

A wide variety of other key political players also find C-SPAN useful for gathering information and monitoring the political process. Members of Congress regularly monitor floor debate both from their offices and from their homes (see chapter 8). When problems keep members away from the chamber, C-SPAN is there. Senator Christopher "Kit" Bond (R-MO), confined to bed after neck surgery, commented, "Thanks to C-SPAN coverage of committee hearings and Senate floor action, I was able to keep up with Senate business."[58] C-SPAN surveys of House and Senate offices indicate that members of Congress are regular viewers. As one staff member put it: "My Member keeps it on in the background, if he sees something interesting . . . he'll want to turn it up and find out exactly what is going on [on] the floor. Announcements, floor scheduling . . . he wants to get that kind of information."[59]

Only a portion of C-SPAN's signal is seen on the Hill. Fifteen minutes before each House or Senate session, the congressional broadcasting system shifts to a direct feed from the floor. Until 1995, that did not make much difference, since C-SPAN's floor coverage was largely unadorned. About the only things that Hill viewers missed were the classical music that used to be played during votes and the text "crawls" that were placed at the bottom of the screen to explain what was going on. Beginning in 1995, however, C-

SPAN began to use the voting period to impart more information. "Squeeze" boxes placed in the corner of the screen during votes are used to replay segments of debate or live interviews with members and journalists. Viewers at home now get more context, which viewers on the Hill are not aware of.

C-SPAN is obviously interested in its elite audience. Before and after the 1992 election, C-SPAN polled new and departing members of Congress about their C-SPAN viewing habits. High percentages of each group reported watching C-SPAN. A number of new members indicated that watching C-SPAN influenced their decision to run and that they gathered important information about their opponents and campaign strategies from C-SPAN programming. Both new and retiring members highly praised C-SPAN for informing the public. A number of members indicated that C-SPAN coverage had changed their positions or perspectives on a public issue.[60]

C-SPAN has become an invaluable tool and source of information for congressional staffs. It ranks third, behind the *Washington Post* and CNN, as a source used every day. The over-time pattern indicates an increased dependence on C-SPAN (see table 4).

Table 4

Information Sources Used Every Day by Top Congressional Aides

	1991	1992	1994
Washington Post	90	87	90
CNN	69	77	70
C-SPAN	58	66	62
National Public Radio	31	32	na
New York Times	30	36	na
Wall Street Journal	30	40	na

Source: Surveys of over 380 senior aides in House offices conducted by Fleishman Hillard Inc., Washington, D.C., and reported in press releases and personal communications with the authors.

For some congressional staff members, C-SPAN is "background noise" running constantly in their offices. Staffers pay attention when a key debate or vote is in progress. For other staff, C-SPAN is an absolute necessity. One staff member whose job it is to cover the floor pointed out: "C-SPAN is my life. . . . I have three television[s] . . . I watch constantly. . . . It's almost addictive."[61]

Competing Political Elites

Lobbyists and journalists report an increasing use of C-SPAN for monitoring the political process. During the early C-SPAN era, when Washington, D.C., lacked cable service, lobbying organizations dominated the list of subscribers to Capitol Connection (see chapter 1). Today C-SPAN plays continuously in the background in most lobbying offices. As one trade association lobbyist pointed out: "I sit in my office every morning, watching C-SPAN's early-morning guests, absorb the headlines that callers point out, and listen to Speaker Gingrich's televised press conference to pick up the hot news and the buzzwords for the day. If the theme is competitiveness, I frame the interests of my association in those terms during my meetings with members that day. If the theme is opportunity, I am sure to work that into my spiel to members. C-SPAN allows me to more effectively play the inside-the-beltway game."[62]

For journalists, becoming part of the C-SPAN audience allows them to expand their physical reach for stories. C-SPAN defies the limits of geography in two ways. Journalists outside of Washington now have access to many events that only the "inside the beltway crowd" could cover previously. Additionally, C-SPAN "transports" reporters to events that even the most liberal expense accounts and free-time schedules would not permit. The journalist's problem is greatest during campaigns, when multiple events happen at great distances apart. Reporter Thomas Southwick argued that the media use C-SPAN programming to cover campaign events that they cannot attend, giving them a full picture of the candidates' performances in a variety of settings. It was clear in 1984 that "far more reporters followed the convention proceedings via C-SPAN than via personal reporting from the convention floor."[63] Journalist Jeff Greenfield talked with pride about how he was going to cover the Senate campaigns in several states with no travel expenses. "I am going to stay at home and turn on my television set. Specifically, I'm going to watch . . . C-SPAN."[64]

Other political elites outside of Washington are also regular C-SPAN viewers. Governors' offices use C-SPAN to monitor events of interest to their states. Surveys of party activists, such as participants in the Iowa party caucuses, indicate that they are much more avid viewers of C-SPAN than is the population as a whole.[65] Similar results are evident for delegates to the national party conventions.[66] A national survey of political elites (legislators, governors, party chairs, newspaper editors, CEOs of major companies, and college presidents) indicates not only that they watch C-SPAN but also, and more important, that almost one-third credit C-SPAN with causing them to change their minds on a public issue.[67] The lesson seems clear: above and beyond providing access to the public's thinking, C-SPAN serves as a significant vehicle for informing key American political decision makers.

The Glitterati Contingent

The line between public affairs and entertainment is blurred in American society. Entertainment figures regularly endorse political candidates and causes while a smaller contingent of entertainment figures have made the switch from the sound stage and studio to the legislative chambers and meeting rooms. It should be no surprise that C-SPAN can count among its loyal viewers several people who have made a name for themselves outside of politics. Despite the reasonableness of this assumption, both C-SPAN and the media express considerable interest in stars with a bent toward C-SPAN. The late counterculture rock musician Frank Zappa was prominently featured on the cover and in a chapter of *America's Town Hall*. Zappa referred to C-SPAN as "the best television available" and chafed at the suggestion that a rock star could not be interested in politics and the future of the nation.[68] Singer Barbra Streisand is known for her public political activity, but she admitted that politics pervades her private life as well. She stated: "My favorite show is C-SPAN. That's what I watch, C-SPAN 1 and C-SPAN 2. When I'm not watching CNN."[69] It is not clear how general such interest is, yet C-SPAN and the media make a great deal of the fact that C-SPAN, by counting some of the "beautiful people" among its viewers, appeals to not only the public affairs nerds.

The International Audience

Although it might reasonably be assumed that interest in C-SPAN is an "American thing," an important international audience is developing. Selected C-SPAN programming is provided by the U.S. Information Agency's Worldnet. Given the limited distribution and lack of international surveys, the size and composition of the international audience is even more of a mystery than that of the U.S. audience. When C-SPAN opens its phones after major political events, a significant number of international callers get through. Numerous letters indicate that the impact of C-SPAN on the political outlook of foreign viewers—especially those in nondemocratic countries—may be even more significant than any impact on individual American viewers.

The C-SPAN offices were abuzz about a young Chinese engineer in Hang-chou who found C-SPAN while fiddling with the satellite dish on the roof of the institute where he worked. He wrote, in broken English but crystal-clear meaning, "Through C-SPAN, I have learned the devoted challenging of your people, the free debate and vote in your presidential selection, your education system and the power of your people."[70]

Above and beyond individual citizens, government bureaucrats and for-eign leaders use C-SPAN to monitor U.S. politics. Senator John McCain (R-AZ) discovered this firsthand. During a meeting, Egyptian President Hosni Mubarak told McCain, "Sometimes you and the Senate get very aggravated." Even though they had not previously met, McCain got the dis-tinct feeling that because of C-SPAN, Mubarak "acted as if he knew" the U.S. representative.[71] To the degree that one believes understanding leads to cooperation, C-SPAN has the potential for contributing to that cooperation.

C-SPAN Reacts to Its Audience

Although C-SPAN abhors "counting eyeballs" and refuses to adjust its pro-gramming for the sole purpose of becoming more popular, it would like to expand viewership on its own terms. In 1988, C-SPAN conducted a num-ber of focus-group interviews designed to assess viewers' understanding of the network and their suggestions for its improvement. The focus groups included Capitol Hill staff, political professionals, and average viewers. Focus-group suggestions have been used to make marginal changes in C-SPAN policies and programming while retaining its basic philosophy (see chapter 3).

Although explicitly refusing to count its audience, C-SPAN has carried out a number of audience-building initiatives. Relying on the assumption that people would tune in "if they only knew about us," generic advertising of the network and promotion of specific programs were initiated. The publication of *America's Town Hall* in 1988 garnered significant publicity for the kind of people who watched C-SPAN and attempted to "make it the 'in' thing to do." The creation of an educational program, and initiatives such as the C-SPAN bus, were designed primarily to promote C-SPAN usage (see chapter 4).

The Audience Responds to C-SPAN

In the early years, the question "Is anyone watching?" was perfectly legiti-mate. We now know that a significant number of people watch C-SPAN at any given time. The remaining question is, "So what?" If C-SPAN is nothing more than niche market entertainment for those who see politics as the "best show in town," its potential for impact is greatly diminished. In fact, there is the danger that vicarious political participation could replace actual participation. On the other hand, if C-SPAN changes political outlooks and behavior, it has the potential for significant influence on the very system it attempts to highlight.

There are two ways to answer the "So what?" question. In a variety of discrete ways, it is possible to discover specific instances of the C-SPAN audience taking individual action. To the degree that C-SPAN activates individual members of its audience, it may well stimulate increased levels of political involvement or create new vehicles for political participation. Such participation has the potential for changing the behavior of political decision makers. More broadly, we will look at evidence that indirectly indicates that some precursors of political outlooks were either caused or reinforced by C-SPAN viewing.

POLITICAL ACTION STEMMING FROM C-SPAN VIEWING

The evidence that C-SPAN programming hits a responsive chord is abundant. The telephone lines light up when call-in programs begin. After a gushy "thank goodness for C-SPAN," many callers complain that they have been trying to get through for weeks. After the 1988 presidential debates, AT&T recorded over 219,000 attempts to connect to the C-SPAN call-in number; 108 callers were successful in reaching the program.[72] Despite such dramatic numbers, C-SPAN surveys regularly indicate that less than 5 percent of C-SPAN viewers have ever tried to participate in a call-in program.

In an effort to increase the number of viewers able to participate in call-ins, C-SPAN initiated a policy limiting callers to one call every thirty days. The policy was intended to work on the honor system. One viewer got so agitated at regular callers who broke the policy that he began recording repeat offenders. His edited tape of "evidence" was used to embolden the call-in hosts and those screening the calls to cut off viewers not abiding by the policy.[73] Both the existence of the problem and the call-in sleuth's response indicate audience commitment.

Call-in guests indicate that they "always get feedback from C-SPAN." Journalists are amazed when they get more letters after a C-SPAN appearance than after a major network appearance. As Charles Bierbauer, of CNN, sees it, the outpouring of letters "says something about [the C-SPAN] audience; sophisticated, involved, aggressive kinds of people . . . and it may be they're an older audience that has time to sit and write letters."[74] During a recent *Journalists' Roundtable*, political commentator Morton Kondracke posed what he thought was a rhetorical inquiry: "I have a theory about [why people are mad at Washington], but please, you tell me." He was not prepared for the responses. He noted, "By phone, fax and mail, a small avalanche of over fifty messages descended."[75] The usual number of letters tends to be in the dozens rather than the thousands, perhaps serving more as evidence of how few letters public officials get from appearances on other networks.

Not all callers want to be on the air. C-SPAN's viewer services division handles hundreds of calls each week from viewers making suggestions. Increasingly, members of the audience are using e-mail and fax machines to make their views known to C-SPAN. The lack of a fixed schedule is a continuing frustration for viewers. Much of the time of employees in the viewer services department is spent explaining network policies, such as the primacy of House and Senate sessions and the desire to cover a broad base of hearings (see chapter 4), policies that make it difficult to establish a long-range schedule. Over the years C-SPAN has reacted to the complaint about scheduling information with a variety of techniques. *C-SPAN Update* was published for a number of years and included as much scheduling information as possible. After subscribers complained about late delivery, each issue carried a statement: "Postmaster: Time-sensitive material. Please deliver by cover date." The call-in schedule hotline was instituted in the late 1980s to provide more up-to-date information. Between 1993 and 1995 it received over 350 calls per day.

With the arrival of new technology, C-SPAN saw the opportunity for providing more-immediate scheduling information. C-SPAN scheduling information is now provided in almost "real time" on the Internet and through a variety of commercial services. Approximately three to ten thousand individuals tap into C-SPAN's computerized scheduling information every day. Some viewers have taken the scheduling information problem into their own hands. The Richland, Wisconsin, chapter of the American Association of Retired Persons initiated a telephone tree to alert members when issues of concern are being discussed on C-SPAN.[76]

The loyalty of the C-SPAN audience is legion. Detroit media critic Jim McFarlin once asked, "Have you ever wondered who watches C-SPAN?" In reply, numerous viewers wrote in to question his implied criticism of "their" network. McFarlin wrote a rejoinder admitting his mistake at being so flippant.[77]

C-SPAN is regularly surprised by its audience's interest in network promotional items and products. When *Booknotes* offered a free bookmark, they expected one thousand requests but received over ten thousand. Surveys indicate that *Booknotes* is one of the least-watched programs, but it draws a well-educated and loyal audience. After the 1994 congressional elections, C-SPAN's 1995 *U.S. Congress Handbook* sold over fourteen thousand copies in the first five months.

There is considerable evidence that members of the C-SPAN audience do more than passively view the network and that they do not limit their political activity to interacting with C-SPAN and its guests. C-SPAN viewers are active in political behaviors such as voting, joining political organizations, and communicating with elected officials. As is discussed in more detail in chapter 8, political initiatives of the C-SPAN audience have permeated the consciousness of political decision makers. Members of Congress regularly report calls and letters from constituents activated by a C-SPAN program. Some policy debates that in the past would have been carried out within the dark and narrow confines of the Capitol corridors have been forced into the sunlight by C-SPAN coverage. The change of venue has affected the issues included for consideration, the thought process of decision makers, and the eventual outcomes of policy battles.

SETTING THE STAGE FOR FUTURE ACTIVATION: LONG-TERM POLITICAL PERSPECTIVES OF THE C-SPAN AUDIENCE

Understanding the potential impact of the C-SPAN audience on the future of the American political process involves more than assessing the discrete and short-term political activation of individuals. The C-SPAN audience harbors unique and significant long-term orientations to the American political process, orientations that have the potential for predicting the individual political behavior that members of the audience will favor in the future. Even more important, to the degree that members of the C-SPAN audience reflect a vanguard of opinion, understanding their outlooks on politics may well help us understand the broader population. If C-SPAN affects only a unique set of individuals with a high degree of interest in public—and particularly congressional—affairs, it is still significant. But if the C-SPAN audience is the tip of the iceberg, reflecting outlooks acquired earlier because of enhanced public affairs information, the growing accessibility of pubic affairs information could make the general population resemble the C-SPAN audience in terms of political outlook. In other words, the political perspectives of the C-SPAN audience may well be a precursor of the political conventional wisdom on which the broader society will base its behavior in the future. In either case, the starting point is the orientation of the C-SPAN audience to the political process and to the institutions that compose that process.

Political theory and conventional wisdom generally indicate that interest in politics and access to extensive information are, without qualification, good characteristics in a democratic society. Few would argue—at least publicly—for a less interested and less aware group of citizens. The move to televise Congress and the motivation for creating C-SPAN shared the same assumption: that increased public affairs information would improve the American political process. Both initiatives assumed that politicians could explain their behavior and that the public had the ability to understand the explanations. Most observers felt that public expectations and the behavior of politicians were largely in sync and that only minor revisions by both would be necessary for each to prosper.

Support for Congress

If the assumption of members of Congress in televising congressional proceedings was "to know us is to love us," the data supporting this premise is likely to be rather disconcerting. Almost from the beginning, C-SPAN viewers have been somewhat more critical of Congress and the political process than has the public as a whole. It is impossible to determine whether C-SPAN draws more disaffected people or whether it creates them. Most likely, both processes are at work.

By the early 1990s, public dissatisfaction with Congress emerged in figures in all public-opinion polls, figures that had been considerably more positive only a few years earlier. C-SPAN viewers began to view Congress in a more negative light earlier than the general population. In 1991, 37 percent of the American public felt Congress had been a "good influence" on the way things were going in the United States. Only 30 percent of regular C-SPAN viewers that year held the same opinion, making them the most cynical evaluators of Congress among regular viewers of other specialized media sources.[78]

The pattern of cynicism among C-SPAN viewers and nonviewers tends to vary according to the poll one uses. It is clear from all polls that C-SPAN viewers are considerably more likely to have an opinion of Congress than are nonviewers.[79] The general pattern of poll results heartily challenges the 1970s conclusion that there is a clear correlation between knowledge of Congress and public support for the institution.[80]

In preparation for a 1994 House Democratic leadership retreat, pollster Mark Mellman was asked to conduct a detailed survey on public attitudes toward Congress. He found that 46 percent of the national sample had watched C-SPAN. Those viewers were much more negative toward Congress than were those who had not watched the network. Viewers also seemed to be aware of C-SPAN's impact on their outlooks. Of those who had watched, more than twice as many (15 percent versus 6 percent) felt that what they had seen had made them "less favorable toward Congress," whereas 22 percent of the respondents perceived no effect, and 57 percent did not know if their attitudes had changed.[81] Overall, the results of the poll indicated significant dissatisfaction with Congress and particularly with the sitting Democratic leadership. Mellman concluded, "To watch Congress is not to love it . . . people see partisan debate and confuse it with partisan bickering; they see the legislative process and confuse it with inaction."[82] Asked to participate in the meeting, Brian Lamb sat a bit uncomfortably as his network was indirectly accused with causing some of the Democrats' problems. Democratic leaders were described as "surprised and concerned" about the findings, and one observer noted that the "air was reverberating" from the briefing.[83] Attendees left the room with a pretty good idea that the 1994 elections would be a challenge for Democratic candidates. In the end, the poll had tapped real public concern. All thirty-two House incumbents defeated in the 1994 general election were Democrats, and Republicans took over both the House and the Senate.

Participants such as Representative J. Roy Rowland (D-GA) felt that televising Congress may have hurt the institution's image. He argued: "Some of the things that go on have to be awfully disturbing to people . . . some of the petty arguments that people get into. I think that detracts from the stature of the institution."[84] Brian Lamb reported, "I've been told personally by high-ranking members of Congress that they made a mistake ever opening up the process to television."[85] Other members attributed the defeat of the Democrats in 1994 partially to the presence of the cameras and the inability of Congress to explain its complex procedures and seemingly damaging norms of partisan behavior. Such opinions were not often expressed publicly, however, since attacking openness in the policy process and taking on C-SPAN are like questioning motherhood and apple pie.

Most comments regarding C-SPAN from practitioners and the media are positive, pointing out its contribution to a better-informed population. CBS newsman Charles Kuralt praised C-SPAN in 1987: "Don't go looking for the country bumpkin in the country. He has been watching the Iran-Contra hearings on C-SPAN, and he is likely to be better informed than his city cousins. In a country that depends for its life on an informed citizenry, this is all to the good."[86]

Criticism of C-SPAN emphasized the shortcomings of Congress in properly dealing with the new media age. Some observers argued that Congress needs to clean up its act and allow more C-SPAN rather than less. American Enterprise Institute scholar Norman Ornstein proposed a Congress that would make its policy process more understandable, and he supported the more aggressive use of C-SPAN by party leaders both in general and specifically to explain ongoing votes as they happen.[87] As we will see in chapter 7, C-SPAN itself and the new Republican leadership considered this issue after the 1994 elections.

The C-SPAN Audience and Citizenship

Far from being simply cynical naysayers, members of the C-SPAN audience exhibit a sophisticated mix of enlightened criticism and tempered hope. According to a detailed analysis, regular C-SPAN viewers were described as "vehemently anti-Washington, yet at the same time . . . strikingly non-cynical and deeply convinced of the importance of one individual's participation in the American democratic process despite a deep discontent with the status quo."[88]

C-SPAN viewers hope to use watching C-SPAN as a tool for empowerment. On standard questions dealing with political alienation and efficacy (the utility of becoming involved in politics), regular C-SPAN viewers in 1992 revealed themselves as more positive about politics, the political process, and their potential for having an impact. At the same time, they wanted to change the current leadership by "throwing the rascals out." Regular C-SPAN viewers were seemingly inconsistent about the role of government in society. They were willing to support an active involvement in social policies, but they generally felt that the government controls too much of our lives (see table 5). There is little evidence that this was purely a partisan re-

Table 5

Political Alienation and Efficacy of C-SPAN Viewers and the General Public in 1992

	C-SPAN REGULAR VIEWERS	TOTAL POPULATION
EFFICACY AND ALIENATION		
(Percentages indicate responses reflecting less political alienation and greater political efficacy. Such responses are most supportive of a democratic system.)		
Efficacy "Those like me don't have any say about what government does."	56% Disagree	49% Disagree
"I don't want to involve myself with politics."	93% Disagree	74% Disagree
Politicians "Most elected officials care about what people like me think."	46% Agree	36% Agree
"I sometimes feel it doesn't matter that much who gets elected."	96% Disagree	63% Disagree
THE DESIRE FOR AN ACTIVIST GOVERNMENT		
"Government should help needy people even if it means going deeper in debt."	61% Agree	53% Agree
"The federal government controls too much of our daily lives."	70% Agree	64% Agree
NEED FOR POLITICAL CHANGE		
"It is time for politicians to step aside and make room for new leaders."	51% Completely Agree	38% Completely Agree
"We need new people in Washington even if they are not experienced."	37% Completely Agree	20% Completely Agree

Source: Times Mirror Center for the People and the Press, "Profile of the Regular C-SPAN Viewer," press release, n.d. Based on a 1992 national survey of 3,517 adults.

sponse to Democratic control of Congress, since the C-SPAN audience at that time included a slightly higher percentage of Democrats than did the population as a whole.

Conclusion

Awareness of C-SPAN and of the size of its audience has grown dramatically over the last two decades. The audience remains a niche audience of upscale viewers with higher-than-average education, income, and interest in politics. Beyond the generalizations, the audience represents a relatively broad spectrum of Americans—in many cases broader than the audiences for comparable upscale news outlets.

Although numbers count in business, media, and politics, the size of the C-SPAN audience may well be less important than its composition and its proclivity to act on its political beliefs. The C-SPAN viewership counts among its members an elite group of policymakers and staff who depend on the network for substantive, monitoring, and political-scheduling information.

Regular C-SPAN viewers are not content to simply watch politics happen. They inform themselves through a wide variety of sources and participate in the political process in both traditional and nontraditional ways. In many aspects, C-SPAN viewers support James Madison's plea: "A popular Government without popular information, or the means of acquiring it, is but a Prologue to a Farce or a Tragedy; or, perhaps both. Knowledge will forever govern ignorance; and a people who mean to be their own Governors, must arm themselves with the power which knowledge gives."[89]

CHAPTER SEVEN

C-SPAN's Impact

A View from the Inside

In early January 1995, Representative Newt Gingrich (R-GA) stood in the U.S. House chamber and addressed both his colleagues and the American public with words of assurance outlining his vision of America. His words were the culmination of years of effort to reach his personal goal of becoming Speaker of the House of Representatives. Throughout his political career he had effectively used words to incite and excite. Within a few months, many of his words would carry new weight as they were transformed from musings and platitudes into operative tools such as rules and laws. A few years earlier Speaker Jim Wright (D-TX) had faced another set of microphones and, in carefully framed words, announced that he was resigning from Congress. His career would be cut short by charges that he had misused his office staff to write a book and had been granted a highly lucrative publishing deal because of his position. What ties Gingrich and Wright together is the role C-SPAN played in establishing their situations and in amplifying the potency of their words. Their experiences reflect only one category of the potential impact that C-SPAN has on the American political scene.

The Place of C-SPAN in the American Political Process

"Word purveying isn't the only game in town, but it is surely Washington's major industry, in the same way that in Detroit people make cars and in New York people buy and sell stocks and in Los Angeles people wait on tables and dream of being a movie star. Words are what we make."[1] In democratic politics, the making of words is only half the game. Unlike the solitary poet who could be satisfied with creating the perfect sonnet whether or not anyone else read it, politicians transmit their words to others so that the words will be acted on. Throughout history, politicians have used various technologies to get their words heard by larger and larger audiences. In earlier years, the most effective new technologies involved bringing the speaker to the audience or the audience to the speaker. Aristotle argued that democracies must be small enough that no citizen would have to travel more than one day to participate in the dialogue. Changes in transportation technology expanded the ability of speakers to efficiently go to their audiences. The carriage rides from town to town for torchlight parades were superseded by the whistlestop campaign tour by railroad, which was superseded by the airport rally with the availability of jet travel. Political leaders with the foresight and stamina to utilize the new technologies often gained power at the expense of those opponents unwilling to take advantage of new technological opportunities.

The next wave of technology made physical distance less relevant and emphasized increasing the impact of the message by expanding the reach of the messenger. While William Jennings Bryan was exhausting himself on whistlestop tours, William McKinley carried out his famous 1896 "front porch campaign" with a telephone beside his chair. He used this tool to arrange visits by political leaders and to assiduously manage his campaign operatives around the country. Franklin Roosevelt effectively used radio to reach the public in the 1930s and 1940s. By the 1950s, television became the technology of choice for most politicians. If C-SPAN had been available, successful politicians such as McKinley and Franklin Roosevelt probably would have used it. For modern politicians, C-SPAN offers a new venue and a unique format for getting their words out.

Careful analysis of previous technological changes indicates two basic principles concerning the impact of technology on politics. First, with each technological change, political winners can usually be distinguished from political losers by their foresight and skill in using the new technologies. Second, technology is not neutral. The form, content, providers, and recipients of the message can all be affected by the technology involved.

Assessing the Impact of C-SPAN

The launching of C-SPAN in 1979 might be seen as simply a footnote in the progression of technology applications associated with politics, but time has proven a range of implications much broader than anyone anticipated. C-SPAN is the contemporary "megaphone," spreading the word of political activists to the segments of the population more oriented toward public affairs. Assessing the impact of technology, particularly in a highly volatile realm such as politics, is risky at best. A number of impediments and pitfalls immediately confront the analyst:

- Technologies do not appear on the scene in isolation. They are often only part of a series of simultaneous changes. Other changes may either reinforce or undermine the change associated with a new technology. There is always the danger of becoming a "technological determinist," associating every change with a particular technology and denying alternative causes.
- Technologies and the organizations and individuals they affect interact, rather than exist in a clear cause-and-effect relationship. "Technologies do not 'impact' on organizations like two ships colliding at sea. Instead, technological innovations are filtered through the organization's traditions, constraints and resources."[2] Organizations and individuals adapt to new technology, and at the same time the technologies are forced to adapt to the organizations and individuals.
- "Intention" is not the same as "impact." The goals of technological inventors or innovators may have little to do with the actual effect. The process becomes complicated as stated goals become the "conventional wisdom" concerning probable impact, which may well cloud the evidence of actual impact. Often the inventors or prime proponents of new technologies have the most distorted view of

potential uses and consequences. Alexander Graham Bell saw the telephone as a one-to-many communication device for broadcasting messages, and Guglielmo Marconi thought that radio would be used in a one-on-one mode.[3]

- Empirical evidence of impact is often either unavailable or relatively insignificant. The most important effects of technology often involve subtle changes in attitude or behavior, changes impossible to capture in measurable and countable form.

Considering all these caveats, the intention of this chapter is to explore some of the key influences of C-SPAN on American politics. The evidence involves a mix of empirical and impressionistic data. It relies on official government documents, public-opinion poll data, searches of the academic and popular literature, and interviews with key participants.

C-SPAN's Impact on Congress

The decision to allow televised coverage of House and Senate proceedings generated a great deal of debate both among members of those chambers and among the media. Unlike some technological changes, such as the introduction of computers into Congress, House members in the late 1970s had some personal experience with television and thought they could predict some of the consequences.[4] The initial decision to televise House proceedings did not anticipate a full-time cable channel such as C-SPAN, so most of the projections emphasized the impact of televised proceedings per se. Opponents worried that members would be tempted to "grandstand" and speak to the public rather than pay attention to their duties. Proponents foresaw greater visibility for the House and the ability to get their message across without having to go through the established commercial media.

In one sense it is hard to distinguish the impact of C-SPAN in isolation from the influence of the televised House proceedings. Initially, C-SPAN was little more than a tool for getting gavel-to-gavel coverage to a wider audience. Since the networks showed little enthusiasm about using the feed from the House floor for their newscasts, C-SPAN made televising the chamber significant. Over the years, C-SPAN expanded its programming well beyond coverage of the House floor, and today a discussion of C-SPAN subsumes the more narrow concern over the impact of the cameras.

By the time members of the Senate got around to seriously discussing tele-

C-SPAN, Not MTV

The U.S. Senate Goes on Television

Most observers expected considerable grandstanding by members of Congress once the cameras arrived. (Cartoon by Bob Englehart, *Hartford Courant*, reprinted by permission.)

vising chamber proceedings, they had seven years of House experience from which to draw. Although the House experience and the technological advances allowing for low-light cameras mitigated some of the fears, opponents in the Senate took the typical tack of touting their chamber's uniqueness. They feared that television would upset the Senate's singular traditions of unlimited speeches, parliamentary delays, and an undecipherable schedule. They suspected that their colleagues would be uncontrollably drawn to the cameras and drawn away from the committee rooms where most of the work is done. Proponents took a more institutional view, expressing fear that in the television age, the Senate would become insignificant without its own presence on the dial.

Thus, two categories of potential consequences pervade the discussion of C-SPAN's impact on Congress as an institution. One focuses on the behavior of the members and the resulting impact on House and Senate proceedings. The other takes a broader view, looking at the impact on the visibility and credibility of Congress.

IMPACT ON FLOOR PROCEEDINGS

Much of the debate over televising both the House and the Senate revolved around the potentially significant effect that the presence of the cameras would have on floor proceedings. From general fears of upsetting the ambiance of the chamber to more specific fears of encouraging "grandstanding" by members, extending the length of debate, denigrating the quality of debate, increasing partisan voting, and promoting floor attendance while adversely affecting committee attendance, opponents painted a relatively dark picture of a televised chamber. Nevertheless, both formal studies and informal evaluations of floor behavior tend to conclude that television has "brought no dramatic changes to floor activities or members' behavior."[5]

General Fears

Perhaps the most solid evidence that television coverage failed to undermine the procedures of Congress stems from the fact that most initial opponents have come around to accepting—if not strongly endorsing—the consequences.

Senator Bennett Johnston (D-LA), one of the most vocal initial critics, had changed his mind within a year. He concluded: "It has worked well.

Some of the fears that I and others had have not materialized. I think Senate TV has been a success."[6] Senator Jesse Helms (R-NC), whose ideological views differ significantly from those of Senator Johnston, nevertheless came to a similar conclusion: "I voted against television in the Senate, because I feared it would create some prima donnas when the Senate already had enough of those . . . I now welcome the great work done by C-SPAN."[7] House Minority Leader Robert Michel (R-IL) used the tenth anniversary of C-SPAN to present his turnaround by saying, "My reluctance earlier on . . . has changed because quite frankly I get the impression that members after they are here for a while, really aren't aware that they're being covered . . . it's a good educational thing for the American people . . . it's been good for the institution."[8]

Grandstanding

Defining "grandstanding" is difficult. One person's superfluous comments to the public form another person's significant speech on public policy. No one should be surprised that much of the discussion on the House and Senate floor is more for public exposure than for the edification of colleagues. Journalist Elaine Povich pointed out: "One senator, who shall remain nameless, came to the Senate floor obviously prepared to make a very detailed speech—he brought his charts and his graphs and everything. . . . Well the Senate business of that day had changed between the time he left his office and when he got to the Senate floor, and they were no longer discussing catastrophic health insurance. However, this Senator, having come so audiovisually prepared to do this for the people out there, insisted and got a special rule so that he should go ahead."[9]

In the era before C-SPAN, that senator might simply have walked away without speaking, but the availability of an audience and his obvious preparation lured him into making a speech that was irrelevant to the Senate business at hand. Such behavior is perceived as rare, however. Former Speaker Tom Foley (D-WA) faced up to the issue by stating: "It was a concern originally that members of Congress were going to be distracted by the coverage and were going to be kind of playing to the cameras rather than carrying on debate. I don't think that's happened very much. It isn't a problem."[10]

The fear that members might be intimidated by the cameras has also not been borne out. As C-SPAN's Bruce Collins sees it, "Judging from some of the crazy things these guys say on the floor, they have *got* to be oblivious to the cameras."[11]

Length of Debate

Senator Russell Long (D-LA), an opponent to the end, argued, "The greatest surplus commodity we have in Congress are speeches that need never have been made, speeches that fail to improve on silence."[12] Such concern for saving time and breath may well have carried more weight if it had not been made by a senator who was filibustering the resolution that would have allowed television in the Senate. After only a few months' experience with the cameras, Speaker Tip O'Neill (D-MA) stated that he had "made a terrible mistake" in authorizing live television coverage of the House. He was concerned that debate was droning on interminably and that members were using tapes of their speeches to increase their exposure back in their districts.[13] Projections and initial impressions aside, the empirical evidence does not back up the assertion that the length of debate has increased.

Graph 10

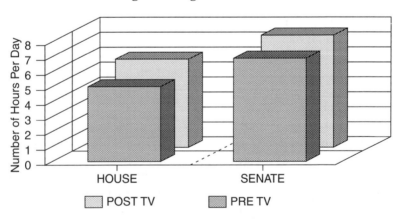

Length of Congressional Sessions

The length of congressional sessions increased only slightly after the arrival of cameras. *Source:* Norman J. Ornstein, Thomas E. Mann, and Michael J. Malbin, *Vital Statistics on Congress, 1995–1996* (Washington, D.C.: Congressional Quarterly, 1996), p. 158.

Proving that the presence of television cameras increased the frequency and length of speeches is difficult, since numerous other factors influence when and how often members take their arguments to the floor. During the early period of television in the House, a survey of members indicated that while only a few believed that television had prompted them to speak on the floor, 77 percent felt that it had influenced their colleagues to do so.[14] In the aggregate, the length of congressional sessions did not significantly increase with the presence of television cameras (see graph 10).

The major exception to discovering a causal relationship between the arrival of television and length of floor proceedings is the increased number of short speeches allowed at the beginning of each day's legislative session and the inclusion of Special Orders speeches at the end of the legislative day in the House. A careful analysis of floor behavior during the test period in the Senate concluded that the only significant changes in senators' behavior were a dramatic increase in the number of Special Order speeches (up by 250 percent) and an increase in the total time spent on such speeches, even though the maximum time for each speech was reduced from fifteen minutes to five minutes. A longer-term evaluation of the House and Senate statistics bears out the dramatic increase in such speeches, an increase associated with the televising of chamber proceedings (see graph 11). All other changes in measurable floor behavior—such as numbers of amendments, quorum calls, and time in session—could be explained by factors other than television coverage.[15]

Quality of Debate

Senator John Danforth (R-MO) asserted that with television, the Senate would be "propelled, now with a mighty spurt forward, to more frantic action, to the one-liners, to oversimplifications."[16] From his unique experience as chairman of the Speaker's Advisory Committee on Broadcasting, Representative Charlie Rose (D-NC) asserted that television had improved the quality of House debate: "If you have very little to say . . . television picks that up far quicker than sometimes does the printed page. . . . I find members more precise and more cautious, just as we are all very cautious when we were in an interview with our local television stations. . . . And I think we're getting better speeches."[17] When debates were recorded only in print, members could use the right to "revise and extend" their remarks without fear of embarrassment; the inability to revise a videotape, however, encourages members to think through their comments more carefully.[18]

Graph 11

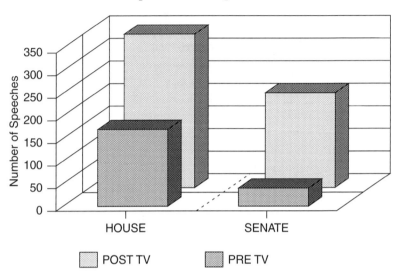

Special Orders Speeches

The arrival of television spawned a dramatic increase in Special Orders speeches in both the House and the Senate. *Source:* Congressional Research Service. House figures compare the month of March in 1977 and 1981. Senate figures compare the first fifteen days of June in 1984 and 1988.

Scientifically assessing the quality of debate is difficult, if not impossible. No one formally tracks the high and low points in floor debate. There is no assurance that we could agree on the criteria. But most observers of Congress, both inside and out, credit the televising of Congress with resulting in higher-quality debate. According to Norman Ornstein, of the American Enterprise Institute, the presence of C-SPAN has "improved the quality of deliberations. . . . Before, there was a sloppiness about what went on the floor. But now they know somebody is watching, so members do more homework."[19] As one member put it, "No one wants to look like a fool when the cameras are on."[20] Senator Robert Byrd (D-WV), the self-appointed in-house historian of the Senate, is sensitive to the Senate's traditions and is well versed in its development. He is quite satisfied with television in the Senate, feeling not only that the initial fears were unfounded but also that positive consequences emerged.

He concluded: "It has worked very well. Contrary to some concerns that senators would play to the gallery, that really hasn't come about . . . speeches became shorter, I believe, and I think their speeches were better prepared."[21]

If one of the criterion of a good debate is to efficiently disseminate information, television may be credited with a significant change in debate style. With the coming of C-SPAN, members have applied the dictum "A picture is worth a thousand words." For example, in 1983, Representative Silvio Conte (R-CT) donned a pig nose to protest the pork barrel legislation of his colleagues, who "were greedily feeding at the government trough."[22] Jim Nussle (R-IA) wore a brown-paper bag over his head to express his embarrassment over the House check-bouncing scandal in 1991. The stunt caught the attention of party leader Newt Gingrich, put Nussle on the fast track for a party leadership position, and made Nussle head of Gingrich's 1994 transition team. Dick Armey (R-TX) unveiled a Byzantine chart of President Bill Clinton's health plan, a chart that "reduced it to a visual cacophony of arrows, boxes and fine print."[23] Senator Bob Dole (R-KS) used a similar devastating chart in his response to President Clinton's 1994 State of the Union message. Senator Jesse Helms (R-NC) titillated the C-SPAN audience with graphic examples of projects sponsored by the National Endowment for the Arts that he found to be pornographic. Senator Alfonse D'Amato (R-NY) punctuated his speech on President Clinton's economic package by stabbing a drawing of a "Taxasaurus" with a giant yellow pencil.[24]

As Senator John Glenn (D-OH) concluded, "Since we went on C-SPAN, I think the cost of doing charts for the Senate floor must have gone up 10,000 percent."[25] There are some limits on the use of visual props. Speaker Tip O'Neill ruled Representative Bud Shuster (R-PA) out of order when he tried to bring a rubber Donald Duck to the microphone to protest the "lame duck" session.[26] Senator Ted Stevens (R-AK) asserted that Senate debates were beginning to look like show-and-tell sessions in elementary school. The impact of visual images goes beyond those watching from the floor or on C-SPAN. With C-SPAN serving as the "network of record," clips and pictures of these visuals have ended up on the evening news programs and in major print publications. At times they have become powerful symbols in national policy debates.

C-SPAN has been used to provide an alternative venue for visual material disallowed in other realms. During the Iran-Contra hearings, Lt. Col. Oliver North was denied the right to show a set of slides detailing the Soviet and Cuban presence in Nicaragua. Senator Daniel Inouye (D-HI) angered

Visual Floor Props

Senator Don Nickles (R-OK) takes advantage of visual aides to make
his point. (Photo courtesy of C-SPAN.)

North's supporters by asserting that the slide show would require darken-
ing the Senate hearing room and thereby would cause a security problem.
House Republicans had the fifty-six slides made into posters and showed
them during forty-four one-minute speeches before the start of the legislative
day on July 13, 1987. One congressional aide involved in the presentation
explained: "It was certainly aimed at the cameras so the American people
could see it. If they couldn't see the slides on the House floor, where else
could they see them[?]"[27] To blunt the thrust of the Republican presentation,
twenty-nine Democrats made one-minute speeches, complete with their own
pictures depicting alleged victims of Contra actions.

Although there is little doubt that the use of visual props during floor
speeches was largely spawned by the desire to more effectively communicate with the television audience, the effect of such visualization on members
in the chamber is less clear. Representative Bob Wise (D-WV) is convinced
that the presence on the floor of petitions with eighty-one thousand signa-

tures swung votes on a bill dealing with telephone access charges in 1983. Ed Markey's (D-MA) use of a series of labeled boxes into which he dropped dollar bills was so effective that the floor manager of the bill asked him to repeat the presentation during the final debate on the Universal Telephone Service Act.[28]

Partisanship on the Floor

The political process expects Congress to make tough value choices on issues over which reasonable people honestly disagree. There is no "right" answer to such choices, of course, since values are unprovable preferences. The fact that one person prefers direct federal government involvement in a policy area such as education and another prefers a hands-off approach does not prove that one position is right and the other wrong. Observers of Congress disagree about the degree to which partisanship should affect this process. On the one hand we hear cries for more distinct parties, giving voters clear choices among values. On the other hand, observers exclaim frustration when politicians do not reach an expeditious compromise on issues. Just before the introduction of Senate television, journalist Matthew Matejowsky expressed fear over the potential hardening of partisan positions that television might bring: "If compromise is the grease in the engine of democracy, then television in the Senate chamber would be the rust."[29] Senator William Proxmire (D-WI) expressed a similar fear: "Instead of an institution where sharp differences are ground down and compromised, this floor will become a place where they are sharpened."[30]

Comparing the pretelevision period with the present, many observers feel that debate has "crackled more and more often." In the words of one veteran lawmaker: "The stakes are higher now. We have an audience, and whatever happens on the votes, we want to score with potentially millions of viewer."[31] But evaluating changing partisanship shows one of the difficulties in performing technology assessment: it is difficult to isolate technology as a cause. Although numerous studies have shown an increase in party-based voting in the House and the Senate over the last decade, it is impossible to determine whether televising the chambers led to more partisan debate, which in turn led to more partisan voting. External influences—such as divided government, the nature of issues, or the political strategies of partisan leaders—may well account for both the partisan debate and the voting outcomes.

Attendance

A number of the opponents of Senate television painted a rather negative picture of members who would be unable to ignore the draw of the camera. Senator Danforth stated, "The nature of television is to attract us to it like sugar."[32] Senator John Warner (R-VA) concurred: "It might be that many of us are forced, or compelled for survival purposes . . . to be there in that chamber and neglecting other important work: meeting people in our offices, attending the committees."[33]

There is no solid evidence that members spend more time on the floor because of television; in fact, the ability to monitor chamber proceedings may well lead to less slavish attention to attendance. An early House survey indicated that only 3 percent of the respondents reported spending more time on the floor, while 23 percent felt they spent less and 68 percent saw no difference.[34] After the first year of television in the Senate, some argued, "Attendance on the floor is better; debate is sharper, with fewer endless quorum calls and fewer interminable speeches."[35]

Veteran lobbyist Ralph Nader expressed appreciation for the C-SPAN cameras in committee hearings. He pointed out that when C-SPAN is there, "more Members show up." He continued: "You can talk to an audience. You don't just talk to a bunch of Members of Congress who are giving you glassy stares because they've got their minds on something else."[36]

For assessing attendance, the issue may well be which stage of the legislative process members attend. Precise measurement of attendance occurs only during quorum calls and votes. It is argued that an increasing number of members monitor debate from their offices and dash to the floor only when a quorum call or vote is imminent. Absence from the floor during debate is not formally measured.[37] The visual image of the final vote on Clarence Thomas's nomination to the Supreme Court in 1991 shows what happens when legislators are fully aware that the whole world is watching. The vote was carried live by both C-SPAN and other networks. Unlike a typical vote, during which the senators wander in and out waiting for their name to be called, this vote found virtually all the members of the Senate sitting at their desks, like schoolchildren waiting to be called on.

Some observers have argued that the presence of the cameras may be "pushing debate off the floor . . . to more secluded corners." Former Speaker Tip O'Neill explained that while on camera, members find it more difficult to secure a compromise: "There isn't enough wiggle room. There is little room to make mistakes and too little opportunity to correct them."[38] The retreat to the back rooms may make floor and committee attendance less important.

Floor attendance for official actions is up in both chambers, but there is little evidence that this increase is directly related to the presence of the cameras. Constituent expectations and increased media scrutiny are just as likely to be the cause.

IMPACT ON VISIBILITY AND CREDIBILITY

According to Representative Robert Dornan (R-CA), C-SPAN "is like taking the walls of the House and blowing them out from coast to coast, and expanding the gallery to include everyone."[39]

Chamber Visibility

One of the reasons both the House and the Senate opted for televised coverage was increased visibility in general and in comparison with the presidency in particular. The evidence is not encouraging. In a detailed analysis of evening news programs before the arrival of television in the Senate, it was discovered that the commercial networks used coverage from the House floor only rarely.[40] Journalist Roger Mudd complained, "I think it is a journalistic outrage, to have the material and just ignore it."[41] There is little evidence that the House permanently increased its visibility advantage relative to the Senate by going on the air seven years before its competing chamber. The Senate continued to dominate stories on the evening news, even when only pictures from the House floor were available. Once the Senate came on the air, the Senate domination seemed to increase (see graph 12). From a more aggregate perspective, both the House and the Senate were covered *less* after television arrived in the chambers.[42] The situation is even more dramatic when one looks at coverage of Congress relative to coverage of the presidency. Congress continued to lose ground compared with the White House (see the prologue). The open question is whether Congress would have lost even more without those televised pictures.

For the narrower C-SPAN audience, the network has increased the visibility of Congress. But C-SPAN, which broadcasts House debates gavel-to-gavel, is available in almost twice the number of households as C-SPANII, which covers Senate debates.

Graph 12

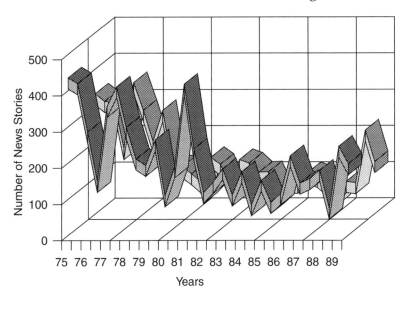

House and Senate Network News Coverage

During the first few years of televised proceedings, when only the House was televised, the attention paid to the House by network television surged past attention paid to the Senate. The Senate began to gain its formerly held ground when it went on the air. *Source:* Calculated by the authors from the *Vanderbilt Television News Index*.

Individual Visibility

In politics, personal recognition is a source of power. Perhaps reflecting the personality bias of network news, the arrival of television narrowed the range of members who received network coverage by focusing attention on a few key leaders who were good news "hooks." With only a limited time to tell a story, the networks focused on the better-known chamber leaders, who did not require a full—and time-consuming—introduction to the audience. Timothy Cook, in his detailed analysis of members' mentions on evening news programs, noted that "party leaders' share of mentions has increased relative to the shares of both committee leaders and non leaders" since the arrival of the television cameras.[43] The news hook for most congressional stories tends to be the well-known and established chamber leader. A 1995 study showed that 5 percent of the House and Senate members—largely chamber leaders—made 50 percent of the news on the three networks.[44]

For some nonleaders, though, C-SPAN does provide an opportunity for greater visibility. This potential is revealed by members' behavior. In the "I don't get no respect" category, very junior Senator Harrison Wofford (D-PA) discovered the impact of C-SPAN. Shortly after being elected in a special election, he complained about repeatedly losing his permanent seat on the floor to senior colleagues "wanting a better angle for C-SPAN's cameras." As he described it, "It was sort of funny, except I would come in to sit in my seat and a page would come by and say, 'No, it's been taken now by Senator Bumpers.' "[45]

Visibility can be a two-edged sword, either enhancing the positive impact of a member's actions or expanding the potential for public humiliation. Shortly after the 1994 elections, Representative Gingrich reflected on the increased scrutiny in the C-SPAN era. Speaking to the Republican Governors' Association, he stated: "I've been reading through [former Republican Speaker] Joe Martin's autobiography, but they didn't have C-SPAN. It was a different world. . . . If you made a mistake back then you just revised and extended and nobody knew it. Now my mother is going to call, 'Now why did you do that?' "[46]

Public Credibility

The old public relations dictum, "I don't care what they say about me in the papers, just so they spell my name right," may work for individuals, but it is not the base on which to build trust in political institutions. Positive evaluations are required to establish the reservoir of legitimacy that institutions need to make tough political decisions.

One of the goals of allowing the television cameras into the House and Senate chambers was to improve the image of Congress. It is now accepted that observing Congress is a right, not a privilege. When changes in camera access to the chambers are discussed, editorials such as the following appear:

> Let's stop and remember whose House this is. It doesn't belong to Speaker
> Tom Foley or any of the temporary residents. It belongs to the American

people, who deserve to see what's really happening. . . . One of the principles that separate the United States from other democracies is the belief in
an "informed public." We are better off with open government—warts and
all—than with coverage that is directed by sensitive incumbent politicians,
complete with special effects. Congress would do well to think about the
words of Abraham Lincoln. . . . "If once you forfeit the confidence of your
fellow citizens, you can never regain their respect and esteem."[47]

Many of the insiders feel that C-SPAN has contributed to respect and esteem for Congress. On the two-hundredth anniversary of Congress, Minority
Whip Newt Gingrich put aside some of his frustrations with Congress and
expressed the belief that the Founders would look relatively favorably on
the conflict and messiness of the contemporary legislative process and recognize that "frustration is the cost of freedom." He went on to assert that the
openness of the legislative process provided by C-SPAN was consistent with
the Founders' desires and was a positive factor for Congress and that "with
the rise of C-SPAN, those who do care about the Congress in fact do have
an opportunity to look at this institution on a daily basis now, and for our
aficionados this is an effective institution and they see how it truly works."[48]
Minority Leader Robert Michel (R-IL) agreed, arguing that C-SPAN had
"made the machinery of the Government understandable and accessible to
viewers across the nation."[49]

Other observers argue that C-SPAN's contribution to legitimacy goes
beyond our borders. We most often think about C-SPAN as providing programming about America for Americans, but its signal reaches around the
world. According to commentator Ralph Baruch, "C-SPAN has made its
mark, not only in this country, where it shows the American public democracy in action, and where it has been quoted and excerpted by both broadcast
networks and stations, but world-wide, with C-SPAN demonstrating to the
nations of the world how open democracy functions."[50] In a similar vein,
President Ronald Reagan used televised coverage of Congress as a selling
point for Western-style democracy as the Communist world was beginning
to collapse. In a speech at Moscow State University he argued, "We Americans make no secret of our belief in freedom. . . . Turn on the television,
and you will see the legislature conducting the business of government right
before the camera."[51]

Despite such rosy evaluations, the empirical evidence raises some serious
questions. Overall public evaluations of Congress have declined recently, de-

spite a dramatic expansion of the potential C-SPAN audience. There is little evidence that C-SPAN coverage or increased knowledge of Congress helped stem that tide. In fact, the opposite conclusion receives more support from the data. An initial argument for televising Congress was the assertion that "to know us is to love us," whereas in reality, "familiarity breeds contempt." As we saw in chapter 6, polls using C-SPAN viewership as a variable indicate that frequent viewers have less respect for Congress than nonviewers.[52] Part of the explanation resides in the fact that members of Congress themselves have used C-SPAN to castigate the institution of which they are a part. Such behavior stimulates and legitimizes "Congress bashing" by the commercial media.[53]

Data concerning C-SPAN viewers is bolstered by similar evidence that people who know more about Congress think less of its performance. A 1994 poll found that 67 percent of those who are most aware of the recent accomplishments of Congress disapprove of its performance, as compared with 51 percent of those with little or no knowledge of what Congress had recently done.[54] Another study indicated that the most politically knowledgeable people make a clear distinction between Congress as an institution and the members of Congress as a group. Highly knowledgeable people are more supportive of the institution and less supportive of its members than are less knowledgeable people.[55] The lesson for Congress seems to be that simply informing the public is not enough. A more informed public may better understand the messiness of the political process and respect the necessity of an institution like Congress, but until members of Congress are perceived as cleaning up their act, overall respect will not increase. Whereas it was once possible for Congress to carry out its business under a cloak of ignorance and bask in the general public acceptance for public institutions, those days are long gone. Creating an aware public may now be a necessary condition for turning the process around. With more awareness of the basic institution, the seed is sown for a new evaluation based on the turnover of membership or on the changed behavior of individual members. Taking their cues from such evidence, Speaker Newt Gingrich and Senate Majority Leader Robert Dole initially took steps in 1995 to expand C-SPAN access to congressional proceedings (see chapter 9). The institutional consequences are not yet known.

Politics is more than a battle between institutions. Institutions are made up of individuals, with their own hopes and goals. C-SPAN has proved to be a resource for some politicians and a detriment to others.

The Making and Breaking of Individual Careers

Events in politics seldom stem from only one event or factor. C-SPAN has become one of those factors that clever politicians use to promote their careers. C-SPAN is accessible to people across the political spectrum, especially those that fall within the mainstream of liberals and conservatives. C-SPAN's commitment to balance forces it to seek out spokespeople for a wide variety of views. There is no question that C-SPAN gets "used" by those with a political agenda, but as Susan Swain, C-SPAN's senior vice-president, sees it: "The good news is that both sides have figured out how to use us. For every Newt Gingrich there is an Al Gore." C-SPAN carefully tracks the partisan associations of its guests. Although there might not be perfect balance on any one program, over time a natural or planned balance emerges.

Like most technology, C-SPAN can be a double-edged sword. An effective set of appearances may create a positive response, but public exposure of foolish or pompous behavior is exacerbated by the publicity that C-SPAN provides.

MAKING CONGRESSIONAL CAREERS

Over the tens of thousands of hours of airtime broadcast by C-SPAN, hundreds of members of Congress have acted in ways that promoted their careers. Whereas some journalists express frustration at being "used" by politicians, C-SPAN's Brian Lamb expresses little concern: "I wish more members would use us. We're here to let members speak directly to their constituents, without being filtered by the media."[56]

Most of the consequences of actions by members of Congress have been limited, affecting only a few viewers or affecting a larger group of viewers in a limited way. Only a few of the most dramatic stories appear below.

Newt Gingrich: "Speaker C-SPAN"

Any discussion of C-SPAN's impact on political careers should include House Speaker Newt Gingrich. In the short form, the story is pretty simple. When asked whether he would be the Republican leader without C-SPAN, Gingrich—normally revealing the verbosity of a former history professor—gave an uncharacteristic and unqualified one-word answer: "No." With some prodding, he went on to explain that C-SPAN challenged the distribution of power in society by giving highly interested individuals access to information that allowed them to participate in a "feedback loop" with policymakers. These policymakers recognized the value of choosing leaders who could use the new technology to the benefit of their policy perspectives. C-SPAN thus accelerated the expansion of participation and the necessity of decision makers to listen to the active public. Thus, "you can't explain Gingrich as Whip without C-SPAN."[57]

A fuller version of the Gingrich story reveals how C-SPAN provided a group of media-savvy House conservatives in the mid-1980s with a method of circumventing the more liberal press and winning a prime-time audience away from the majority Democrats. By carefully scheduling speeches during the Special Orders period at the end of each day's session, Gingrich and his colleagues in the Conservative Opportunity Society (COS) blasted Democratic initiatives and performance, with little or no opposition. In Gingrich's words: "C-SPAN started the year I arrived. It has brought Congress into the living room. It has created a network of several million people who have dramatically higher awareness of how the system works than anybody has before."[58] C-SPAN was the perfect vehicle for a minority faction within the minority party. New technology is often adopted by those who feel disadvantaged by traditional means of communicating. Representative Dan Glickman (D-KS) complimented the COS for its clever use of Special Orders. He explained the Democratic hesitancy: "Being the majority party, I don't think we felt the need to do it. We weren't hungry enough. Mr. Gingrich and his friends were hungry, and they saw an opportunity to get their message across."[59]

If Gingrich's impact had spread no farther than the C-SPAN audience, he would probably still be a backbench member of the minority. Gingrich's genius lay in using C-SPAN to "get a bounce into the news."[60] The mainstream networks picked up the stories and sometimes the videos of him and his colleagues castigating the Democrats.

Perhaps as the ultimate unintentional celebration of C-SPAN's tenth anniversary in the spring of 1989, Republicans chose Newt Gingrich, the "Son of C-SPAN," as their second-ranking party leader. According to most observers: "He had neither the seniority nor the legislative record once thought necessary for such a post. But he did have the fame."[61] On a C-SPAN call-in show, Gingrich himself validated the assumption that he had used C-SPAN for his own purposes. He explained: "I'm the first leader of the C-SPAN generation. . . . In Washington, if you're in the newspaper every day and on TV often enough then you must be important."[62] He pointed out that the characteristics of effective leaders "are not fixed. Leaders need to adapt to the

> ### *Box 6.*
> ### Newt Gingrich: The Hot Leader on a Cool Medium
>
> The media descriptions of Newt Gingrich represent a combination of dislike and grudging admiration. Many reflect his grand view of history or his direct link with the new media. Characterizations include the following:
>
> "Son of C-SPAN"[a]
> "The scorched-earth strategist with a grand vision"[b]
> "The conservative firebrand"[c]
> "Motormouth run amok [who] manipulates the media"[d]
> "A quick-draw artist of the television generation"[e]
> "Bomb thrower"[f]
>
> [a]Used dozens of times in major publications (LEXIS-NEXIS database).
> [b]Michael Doyle, "Will New Level of Influence Make Newt Gingrich More Engaging?," *Houston Post*, November 10, 1994, p. A22.
> [c]AP Worldstream, November 8, 1994.
> [d]Martin Schram, "The Media Performs as Gingrich's Dupes," *Newsday*. December 7, 1994, p. A35.
> [e]Ronald Elving, "Politics of Congress in the Age of TV," *Congressional Quarterly Weekly Report* 47 (April 1, 1989): 722.
> [f]Used over eighty times in major U.S. newspapers in association with Gingrich between 1985 and 1994 (LEXIS-NEXIS database).

tenor of the times and the available technology."[63] For Gingrich, that new technology was C-SPAN.

Few politicians have come to high political office with more political baggage than Newt Gingrich. The mainline media seems to have a love-hate relationship with him (see box 6). They grudgingly respect him because of his effective use of C-SPAN but are miffed at his success in circumventing their control over public information. After the Republicans won a majority in the House in 1994 and Gingrich was tapped as Speaker, profiles of Gingrich invariably mentioned his pioneering use of C-SPAN.

The personally polite Gingrich probably does not much care what the media calls him, just so long as those in power call him "Mr. Speaker." Cognizant of the debt he owed to C-SPAN, Gingrich granted Brian Lamb an in-depth interview just before assuming the speakership and responded fa-

Gingrich on the Floor

Newt Gingrich became a familiar figure on the House floor even before becoming Speaker of the House. (Photo courtesy of C-SPAN.)

vorably to C-SPAN's initiative to open Congress more to the public. The daily leadership press conferences were opened to C-SPAN cameras, committee coverage was expanded, and talks were initiated to give C-SPAN more control over the coverage of floor debate.

DICK ARMEY: "C-SPAN ACTIVIST"

C-SPAN coverage may well stimulate political activity, either by raising an individual's level of disgust or by giving an individual enough information to become a player in the political process. As of 1989, political scientist Norman Ornstein indicated that he knew of at least a half dozen members of Congress who admitted they had run for public office because of what they had seen on C-SPAN. Some were confronted with issues they wanted to help solve; others took a look and said, "I am better than these guys."[64]

The clearest example is current House Majority Leader Dick Armey. This economics professor was watching C-SPAN one night in the early 1980s and remarked to his wife, "Honey, these people sound like a bunch of darned fools." She replied, "Yeah, you could do that." He long credited the experience with his decision to run and with his upset win in his suburban district. After becoming part of the majority party leadership in 1995, Armey began to back away from the story, indicating that he had been "only joking" and that he had thought about running for public office long before.[65] He did

Gingrich Stole Congress

The news media quickly recognized C-SPAN's role in Newt Gingrich's rise to power. (Cartoon by Leo Michael, *Multichannel News*, reprinted by permission.)

admit, though, that C-SPAN had "demystified" Congress for him.[66]

Once in the House, he gained publicity by refusing to establish a second household for a few days each week and instead spending his nights on a cot in the House gymnasium. After being evicted by Speaker Tip O'Neill, he reluctantly retreated to a sofa in his office.[67]

Robert Walker: "Instant Recognition"

The realization of C-SPAN's potency for elected officials is reinforced by anecdotes revealing the celebrity status that appearances on C-SPAN grant. One often-told story involves Robert Walker (R-PA), in the early 1980s a relatively junior member of the minority from Pennsylvania. As one of the early members of Gingrich's COS, Walker took to the floor regularly to chastise the Democratic leadership in Congress. In fact, Walker was speaking to an almost empty chamber during the Special Orders period in 1984 when

Speaker Tip O'Neill ordered the cameras to pan the chamber, in the hopes of embarrassing Walker and his colleagues. Walker handled the affair with aplomb, referring to the change of camera angles as "another example of Democratic party duplicity." Clips of the event flooded the mainline media, and Walker's picture showed up in newspapers all over the country. Until that point, few would have put Walker on the same level of recognition as fellow Congressman and former Buffalo Bills quarterback Jack Kemp (R-NY). Kemp himself revealed the power of C-SPAN by telling a story about walking on a beach in Puerto Rico. "A breathless tourist squealed and ran up to him saying, 'You're Jack Kemp, aren't you?' Pleased at the recognition, Kemp straightened his posture, smoothed his hair and said, 'Why yes, I am.' The star-struck tourist asked, 'Do you know Bob Walker personally?' "[68] With a bit of tongue-in-cheek humor, Speaker O'Neill began calling on Walker by saying, "The chair recognizes the media star from Pennsylvania."[69]

Bob Dornan: "Not Bombing Out on TV"

When conservative Republican Representative Bob Dornan, from California, announced his candidacy for the presidency in 1996, no one gave him any chance of winning, but few were surprised. Bob Dornan's political style was built on surprise and outrageous activity. He gained the nickname "B-1 Bob" for his strong support of the B-1 bomber and increased defense spending, but for many the name better represented the bombs that he dropped during congressional debate, which resulted in collateral damage through C-SPAN broadcasts. Breaking the norms of civility and talking beyond his colleagues to the public, Dornan called a San Francisco Democrat "the gentlelady from Babylon by the Bay," charged liberal Democrat Tom Downey (D-NY) with being a "draft-dodging wimp,"[70] and labeled southern California Democratic campaign contributors the "coke-snorting, wife-swapping, baby-born-out-of-wedlock, radical Hollywood left."[71] Dornan's weapon of choice was the Special Orders speech at the end of each legislative day. He used his previous experience as an actor and talk-show host to vigorously spread his "verbal napalm."[72]

In 1992, Dornan's attacks on presidential candidate Bill Clinton as a "pot smoking, draft dodging philanderer" were broadcast in full on C-SPAN and helped build Dornan a national following (see chapter 8).[73] The speeches prompted C-SPAN viewers to deluge news organizations with calls, asking why they were covering up the "true" stories of Clinton's trips to Moscow and antiwar activities. Dornan used the public uproar to add legitimacy to his claims and later leveraged a White House meeting on the topic.[74]

Dornan's effectiveness on C-SPAN caught the attention of talk-show entrepreneur Rush Limbaugh, leading to Dornan's being cast in a new role: as Limbaugh's regular vacation replacement. Without C-SPAN, Dornan's national exposure on radio and his presidential bid lack any explanation.

Al Gore: "The Glory of Opening Night"

C-SPAN has not granted visibility only to Republicans. Vice-President Gore's fascination with new technology and the emerging information highway is not a newfound captivation. In March 1979, Representative Al Gore (D-TN) was the first person to stand in front of the cameras to give a speech, inaugurating a new era of television's relationship with Congress. He stated: "Television will change this institution, Mr. Speaker, just as it has changed the executive branch, but the good will far outweigh the bad. From this day forward, every Member of this body must ask himself or herself how many Americans are listening. . . . The marriage of this medium and our open debate have the potential, Mr. Speaker, to revitalize representative democracy."[75] He repeated his first-in-line performance seven years later during the first televised floor debate in the Senate. "The marriage of television and free debate in the Senate will be of benefit to American citizens. . . . It will bring changes in this institution. But whatever changes are due to the feeling that millions of Americans are paying attention to what goes on on the floor of the Senate will, in the final analysis, be good and positive changes."[76]

As a House member, Gore became an avid C-SPAN viewer. One night John Evans, president of the Arlington, Virginia, cable company, got a call from the somewhat agitated congressman. Gore was at home eating dinner and watching C-SPAN to determine when he would have to go back to vote. He was mad that the signal had gone down. It took the cable company about twenty minutes to fix the malfunctioning equipment, and Evans immediately called to apologize to the congressman.[77]

George Mitchell and Tom Daschle: "The New Era Leaders"

The choice of Senator George Mitchell (D-ME) as majority leader in 1988 over Senators Bennett Johnston and Daniel Inouye was widely reported as being based on the fact that the Democrats were looking for "an effective, articulate spokesman . . . more attuned to the television age."[78] Despite being "hardly a blow-dried, made for television politician . . . his solid manner comes across well on television."[79] Although Democrats were interested in network television skills, they recognized that their leader would also have to portray their party well under the continuing eye of C-SPAN. Mitchell "may not have vaulted over two more senior rivals . . . in an age less attuned to media presentation."[80]

Robert Byrd, Mitchell's predecessor who had been elected before the cameras were installed, was quite honest when he said: "I don't believe that I would have been chosen as a congressional leader with C-SPAN. My dour demeanor and pompadour don't fit TV. My skills are elsewhere from forming ten second sound bites." Once the cameras came, Byrd learned to use television and viewed it as an "additional weapon," one that he "could mold" to his own will. Byrd expressed a lot of respect for television and its ability to separate prophets from windbags and demagogues. In his view, "You can't fool that [television camera] eye . . . it sees deeply in the window of the soul."[81]

When the next round of Democratic leadership selections came, with Mitchell's retirement decision in 1994, the two front-runners were Jim Sasser (D-TN) and Tom Daschle (D-SD). "Both men have mastered television, the prerequisite for a modern leader."[82] With Sasser's unexpected defeat and the Democrats' loss of the majority in 1994, the Democratic leadership race for minority leader pitted Daschle and Christopher Dodd (D-CT), another television-age politician. Dodd went public with his quest; Daschle refused television interviews and eventually won, using the insider's strategy. Since both candidates were known for their television skills, ability in front of the camera failed to materialize as an issue.

AFFECTING THE PRESIDENTIAL CONTEST

If presidential contests were theatrical plays, C-SPAN would play a leading role in the first act, setting the context and sorting out the stars from the bit players. By the second act, C-SPAN's role would fade as the other media picked up established stars, some of whom owed their status to C-SPAN. By the third act, C-SPAN would be presenting soliloquies to a small segment of the audience, with the established stars seldom deigning to speak on the C-SPAN stage.

Ross Perot: "The Making of a Candidate"

The story of Ross Perot's 1992 presidential campaign has largely focused on the public kickoff on the *Larry King Live* show, where Perot "permitted King to extract from him a pledge that he would run for president if supporters would qualify his name for the ballot in all 50 states."[83]

The implication later made by many observers, including King himself, was that the Perot phenomenon sprang from that one program.[84] In reality, Perot had been assiduously using a variety of talk shows and alternative media to circumvent the conventional media, which paid him little attention.[85] C-SPAN played an early and important part in Perot's media mix. Perot had appeared on C-SPAN numerous times before February 20, 1992, when he made his announcement on Larry King's show. As Brian Lamb remembered it, "Viewers reacted enthusiastically to each appearance . . . people called, people reacted. . . . When there was an opportunity to buy transcripts . . . they bought more than they bought from anybody else."[86] Appearances by Perot on C-SPAN activated a number of key supporters. One Massachusetts supporter remembered "picking at the remote control" when he "landed on C-SPAN's rebroadcast of the undeclared candidate's speech to the National Press Club."[87] *Newsweek* concluded, "The grass roots insurgency . . . has been fueled by scores of C-SPAN epiphanies."[88]

Once Perot established a base and got a "kick start" on publicity, the conventional media picked up his story. Between May and July 1992, his coverage exceeded that of Bill Clinton's candidacy. Part of Perot's later problems with the media stemmed from C-SPAN. After a particularly heated appearance on *Meet the Press*, Perot vowed not to be interviewed. He relented a few weeks later, allowing Brian Lamb to interview him for one and one-half hours in his Texas offices. Lamb attempted to walk the fine line between C-SPAN's objective approach to interviewing and the need to get Perot to answer key charges that he had avoided. Without making any new charges, Lamb simply asked Perot to respond to a series of charges made in the other media. Perot's answers on C-SPAN seemed either inconsistent or at odds with other evidence, and the C-SPAN interview served as the basis for a spate of negative stories in the commercial media. With the media hitting so close to home, Perot began to think about dropping out. After his departure from the race in July, his coverage decreased precipitously, and he was doomed to third place in terms of coverage even after he reentered the race late in the campaign.[89]

Ross Perot on C-SPAN

Ross Perot was a regular guest on C-SPAN even before beginning his presidential bid. (Photo courtesy of C-SPAN.)

Paul Tsongas: "Keeping a Candidacy Alive"

Former Senator Paul Tsongas (D-MA) credited C-SPAN with giving his 1992 presidential bid a major boost. In response to questions about his health after his bouts with cancer, Tsongas allowed C-SPAN to film his vigorous morning swim. The image of Tsongas in his Speedo swimsuit became a symbol of his campaign through repeated airings on C-SPAN and rebroadcasts of the "sound-bite version" of the event on other networks. Tsongas gave the C-SPAN crew and the viewers a start when he began stripping off his swimsuit in front of the cameras. People gasped that C-SPAN was about to be X-rated until they realized that he had two suits on and that he was simply trying to get the drawstrings straightened out. After the segment aired he said, "No matter where I went around the country, people . . . would always come up to me and I was shocked at how much they knew about me, particularly how I looked in a Speedo bathing suit."[90]

More seriously, Tsongas concluded, "For someone who was so under-funded, C-SPAN was a way—given how the networks cover the races—for someone who was interested in politics to hear a candidate at length and arrive at their own judgments."[91] Tsongas asserted: "The media in 1991 did not take my campaign seriously . . . C-SPAN did. . . . As we traveled through 1991 in particular, people had watched C-SPAN and knew what I stood for."[92] Although Tsongas's front-runner status before the New Hampshire primary faded as the campaign progressed, C-SPAN's cameras helped keep him in the race by reducing the concerns over his health and giving him a vehicle for spreading his substantive message.

The advantage of C-SPAN's commitment to carry a story through is also exemplified by its coverage of the end of the Tsongas campaign. In the plastic world of blow-dry candidates whose every move is managed, there is something very appealing about unexpected human insights. C-SPAN is often positioned to capture such moments. In 1992, the networks showed only a short clip from Paul Tsongas's resignation speech. In C-SPAN's coverage of the entire event, his wife leaned over and planted a kiss on his cheek, leaving a lipstick imprint that remained for the rest of the event. This version gave us a more in-depth feel for Tsongas and his wife.[93]

Bill Clinton: "The Long-Term Strategy"

Bill Clinton's use of C-SPAN started well before he became an announced candidate for president. After watching Brian Lamb interview Governor Bill Clinton, Mike Gauldin, Clinton's press secretary, came to the conclusion that C-SPAN could be useful. From then on he targeted C-SPAN, informing the network of upcoming meetings at which Clinton would be speaking. It was not only an attempt to get Clinton known to the public but also "a good method of 'shooting over the wall' to reach the political elite."[94] As Gauldin explained it:

> Clinton's strength was his substantiveness, his familiarity with the issues, his ideas for the future; now these are not things that can be communicated in a ninety-second sound bite. . . . We always paid special attention to C-SPAN, because on C-SPAN the viewer got the whole speech. . . . I would consciously call up C-SPAN and tell them what he was doing . . . and whenever we were at conferences, we always tried to make time to do the call-in programs. . . . Our strategy was to get him into forums where he could get his message to decision-makers. . . . We recognized that C-SPAN

provided us with a forum perfect for Bill Clinton, so we tried to make use of it wherever possible.[95]

As founder of the Democratic Leadership Council and an active participant in various governors' councils, Clinton had a lot of potential for coverage. His public exposure marked him as an up-and-coming political force.[96] And Bill Clinton was always an eager guest. At the National Governors' Conference in Traverse City, Michigan, in the mid-1980s, C-SPAN's evening call-in guest canceled at the last minute. Even though Clinton had already done the morning call-in, the C-SPAN staff was desperate. A call to Clinton's hotel room found him in the shower. A staff member accepted the invitation almost immediately. Within five minutes a wet-haired Clinton arrived at the makeshift studio and went on the air immediately.

Between 1987 and his nomination in 1992, Clinton appeared as a featured participant on 175 C-SPAN programs. His initial foray into presidential politics in 1987, at a speech before the New Hampshire State Democratic Party, was covered in full by C-SPAN. The increased visibility helped garner him the keynote speaker's position for the 1988 Democratic National Convention. Again C-SPAN showed the entire speech. Unfortunately for Clinton, however, gavel-to-gavel coverage of that extraordinarily long speech gained him the reputation as someone who knew how to begin a good speech but not how to end one.

As he began gearing up for the 1992 race, Bill Clinton made a number of speeches that were covered fully only by C-SPAN. Campaign director Frank Greer commented:

Through the early days, when we were thinking about running, we realized the power of C-SPAN. . . . Whenever he gave a speech, we would start getting letters and calls in Little Rock from people who had seen the speeches. The only way they were getting this was C-SPAN. . . . Democratic activists were making up their own minds, and they were able to see in totality his message, as opposed to getting sound bites and labels. [President Clinton] really believes that C-SPAN did transform the coverage of the media, and that it provided him with a forum and access to people that he never would have had otherwise. He would probably say that C-SPAN was a critical part of his ability to communicate, educate, and finally be elected President.[97]

Early in the race for the presidency, Clinton realized that his most important audience was not the general public. He first had to convince the party activists who would be voting in primaries and attending conventions. A disproportionate percentage of this group had "been watching C-SPAN for 10 years and [had] been watching Bill Clinton . . . for 10 years."[98]

Frank Greer was so impressed with Clinton's performances on C-SPAN that he adopted the C-SPAN approach during the New Hampshire primary. Greer said: "We bought thirty minutes of time—on commercial stations . . . it was unscreened phone calls, just like C-SPAN. . . . As the campaign progressed, you saw the networks—especially the morning shows—following the format of C-SPAN. . . . I really look to Brian Lamb as a visionary in the world of media. . . . I think he transformed the whole character of the campaign."[99]

After the election, Clinton invited the C-SPAN cameras to his day-long economic conference. As observers noted, Clinton seemed "anxious to be the C-SPAN president, displaying his knowledge of arcane details before a national television audience."[100] Inauguration day allowed C-SPAN to show another typical "C-SPAN moment." Turning the cameras on and letting them roll while the president-elect waited for his family in front of Blair House, C-SPAN presented its viewers with a tableau familiar to millions of American couples. In his role as husband and father, Clinton "impatiently yelled up to [his wife] and Chelsea to hurry up because people [were] waiting. A few moments later, Hillary Clinton came down the stairs glowering."[101]

Once in the White House, Clinton gave C-SPAN unprecedented access to White House social events, policy conferences, and one-on-one interviews. The day after the inauguration, the C-SPAN cameras captured hours of "up close and personal" time as the president and Mrs. Clinton greeted thousands of supporters in a receiving line.

Phil Gramm: "Presidential Preemption"

As the 1996 presidential contest approached, Senator Phil Gramm (R-TX) used C-SPAN to steal the thunder from both President Clinton and Gramm's opponent in the Republican nomination race, Senator Bob Dole. Journalist Jack Germond described Gramm's strategy of using immediate access to C-SPAN to upstage both Dole and Clinton by preempting their announcements with policy initiatives of his own. This was particularly evident when he announced his plan for a middle-class tax cut in December 1994 just hours before the president was to address the nation on that same issue. Germond concluded, "If there were not a C-SPAN, would there be a Phil Gramm?"[102]

Backup Players and Surrogates: "Hoisting Trial Balloons"

Politics is a risky business. Winning and losing often depend on who makes the fewest mistakes, as opposed to who makes the best decisions. Politicians have long floated trial balloons to determine public reaction. If an idea is confronted with derision or apathy, it is quickly dropped. If it captures the fancy of those they want to impress, it is restated with more force and detail. Often the principal players in a presidential contest keep their distance from trial balloons by allowing a surrogate to first float them. During the 1988 campaign, the Republican theme of using furloughed murderer Willie Horton as an example of Michael Dukakis's softness on crime was first tested on the House floor by a group of Republican members. A similar approach was used for the issue of the American flag as a symbol of patriotism. George Bush's media adviser noticed the public reaction to floor speeches and knew he had a couple of issues that were winners.[103]

In 1992, speeches on the House floor were used to test the character issue against presidential candidate Bill Clinton. Day after day, members such as Representative Bob Dornan titillated the C-SPAN audience with detailed itineraries and speculations concerning Bill Clinton's foreign anti-Vietnam travels. Dornan's Special Orders comments on Clinton's trip to Moscow were not immediately picked up by the traditional media. It was Larry King who eventually asked Clinton about the trip, citing Dornan as his source. The next day Phil Donahue repeated a similar question. The notoriety was enough to get Dornan a meeting with President George Bush and his campaign manager, Jim Baker, at the White House. A few days later, when Bush was on Larry King's show, he was asked to comment on Clinton's trip. Bush tried to use the issue while remaining above petty name-calling, but to many listeners he "sounded as if he were reading Dornan's notes."[104] Bush's willingness to use innuendo, and media revelations about Dornan's White House visit, probably brought the Bush campaign more political embarrassment than benefit.[105]

BUILDING OTHER CAREERS

C-SPAN has provided exposure to nonpoliticians also. Journalists such as Wolf Blitzer (*Jerusalem Post*, CNN) and Steve Roberts (*New York Times*, *U.S. News and World Report*) found that C-SPAN helped them build a national reputation and furthered their careers. The clearest example of an impact on a journalist's career was Larry King. In 1983, C-SPAN simulcast his talk show. King had assumed that radio was the only appropriate medium for the kind of show he did. He was flabbergasted at the audience response to the televised version. For months, people he met in his travels mentioned the television coverage. King noted, "I never got a reaction like that to my radio program."[106] King concluded that the format might work on television. A few years later he became a regular on CNN.

Larry King on C-SPAN

C-SPAN cameras capture Larry King's radio program for the first time on television. (Photo courtesy of C-SPAN.)

The personal stories could go on. C-SPAN offers significant resources for politicians and others to promote their careers. The potential is not unlimited, however. C-SPAN can expose weaknesses as well as magnify strengths.

CAREER BREAKING IN THE C-SPAN ERA

Technology is often a double-edged sword, encompassing a possible benefit with a commensurate danger. Whereas C-SPAN has clearly helped make some political careers, it has undermined others. The potential advantage of positive exposure carries with it the potential risk of negative revelations. Since the public tends to believe that the "camera does not lie," events caught and disseminated on video have a greater potential impact than stories spread by word of mouth or reported solely in print.

Tip O'Neill and Jim Wright: "Wounding Shots"

Speakers Tip O'Neill and Jim Wright ran afoul of the cameras in different ways. During the summer of 1984, Speaker O'Neill became increasingly agitated at the COS and Newt Gingrich's effective use of Special Orders speeches to an empty chamber at the end of each legislative day to pillory Democratic policy positions and leadership. O'Neill countered by having the cameras pan the empty chamber, and he verbally struck out at the offenders. The incident resulted in Speaker O'Neill's words being stricken from the *Congressional Record* (see chapter 2). Many insiders felt that being formally reprimanded by his own chamber was such an indignity for the Speaker that it led to O'Neill's initial thoughts about retiring. Publicly, O'Neill insisted that allowing the television cameras into the House was "one of the best decisions" he ever made.[107]

C-SPAN's association with Speaker Wright's departure was more indirect. Newt Gingrich used Special Orders speeches to attack Wright's policies and leadership, as well as his "sweetheart" book deal giving Wright a royalty percentage unheard of for even the best-known authors. Initially viewed as partisan nitpicking, Gingrich's charges on C-SPAN caught the attention of the mainstream media and eventually led to a referral of the matter to the House Ethics Committee. From that point on, Wright's battle for his political life occurred on television. Democratic colleagues began to fear that with the extensive publicity, Wright's future might not be the only thing on the line— the image of the Democratic Party and even public respect for Congress were also on the line. After the Ethics Committee expanded the scope of the investigation and Wright realized that he had lost his effectiveness as Speaker, he took to the floor in May 1989 to make an emotional speech, covered by C-SPAN, in which he announced his resignation.[108] Clearly C-SPAN did not cause Wright's demise, but the network's coverage established more favorable conditions for effectively publicizing the charges and probably sped up the process.

David Durenberger, Bob Kerrey, and Joe Biden: "Glancing Blows"

C-SPAN's strategy of keeping the mikes on during events and of wiring politicians during entire days of campaigning can be a recipe for disaster—or at least significant embarrassment. Politicians accustomed to formal television or radio programming, with red lights to indicate when they are "on the air," can have a rude awakening with C-SPAN's approach of continuous broadcasting. Senator David Durenberger (R-MN) found this out the hard way. During a break in a radio talk show being covered by C-SPAN, the hostess tried to loosen up the divorced senator with a few questions about his personal life. When asked about women in his life, he mentioned a divorced woman with two children, assuming that the mikes were dead. Instead, this personal tidbit was broadcast to a national audience.[109] The embarrassment probably did not do too much damage to his waning political career, but it did complicate his personal life.

The stories of Senators Bob Kerrey (D-NE) and Joe Biden (D-DE), told in chapter 5, fall into the same category. Whether telling an off-color joke or embellishing one's record, politicians can easily discover the danger of live mikes, which do not allow one to deny lapses in judgment.

Promoting Political Causes

C-SPAN offers an alternative venue for setting the national policy agenda. The historical commitment to broadcasting House and Senate sessions in their entirety and the philosophical commitment to balance and breadth in covering political causes create significant opportunities for a wide range of voices to be heard. The commercial networks, with their limited news allocation for public affairs programming and their need to build ratings by satisfying a broad audience, tend to focus their efforts on mainstream sources. Relieved of much of that pressure, C-SPAN "is a godsend for those who want to see not only Congress, but any number of other organizations—too far off the political spectrum to get news coverage—discussing strategies on social and political issues. Everything from the Rainbow Coalition to Operation Rescue, from the Ku Klux Klan to the Communist party is likely to be covered by this little odd network."[110]

As minority leader, Robert Dole was often frustrated with the media. He used the presence of C-SPAN as a way of chiding the media and circumventing it. He commented: "I know the liberal media does not have any interest in this. But I want a record made somewhere. . . . But in the event somebody might be watching on C-SPAN . . . or might read the *Congressional Record,* I want them to understand precisely what happened."[111]

C-SPAN helps establish the legitimacy of issues for the public agenda. As political commentator Murray Kempton concluded, "A subject cannot be taken for serious . . . until it qualifies for discussion on C-SPAN."[112] The premier example of using C-SPAN to help set the agenda must be Newt Gingrich and the COS. Although Gingrich and the COS are often credited with "discovering" Special Orders speeches, "the idea of using floor speeches to capture a TV audience actually started" in 1983 with Bill Alexander (D-AR). According to one account: "Alexander's strategy was to notify TV stations around the country when a Democrat makes a speech on a topical issue. The local stations tape the brief speech and run it on their news shows. Alexander said Democrats need to learn how to use television to compete with Reagan."[113] What Gingrich and company did was expand the strategy and use it with a vengeance. COS members orchestrated their attacks, alerted their supporters in the public, and garnered commercial media attention. When asked about speaking to an empty chamber, COS strategist Robert Walker commented, "With C-SPAN broadcasting the signal, we never thought of it as an empty chamber."[114]

Most Democrats initially missed the significance, dismissing COS efforts. Majority Leader Jim Wright said: "If a fellow wants to waste the time of the television audience with bombast, the rules permit it. But I suggest that the public knows it's as phony as a $3 bill."[115]

COS recognized that it had "struck public relations gold—hours of free TV time enabling its members to reach beyond their local constituencies to the entire nation."[116] When asked about a 1984 audience estimated at 200,000, Gingrich commented: "That's not a bad crowd. This is the beginning of the ability to have a nationwide town meeting."[117] As the minority in Congress, Republicans during the 1980s clearly recognized C-SPAN as a vehicle for circumventing the commercial media, capturing the attention of a segment of the public, and indirectly being heard *within* Congress. Jon Kyl (R-AZ) commented: "We Republicans very much appreciate C-SPAN covering our speeches, because this is one of the few ways that we can get our message to the public, and through them to the Democratic majority. . . . The Democrats have adopted rules that refuse to allow debate or amendments. . . . We would prefer to address our views and proposed amendments to the full House, but they have made this impossible, and our only alternative is to address them to the Nation, for which we thank C-SPAN."[118]

Gingrich and his colleagues continued to use C-SPAN in a variety of ways. With the growth of talk radio and real-time politics, the immediacy of C-SPAN added a new dimension. When Gingrich was pushing a jobs bill, the Democratic leadership would not allow it to be debated. Gingrich called Rush Limbaugh's program director and urged him to turn on C-SPAN to see what was happening. Limbaugh went on the air with the story. The phones on Capitol Hill lit up to the point that in many offices, the phones could not be used. C-SPAN had become the vehicle for informing Limbaugh and, indirectly, getting the story out.[119]

The success of COS and Gingrich was not immediate, but they were using C-SPAN to build a constituency, recruit candidates, try out ideas, and contribute to the public agenda. Many of the Republican farm-team members were avid C-SPAN viewers. The Republican efforts came together in 1994 when over three hundred Republican congressional candidates participated in a signing ceremony, broadcast by C-SPAN, for the ten-point "Contract with America." As discussed earlier, Newt Gingrich challenged Americans to read the contract and then "watch C-SPAN after the election to see" the Republicans "carry it out." C-SPAN political reporter Steve Scully argued: "The 1994 election hinged on the perception that nothing was working in Congress. . . . How would the public know without C-SPAN?" Scully was not alone in his evaluation. Newt Gingrich concluded: "Without C-SPAN, without talk radio shows, without all the alternative media, I don't think we'd have won. The classic elite media would have distorted our message."[120]

With the avalanche of media attention after the 1994 Republican victory, the Republicans did not need C-SPAN to get their message out, but they rightfully recognized the debt they owed the network that had paid attention to them when they needed it. If imitation is the sincerest form of flattery, the Democrats in Congress—reeling from electoral defeat—paid the Republicans a decided compliment. In early 1995 they decided that a coordinated use of C-SPAN and other news media might be their salvation. Under the leadership of Minority Leader Dick Gephardt (D-MO), the House Democrats initiated a "group approach to defining national themes . . . to positively influence the daily news cycle."[121]

Conclusion

A 1960s television drama started by reminding viewers, "There are thousands of stories in the city." Similarly, assessing the impact of C-SPAN largely involves collecting and evaluating numerous stories, as opposed to analyzing clear empirical evidence. It is hard to find anyone who would argue that American politics as practiced by the insiders today is the same as it was in the era before C-SPAN. At least some of those changes were directly advanced by the presence of C-SPAN itself. Despite being a medium intentionally devised to record public affairs, C-SPAN clearly leads to public exposure, with significant consequences.

CHAPTER EIGHT

C-SPAN's Impact

A View from the Outside

Before my husband's medical problems, we practiced politics firsthand. Until we discovered C-SPAN we were so isolated. Now a whole new world has opened up to us. We schedule our meals around hearings, read the upcoming books on *Booknotes*, and stay up late to watch *Question Time*. We have even begun to write letters to our politicians again. It's our window on the political world.

Authors' interview with C-SPAN viewer

A popular Government, without popular information, or the means of acquiring it, is but a Prologue to a Farce or a Tragedy; or, perhaps both.

James Madison, 1822

Informing the Public

With twenty-four hours of information on two channels, C-SPAN broadcasts a great deal of information. It has been called "a remarkable university of the air."[1] The questions are, Who does it inform, and what does it inform them about?

THE PROMISE FOR DEMOCRACY

C-SPAN supporters of all stripes would endorse Madison's comments and would count him among their number if he were still around. Some observers saw the potential for C-SPAN relatively early. Their thoughts foreshadowed the events that would occur almost a decade later. In 1985, journalist Carol Richards wrote: "C-SPAN is making a difference. It is probably a very small difference right now. But as cable reaches more and more homes, and citizens become accustomed to switching over to C-SPAN every so often to see what their elected officials are up to, politicians will be held to a new standard."[2] Presaging the increased congressional turnover during the 1990s, political correspondent T. R. Reid said in 1986: "C-SPAN will change Congress and cause better people to be elected. . . . When people see the quality of the members and see how dumb some of them are, they will boo them out. They will know what the rascals are up to."[3]

On the positive side, C-SPAN allows competent members of Congress to shine. Charlie Rose (D-NC) chuckles with some pride about the elderly man who approached him during a trip to his district. The man, sticking out his hand, said: "Brother Rose, I been watching the stuff on television of yours. It's sometimes pretty dull. . . . [But] you all ain't half as stupid as the press makes you out to be."[4] Many of C-SPAN's plaudits come from those frustrated that commercial television has not lived up to its potential for informing the public. Pioneer television journalist Edward R. Murrow demonstrated a great deal of foresight about television at a time when other commentators, such as Newton Minow, saw television as simply a "vast wasteland."[5] Murrow's comments in 1958 almost seem to presage services such as C-SPAN: "The instrument can teach, it can illuminate, it can even inspire. But it can do so only to the extent that humans are determined to use it to those ends. Otherwise it is merely lights and wires in a box."[6] But even Minow eventually saw the value of C-SPAN. He commented: "What C-SPAN does is give everybody a seat at the event. . . . [It] fulfills one of the great dreams of the people who invented TV—that you can see an event while it's happening, unvarnished, without commentary, so that people can make up their own minds."[7]

George Orwell's novel *1984* raised the ominous specter of an intrusive government using new technology such as television to watch members of the public. As the year 1984 approached, Wilton Dillon, chairman of the Smithsonian Institution's Orwell Symposium, used a C-SPAN broadcast to remind us that "technology is NOT neutral" and that it can be used to either open or repress society. He singled out C-SPAN as an example of a technological application that opened the political system and provided a link for citizens to watch their government in action.[8] Media critic Tom Shales expressed this idea even more eloquently, with a touch of historical parallelism to a pioneering technology of a previous age: "[C-SPAN] is the driving of the golden spike that unites the people with their government through television. It is a crucial, pivotal moment in the life of the newly emerging wired republic."[9]

Although information transfer is difficult to measure, members of Congress certainly realize the importance of C-SPAN in educating Americans. As one television critic put it, "C-SPAN is the most revolutionary form of television, dispensing the most liberating of all weapons—information."[10] Floor speeches are filled with references to C-SPAN's role in this realm:

> C-SPAN continues to do its excellent work informing the American people of what is being debated (Senator Robert Byrd [D-WV]).

> I think we tend to underestimate the American people. . . . What they are seeing on C-SPAN today is very educational to America (Senator Conrad Burns [R-MT]).

> Here we are in the electronic age with C-SPAN, CNN, and faxes, and computers. . . . We have to rethink what does it mean to be a citizen. . . . How do we organize citizenship for the 21st century? . . . Let me suggest that C-SPAN is an important first step (Representative Newt Gingrich [R-GA]).

> Greater attention is paid to what we are doing here, because of C-SPAN— the more people see debates, the more they get it straight, the more they understand the fundamental difference between the two parties (Senator Phil Gramm [R-TX]).

> C-SPAN has whetted the curiosity of America as far as the running of the country is concerned (House Speaker Tip O'Neill [D-MA]).[11]

The cable industry sees C-SPAN as a jewel in its crown—its primary contribution to democratic society. Cable operator John Evans, one of the network's earliest supporters, waxed philosophical about C-SPAN: "I believe that historians, a hundred years from now, will look back on C-SPAN and see that it has been important in preserving freedoms. . . . Only when you have an informed electorate can the strength of democracy truly come out.

That's what C-SPAN is all about. . . . It's fundamental to the preservation of our society."[12]

C-SPAN viewers are equally enamored. In 1984, C-SPAN sponsored a contest and asked viewers to complete the statement "I watch C-SPAN because . . ." The winner asserted: "[C-SPAN] takes me beyond the tidy *Congressional Record*, past the deletion of the daily press, behind the strangely attractive reporting of the network news, straight to the House and Senate floors, even into the subcommittees where our elected representatives stand directly before us in all their eloquence and inarticulateness, their wisdom and their foolishness, their openness and evasiveness, their glory and disarray. . . . I watch C-SPAN because of its great implicit and active faith in free speech and open deliberation."[13]

DETERMINING THE ACTUAL IMPACT ON VIEWERS

Assessing the impact of C-SPAN on the individual viewer is a daunting task, since viewers use the network in so many ways. For a small number of C-SPAN "junkies," it is "their" network and a primary source for an almost continuous flow of information. Other viewers seek out particular programs.

Most viewers are much more irregular, scanning by C-SPAN as they "channel surf" and dropping in for a few minutes when they see something of interest. For many viewers this is a positive experience, providing them with new perspectives and different information. For others, seeing a few minutes of a member of Congress addressing an empty chamber or getting excited about a procedural rule feeds existing biases; the viewer changes channels, muttering, "What can you expect from those duplicitous bozos?"

For the less frequent viewer, or even for people who have never watched C-SPAN, its existence is still important. Their experience is something like driving by the public library every day and seldom going in. For them, the importance of the public library does not lie in frequency of use but rather in symbolism: its very existence represents free information. Its availability implies that one can get the information one needs when it is needed. Seeing C-SPAN as one surfs by gives a viewer some comfort that the system is open and aboveboard and that the information is there if the viewer really needs it.

C-SPAN AND THE INFORMED CITIZEN

By opening the door on Washington, C-SPAN has changed the role of informed citizens in the policy process. It is "rapidly redefining inside Washington's relationship with the vast territory beyond the beltway."[14] In Brian Lamb's view: "It used to be the president proposes, Congress disposes and the public was not involved. Now the public is involved in every step of the way."[15] A number of discrete incidents reveal the new power of C-SPAN to change public perspectives. After President Bill Clinton's attorney general nominee, Zoe E. Baird, admitted she had hired an illegal alien, the inside-the-beltway reaction was "no big deal." Lamb noted: "There's always been a difference in the way people in Washington think about an issue and the way it's done in the rest of the country. What Washington will tolerate, more and more, the rest of the country will not."[16]

Among the Washington establishment, it was assumed that the hiring of illegal aliens was commonplace and that most Americans would not be concerned. The public had a harder time identifying with a wealthy corporate lawyer who cut costs by hiring an illegal day-care worker. Unlike the commercial networks, C-SPAN broadcast the Baird hearings in their entirety. "America listened, and it did not like her words or the tone of her voice. Callers to radio talk shows, senators, newspapers and C-SPAN itself said they detected arrogance."[17]

One by one, Baird's supporters in the Senate faded away in the wake of public discomfort. She withdrew, and the visibility of her case established a new criterion for presidential appointments in the post-"nannygate" era. C-SPAN did not kill Baird's chances per se; it simply opened up the process and allowed her to self-destruct in public. In Washington politics, a public battle often has a much different outcome than one carried on behind closed doors.

The C-SPAN coverage of the debates on Clinton's 1993 economic stimulus package undercut the momentum following his election. Senator Hank Brown (R-CO) used C-SPAN to "make the argument around the country that the 'stimulus' package would finance a lot of 'pork barrel' projects of marginal value."[18] The criticism hit its mark, and a scaled-back version of the package only barely passed, with no Republican votes. C-SPAN captured some of the last-minute vote-switching and agonizing statements of members pressured by the White House to support the package. A number of Democratic Party defectors expressed their frustration about having to support a program they could not fully accept. Many of the Democratic supporters found themselves in political trouble in the 1994 elections. Of the Democratic House incumbents who lost in 1994, 89 percent had voted for the package.

With the dramatic shift in party fortunes after the 1994 elections, a number of long-shot candidates found themselves thrust into office and in need of some information on the congressional process. They often turned to C-SPAN for a quick, unvarnished education on how our political system works. Other new members had more time to prepare. After winning the primary and facing little opposition in the general election, Houston City Councilwoman Sheilah Jackson Lee (D-TX) commented, "I've been keeping C-SPAN on like 24 hours a day, trying to get a sense of it."[19]

A Cautionary Note

Although questioning the value of becoming informed is a bit like attacking motherhood, it is important to honestly evaluate some of the potential problems that C-SPAN's style of informing could exacerbate. Not all observers are optimistic about the impact of programming such as C-SPAN's on American democracy. Raw data may confuse as much as edify. C-SPAN is part of the larger talk-show democracy revolution. Even though many observers point out that C-SPAN's call-in format is different from and more responsible than other talk-show venues, some potential dangers apply to C-SPAN as well.

Raw Data and Dumping the Whole Load

C-SPAN emphasizes raw, gavel-to-gavel programming and over-time balance. Much of what C-SPAN presents is raw data without context. Hearings, speeches, and policy conferences of related public policy issues are often separated by days or weeks without any linkage. To get a balanced picture from C-SPAN, one must watch the network regularly over a long period of time. C-SPAN's mixing of programs is like making successive batches of stew. If any one ingredient is used on a particular day, the entire amount of the ingredient has to go into that day's batch. There is no such thing as a little salt. Today's salty batch will be compensated for by tomorrow's batch, which may be dominated by the taste of bay leaves. Viewers are forced to overdose on one topic or viewpoint one day, in the hopes of receiving an antidote the next day.

Naive viewers might conclude they are privy to everything that is going on; in reality, C-SPAN covers public events that have often been preceded by significant private planning and manipulation. In some cases viewers see the puppets but not the puppeteers. The "window" provided by C-SPAN is larger than the one we once had, but it is still a window and thus limits perspectives. For some events, the window is little more than a knothole in the fence, whereas for others, viewers get the broader picture of a bay window.

In the end, much is expected of C-SPAN viewers, who have to make sense of what they are seeing. C-SPAN increasingly provides context, but its efforts are tempered by the fear of misrepresenting the actual events and undermining its unique mode of coverage.

Talk-Show Democracy

In a 1986 speech to the National Cable Television Association convention, House Majority Leader Jim Wright (D-TX) praised C-SPAN for "democratizing the airwaves" and coined the description of C-SPAN as "America's town hall." If imitation is the sincerest form of flattery, C-SPAN has much to be proud about, since it initiated the national television call-in format well before programs such as *Larry King Live*. Call-in programs have a number of advantages over the more typical sound-bite journalism. From the candidate's perspective, they give a politician enough time to fully develop his or her ideas, and they make the candidate appear "accessible, in touch with everyday concerns of callers and viewers." For voters, call-in programs are "a chance to measure their would-be leaders for themselves, unfiltered. . . . [The shows] humanize candidates, draw out and explore their way of thinking—and expose their weaknesses."[20]

On the negative side, "call-in democracy" has some potential drawbacks. The quality of the questioning varies. "Some candidates and their handlers still think of talk shows as a way to avoid tough press grillings." Also, in the process of humanizing candidates, the call-in show risks "converting them from statesmen into celebrities" or demeaning the rightful and necessary authority of their office.[21] C-SPAN avoids highly personal questions, such as the MTV inquiry as to whether President Bill Clinton preferred boxer or jockey shorts, which add little to our understanding of the presidency. The fact that Bill Clinton felt beholden to answer such a question shows the danger of expected accessibility in the talk-show era.

While C-SPAN's low-key, "just the facts, Ma'am," objective approach of letting both callers and guests say their piece carries a great deal of appeal, it can often allow stupid questions and unresponsive answers to go unchallenged. The neutral, facilitator role of C-SPAN hosts and hostesses may leave outrageous charges and misinformation undisputed. In our society, silence often implies acceptance. The presence of a guest political practitioner at many of these call-in sessions provides some protection but does not eradicate the potential problem.

The new electronic era is marked by an interconnected world with fewer information gatekeepers. Call-in shows on C-SPAN and other networks contribute to real-time news, which spreads like wildfire. Brian Lamb talks about the "wave factor" as news stories spread through the listening and viewing audiences. Shreds of information mentioned on one program are spread by hosts and callers to other programs. With each repetition, the credibility of the message increases. The validity of an idea is based on "who" repeated it and how often it was repeated. This "escalation of validation" builds a wall around misinformation and makes it difficult to challenge. In studying small groups, Irving Janis found similar processes underlying what he called "Groupthink."[22] Information and misinformation can each have their believability enhanced by repetition. Multiple outlets and the speed of electronic information transfer reduce the ability of anyone to check the validity of a story before disseminating it further. In the print era, editors and reporters checked the truthfulness of stories to protect their professional credibility and to defend themselves against lawsuits. The anonymity of call-in participants and the lack of a host who will serve as a credibility check undermine the ability to hold anyone accountable for the spread of misinformation. Apocryphal exaggerations or just plain lies can quickly be elevated to the position of absolute truths.

Spreading misinformation to the relatively well-informed C-SPAN audience is less of a problem than is the next iteration of the information cycle. Callers to subsequent talk shows, ones not on C-SPAN and with a less selective audience, often invoke the mantra of C-SPAN to add credibility to their claims. Callers will begin an argument by saying, "This morning I heard on C-SPAN . . ." Since people often judge the validity of information by its source, such an invocation—often lacking any further specifics on the more proximate source—increases the potential impact of the assertion. Information and comments from morning C-SPAN call-in programs regularly show up on talk shows such as *Talkback Live* (CNN) and *Rush Limbaugh* later in the day. With their larger audiences, the public dialogue can be either improved or degraded, depending on the quality of information or misinformation that C-SPAN provides.

The immediacy of live call-ins feeds the public dialogue with no media filter, allowing misinformation to spread with as much effect as valid information. As media critic Molly Ivins put it: "The trouble with call-in shows is that there is no check on the misinformation they spread. Some earnest nut case calls in and says something preposterous—'I hear Ed Meese has AIDS'—and before you know [it] that rumor has made the circuit of the other call-in shows, and you can't kill the darn thing with a stick."[23]

The danger of call-in politics is exacerbated by the potential of organized call-in campaigns orchestrated by groups or individuals. Far from providing the equivalent of an electronic town meeting where all views are heard, such orchestration ensures that public responses will not be random. Presidential candidate Ross Perot regularly instructed his followers, "See us on C-SPAN." With a stacked audience, the calls often seemed more like a love feast than a critical analysis of policy alternatives.

Societal Fragmentation

Although most of us revel in the expanding base of information resources, a word of caution is warranted. Societies require a common base of language and shared information for members to communicate and interact effectively. Some cultural diversity is possible, even desirable, but a society without some degree of shared perspectives has a precarious future. In the modern era, network television has served, to an extent, as the "cultural glue that certainly has held this democracy together."[24] Cultural defining events, such as the Watergate hearings and the miniseries *Roots*, provide viewers with a common base of knowledge and shape virtually the whole society's views on key public issues.

With the explosion of diverse cable channels providing unique programming and perspectives, society could lose its common points of reference, resulting in a more fragmented society. Fragmentation comes in two types. *Substantive* fragmentation, stemming from personal programming preferences, will lead to people with an in-depth understanding of more and more narrow realms. Simultaneously, *access* fragmentation devolves from the differential ability to pay for cable access. The difference between information "haves" and information "have-nots" in the cable era will be dramatic. C-SPAN is little more than a player in this broader trend toward "cableization," but it may well inordinately contribute to the fragmentation in the realm of public affairs. With the planned expansion of C-SPAN to five or more separate channels (adding international, committee hearings, and state government channels to its current House and Senate channels), even the C-SPAN audience(s) will come away with different information, vocabularies, and understandings.

INCREASING THE TRANSPARENCY OF GOVERNMENT

Proving the potentially broad-range impact of C-SPAN on democratic government is impossible at this time. But even if the optimistic hopes of its supporters or the cautionary predictions of its detractors fail to materialize, C-SPAN does provide its viewers with extensive information on public affairs.

C-SPAN's programming goes well beyond opening the doors of the House and the Senate. C-SPAN has shown its viewers public affairs events traditionally reserved for a very small and select audience. In the process, C-SPAN has not only informed its viewers but also, in some cases, changed the events themselves. Some organizations—realizing the potential impact of cameras—now bar C-SPAN.

Much of what goes on in Washington happens behind closed doors. For decades, events such as the Gridiron Club and the Washington Press Club Dinner served as places where politicians and journalists could let down their hair and parody both their colleagues and the political process. Stories of the happenings at these events filtered out through the print media, but these versions were pale representations of the actual performances. Hearing about Richard Nixon playing the piano or about Bill and Hillary Clinton dressing up as the anti–health care advocates "Harry and Louise" falls into the category of "you just had to be there."

For a number of years, C-SPAN attempted to get access to such events. The Gridiron Club, the "Who's Who" of the Washington press corps, has met annually for over one hundred years to "don funny costumes to roast the President."[25] The event is clearly off-the-record, although summaries of the parodies often show up in newspaper reports. The media, which sponsor the dinners and fight for open access in other realms, bar cameras, despite some embarrassment. Each year C-SPAN requests access and is politely turned down. The organizers of the Washington Press Club Dinner were also not excited about televised coverage of its satire, but they nevertheless decided to allow the C-SPAN coverage. The implications of this change failed to inhibit many of the participants. In 1990, Representative Barbara Boxer (D-CA) seemingly overstepped the boundaries of political correctness when she expressed the somewhat off-color comment that she wondered how men were supposed to feel when they heard "those S&L ads on the radio that say, 'Severe penalties for early withdrawal.' "[26] Representative Claudine Schneider (R-RI) repeated a fictitious conversation with Senator Alan Cranston (D-CA) in which he supposedly asked her "if she knew the difference between a Caesar salad and sex, and she replied 'no,' only to be asked: 'Then what are you doing for lunch tomorrow?' "[27] Columnist Mary McGrory, writing about the dinner, said that "the best lines were too blue to repeat." She noted that if a man had made a woman the butt of such a raunchy joke, "that man would be having a farewell news conference in short order." The C-SPAN phones "burned with enraged calls."[28] With such events now going on-the-record via C-SPAN, both good taste and even the application of political correctness will probably follow. C-SPAN is holding the Fourth Estate accountable, even as the media hold the rest of society accountable.

During election campaigns, C-SPAN offers its viewers political events in their entirety, cinema verité coverage of candidates on the campaign trail, analysis of political advertising, and call-in programs. For those concerned about sound-bite journalism and the dangers of political advertising, the expansion of information sources may have a positive effect. Speaking about the broad range of cable and talk-show programming, of which C-SPAN is a part, Kathleen Hall-Jamieson, media critic and author of *Packaging the President*, argued that the new environment has a "mitigating effect on the candidate's own advertisements. . . . The ads were the primary source of information for the uninvolved viewer until this year, now the uninvolved viewer has another forum."[29]

Keeping the Participants Informed about Other Players

Creating a better-informed public was clearly the primary intention when the cable industry founded C-SPAN, but the public's tool also provides significant information benefits for those who are being monitored. Some of the most ardent viewers of C-SPAN are members of Congress and other members of the Washington establishment. Television sets run constantly in most congressional offices, allowing members and their staffs to monitor ongoing debate both for content and for "heads-up" information as to when the member might be required on the floor for a vote or a quorum call. Even when members are home at night or are back in their districts, they use C-SPAN for both entertainment and edification. As consummate politicians, elected officials can seldom get enough of politics. The impact on the chamber and its members may be even more important than the impact on the public.

Relatively early in the history of House television, Representative Michael Barnes (D-MD) recognized the importance of televised coverage for members themselves, an importance he coined the "Barnes effect." After years in Congress as both a staff member and an elected representative, he concluded: "It used to be that all a representative knew about the vote he was running to cast was 'If it's Tuesday, this must be Fisheries Management . . .' But it's not quite so superficial anymore. It's on TV. If the member is in his office . . . he can leave the television turned on . . . and follow what's going on. . . . I think the case can finally be made that Congress has finally found a way to keep up with itself."[30]

Representative Bob Michel (R-IL) concurred: "[Members] began not only to rely on C-SPAN to portray our work to the Nation, [but] we also began to understand ourselves better through the introspection a lens offers."[31] Senator Daniel Moynihan (R-NY) made the same point on the floor. After pointing out that C-SPAN allows members to see the satire of floor proceedings, he repeated Robert Burns's famous lines:

> O wad some Pow'r the giftie gie us
> To see oursels as others see us![32]

Monitoring allows key players in the political process both to become better informed and to use their time more efficiently. Two types of monitoring stand out, substantive and strategic.

SUBSTANTIVE MONITORING

Knowing what your supporters and opponents are saying is critical in politics. With so many venues for communicating, political leaders often rely on secondhand accounts or simply remain unaware of what is being said outside of earshot. C-SPAN allows decision makers the potential to tune in to a wide variety of political dialogues. Congressional floor debates are sprinkled with comments from members who had been monitoring debates back in their offices and were encouraged to participate. For example Representative Duncan Hunter (R-CA) explained, "As a Member who was doing some other things today and thought I had a few other places I had to be, I saw a piece of this debate on C-SPAN and thought that it was important to come down and participate in it."[33] Representative Fred Upton (R-MI) pointed out, "I was sitting in my office finishing a long week here in session, and I caught some of this debate on C-SPAN, and thought I had to come over."[34]

Monitoring also crosses institutional boundaries. Modern presidents and their staffs follow debates on the Hill and use the information to frame their influence attempts. President Ronald Reagan was a regular C-SPAN viewer who often picked up the phone to call a member of Congress who had just been on C-SPAN. Members of Congress use the network to watch White House press briefings and presidential events not covered by the commercial networks. In 1988, White House National Security Adviser Colin Powell asserted, "[You] would be lost in the West Wing of the White House or anywhere in the senior levels of government these days if you did not have C-SPAN and CNN to tell you what is happening around the country and around the world in real time."[35]

As C-SPAN expands its coverage outside of Washington, greater potential exists for informing decision makers. In a speech before the National Governors' Association, Republican leader Bob Dole (R-KS) commented: "I watch C-SPAN a lot while I'm doing my treadmill. I don't do it very fast, so I get to see a lot of C-SPAN, and so I've listened to a lot of your debate down here and I know what you think of what we are doing."[36]

C-SPAN also changes the linkage between political decision makers and the public. Bob Michel, House Republican leader, recognized the role of C-SPAN in providing a more effective link between Congress and the public. He pointed out, "[The] Washington mentality we so often speak about is waning as C-SPAN has helped to create a community more in touch with Government and frankly, Government more responsive to the public." Later he reaffirmed this view: "Washington is far less isolated from the rest of the country and the world than it once was."[37]

Throughout much of history we have learned to view the media as largely a one-way communication vehicle. Although letters-to-the-editor provide some opportunities for feedback, the time delay reduces their impact. Traditional talk-show programming, involving a host interacting with a set of listeners, collapses the time lag but includes no assurance that political decision makers will become part of the dialogue. C-SPAN call-in programs provide viewers with direct access to decision makers, and there is some evidence that the decision makers are listening. After appearing on C-SPAN call-in programs, members of Congress increasingly report communicating viewers' outlooks to their colleagues during private conversations.[38] Floor debate is sprinkled with references to information generated by C-SPAN call-in programs.

STRATEGIC MONITORING

Politics is more than a process of learning about something; it involves taking meaningful action. Knowing when to act is as important as knowing what to do. On a regular basis, members of Congress and their staffs keep an eye on C-SPAN during sessions to determine what is happening when a vote is about to take place and how much time they have to get to the floor to cast their votes. Some interviews for this book were interrupted when a member looked up at the television and said, "I have got to run to the floor; there is only five minutes left to vote."

For congressional staffs, following the legislative process in Congress is more than an avocation—it is part of their job. Legislative assistants are expected to keep their bosses informed about what is going on so that the legislators can plan their day. This is especially important in the Senate, since its flexible rules allow a member to block or delay action. One staff member called Senate debate "trial by ambush," in which senators offer countless amendments to change or stall legislation.[39] The Majority Leader negotiates amendment initiatives and organizes the floor schedule. His regular comments on upcoming action during debate give senators an idea of what is about to happen. Staff members keep C-SPAN turned on in their offices to monitor debate and to "get all the nuances . . . [of] what's in the Majority Leader's head."[40]

At times the impact of C-SPAN is very clear and direct. In 1987, C-SPAN coverage became an unwitting player in a battle over last-minute vote-switching. According to the rules of the House, members have fifteen minutes to cast their votes using the electronic voting terminals. The Speaker traditionally has the right to keep the terminals open if members are still trying to vote or want to change their position. Toward the end of a vote on a deficit-reduction bill, Representative Jim Chapman (D-TX) returned to his office after voting against the proposal. He glanced at the C-SPAN coverage on his office television and noticed that the vote-total screen reported that the measure was being defeated by one vote. Although opposed to the legislation, Chapman was also a loyal Democrat. He explained, "I could not sit and watch it go down."[41] Chapman dashed back to the floor to change his vote. Although all the time had expired, Speaker Jim Wright, a supporter of the legislation, extended him the courtesy, and the bill passed. C-SPAN viewers were treated to a scene of "rare ferocity" as opponents yelled and booed at Wright's bending of the rules.[42] Lost in the hubbub was the fact that without C-SPAN's coverage, Chapman would never have known that his vote was needed.

The ability to monitor debate from congressional offices may serve the members well but may also change the way they interact. In the pretelevision era, members spent more time on the floor or in the lobbies, waiting for votes to occur. There was more time for interpersonal influence. Now, as they monitor the process from their offices and dash to the floor only for quorum calls or votes, such interaction is less likely. Party leaders feel that absence from the floor makes it more difficult to inform and influence members.[43] Representative Michel noted, "Members now follow the affairs of the House from their offices, rushing to the floor to vote only at the last minute . . . valuable face-to-face contact is gone."[44]

The availability of C-SPAN coverage allows political activists to monitor the comments of adversaries and hold politicians more accountable for what they say. After a Democratic press conference on health care, covered by C-SPAN, Senator Arlen Specter (R-PA) used the C-SPAN coverage to chastise Senator John Rockefeller (D-WV) during floor debate, referring specifically to the press conference that he had seen "repeated on C-SPAN last night."[45] Since Senator Rockefeller was not then present and since Senate rules and traditions limit personal attacks on colleagues, Specter's challenge created a political brouhaha about questioning a member who is not present. Specter was finally allowed to proceed with his comments after Rockefeller's office had been alerted so that it could monitor the debate on C-SPAN.

Monitoring and communicating via C-SPAN can cross institutional boundaries. C-SPAN has been used as a back-channel method of communicating information among key policymakers within the Washington beltway. Serving as House Republican whip, Newt Gingrich saw C-SPAN as a vehicle for communicating with the Reagan White House: "Reagan would watch C-SPAN. If we couldn't get through the staff we would go on C-SPAN. Then he would call us. We spend more time with Reagan on C-SPAN than in his office."[46]

C-SPAN access in the White House changes the ability of the president and his staff to monitor their executive branch colleagues' performance in front of the legislative branch. Rosy secondhand interpretations are replaced by real-time monitoring. "A president can watch his cabinet members and aides testify. Before, he couldn't know how one of his appointees was doing on Capitol Hill. Now he can see them perform."[47] Tom Mann, of the Brookings Institution, described C-SPAN as becoming "a central part of the Washington insider network" and as having a "leveling impact" on the distribution of information within Washington.[48]

It is not only the policymakers who use C-SPAN for strategic monitoring. Lobbyist Gary Hymel stated that C-SPAN cuts his office workload in half. "We don't have to go to the House and try to keep a mental note of what's going on. . . . When a bill is being debated on the House floor . . . we take notes, eat lunch, and drink coffee, and tape it for a permanent record—we can't do any of these things in the House gallery."[49]

Not all of the implications of C-SPAN are positive for lobbyists, who make their living by portraying themselves as ultimate insiders. In one lobbyist's view, the ability of clients to watch political activities was "just another in a series of events" that had "basically taken the mystique away from what's going on in Washington." He explained, "Anything that removes or takes away from the mystique of the Washington lobbyist can ultimately have an impact on the values, the amount of fees that one can justify charging in this business." Since the inception of C-SPAN, some lobbyists have "found themselves caught off guard by clients who monitor floor action from their office televisions." As one lobbyist noted: "When you have a client in Omaha, Nebraska, and you call him up and tell him, 'I just came up from the House floor and this is what happened,' in the old days, that client would be pretty impressed. Now, he calls you and says, 'I just saw this on the House floor on C-SPAN.' "[50]

The presence of the C-SPAN cameras can also draw back the veil of political events and reveal less-than-flattering images. Using its typical, complete form, C-SPAN covered the 1993 gay rights march. While the commercial networks worked to portray the event carefully, "showing little of the scenes many Americans would find disturbing, . . . C-SPAN showed the event in unexpurgated format, including scenes of topless women, homosexual embraces, men in drag and leather, and other images and language not commonly viewed by most Americans."[51] The shock started the C-SPAN phones ringing and resulted in an avalanche of media notoriety for C-SPAN. Most viewers blamed the marchers for going too far and praised C-SPAN for showing the real picture of the event.

C-SPAN and the News Media

Above and beyond informing participants and influencing political decision makers, C-SPAN has changed both the substance and the procedures of those who interpret politics for the public. It is no coincidence that the term "media" appears in the middle of the word "inter*media*ry"—they both come from the same root (*medius*, "to be in the middle"). The news media stand as intermediaries between the public and actual events, picking and choosing which events to report and determining the way in which those events will be interpreted. As an alternative media source, C-SPAN changed the nature and range of available information. Its gavel-to-gavel approach allows viewers to circumvent the selection choices made by the mainstream media.

DEFINING "NEWS"

All media are selective. The word "news" itself should be a tipoff. It is the archaic plural of the word "new," meaning "different" or "out of the ordinary." The headline "Dog Bites Man" is not news, but "Man Bites Dog" may be. In public affairs, it is conflicting public personalities that draw the media like moths to a flame.[52] What distinguishes C-SPAN from most other media is its focus on "long-form" reporting. With its larger "news hole" of twenty-four hours on two channels and its commitment to gavel-to-gavel coverage of political events, C-SPAN avoids sound-bite journalism, which by intention hones in on political personalities in conflict. When C-SPAN covers events other than House and Senate proceedings, it does make journalistic choices concerning which events to cover, but the choices are driven by more than conflict and personalities. Issues of balance and a commitment to covering those events avoided by the other media figure significantly in C-SPAN editorial decisions. With perhaps a bit of exaggeration, Thomas Southwick, then of *Multichannel News*, a cable industry insiders' publication, predicted, "The 1988 election will break the stranglehold of the broadcast networks over the political process and usher in a new era in which the most important conduit of information . . . will be [the] cable television network C-SPAN."[53]

For C-SPAN, news is largely event-driven. The network covers sessions of Congress, presidential speeches, candidate announcements, committee hearings, and public policy conferences. Its emphasis on covering entire events provides the raw material for interested citizens and for other media. In the process, C-SPAN has changed not only the information that is available but also public expectations of news coverage and the strategies of both media seekers and media practitioners.

EXPANDING THE RAW MATERIAL OF RECORDED EVENTS

In the not-so-distant past, news did not exist unless a reporter either was there or could get a reliable firsthand account. Just like the old philosophical question of whether a falling tree in a forest makes any noise if there is no one there to hear it, news is not news unless there is someone there to record and report it. Even if a reporter is on hand, he or she may not perceive the newsworthiness of some portion of an event or may report it differently from another observer. In the past, it was common to hear rumblings about a particularly insightful comment, emotional moment, or outrageous statement, followed by the caveat, "I guess you just had to be there." The uniqueness of C-SPAN lies in its being "there" so predictably and so often. For certain events, such as congressional floor debates and presidential speeches, journalists know that C-SPAN will always "be there."

C-SPAN as the "Medium of Record"

Increasingly, C-SPAN has become the "medium of record" for both official and unofficial public affairs events. It is not so much that C-SPAN *makes* news as the fact that C-SPAN's large "news hole" and eclectic news-gathering approach permanently capture a wide range of events.

After an event is identified as having news value, print journalists and reporters from the mainstream electronic media often turn to C-SPAN tapes to refresh their recollections or to get a feel for an event they were unable to cover. Most often this involves a mad dash down to C-SPAN headquarters to review a tape or a call to C-SPAN to see when a program will be re-aired. C-SPAN provides other media with footage for their programs.

It is true that the existence of C-SPAN is not a necessary condition for clips of congressional floor debate to appear on other television programs, since the networks can receive and tape the floor debate directly. But usually it is much easier for them to go to C-SPAN for the footage. More important, events on the floor become news because the participants have already experienced them. In many cases C-SPAN's dissemination of a story alerts those "in the know" about an interesting event, and the news media then widen the circle and make more people aware. Aside from congressional floor debate and presidential news conferences, C-SPAN cameras are often the only ones at an event or the only ones recording an event in its entirety.

Exposing Duplicity

In recent years, many of the stories directly emanating from C-SPAN coverage or requiring C-SPAN verification have involved exposing examples of members of Congress editing their public remarks before those remarks can become a part of the *Congressional Record*. By long tradition, the House and the Senate allow members to "revise and extend their remarks": cleaning up grammar, correcting factual errors, and inserting supporting arguments. Although some might view such a right as an invitation to duplicity, the rationale for the rules is more benign. Members recognize that both tradition and the courts view the *Congressional Record* as the official account of legislative intent. In administering and interpreting the laws, bureaucrats and judges often go back to the *Congressional Record* for guidance. A speech that is made during the heat of debate and that includes misstatements and garbled syntax is viewed as a "rough draft" of what the member wants in the official record. The floor of Congress is one of the few places where one can say, "Gee, I wish I had said that," and then officially do so. Though the overall rationale sounds reasonable, the "revise and extend" privilege is often used to sanitize the *Congressional Record* of politically embarrassing statements. Reporters use C-SPAN coverage to write stories stating, "This is what they really said, and this is what was recorded." Alerted to such events, network news programs often use C-SPAN clips to authenticate their stories.

Some revision and extension flaps recorded by C-SPAN are both funny and insightful. During a heated exchange on Senate rules, Senator Robert Byrd, a master of the rules, got so exasperated with Senator Alfonse D'Amato (R-NY) that Byrd told him to "shut his own mouth." The official record was changed to read, "The Senator can just . . ."[54] While debating grazing rights, Senator Alan Simpson (R-WY) complained about the "cow poop" that grazing cattle would leave on public lands and then had second thoughts about the public reaction to such an earthy term. At Senator Simpson's request, the official record used the term "cow pies."[55]

Each of these events showed up later in the print and electronic media, often subtly undermining the image of Congress and its members. The C-SPAN record gave reporters "cover" if anyone sought to challenge their recollection. As veteran journalist James Reston almost gleefully put it, "Television, with its daily pictures of what happens in the House and Senate, can identify better than print journalism the political frauds, merely by putting them on the screen and letting them talk."[56]

C-SPAN also captures more extreme revisions of floor behavior. House and Senate rules forbid personal verbal attacks on colleagues. Members going beyond the limit can have their remarks "taken down" (excised from the *Congressional Record*) at the discretion of the chair or by a vote of the chamber. The first and most famous of these events involved Speaker Tip O'Neill when he criticized members of the Conservative Opportunity Society during the 1984 "Camscam" brouhaha. After the 1994 elections and the Republican takeover of the House, partisan animosity flared, and two incidents of words being "taken down" occurred within one week (see box 7).

What in the past might have been simple insider stories to be told around the First Amendment Bar at the National Press Club became national news events as other media, using the televised record, picked up what was said on the floor. In the pretelevision era, a reporter in the House gallery might have been unwilling to write stories without the comfort of a taped backup to verify what had happened. The stories of real people losing their temper or making a mistake are appealing because they fit well with the mainline media's convention of defining congressional news in terms of personalities and conflict.[57]

By creating a permanent record, C-SPAN extends the "shelf life" of potential news stories. Although news is almost by definition transitory, some stories often require time to percolate through the process. Either the right person later hears about an earlier event, or subsequent activities make a seemingly insignificant earlier event into something of importance. The stories of Senators Joseph Biden (D-DE) and Bob Kerry (D-NE), told in chapter 5, are clear examples.

Validating Public Commitments

Knowing that C-SPAN will cover certain events allows politicians to add validity to their words by challenging the public: "Watch what we do, not what we say." During the 1994 campaign, Minority Whip Newt Gingrich continued to follow his long-term political strategy of using C-SPAN for his partisan goals. In announcing the Republican "Contract with America," he

Box 7.
Making News by Recording It

Representative Carrie Meek (D-FL) challenged Speaker Newt Gingrich over a lucrative book-publishing deal by saying: "Exactly who does this speaker really work for? Is it the American people or his New York publishing house?" Even though the Republicans had changed the rules so that the *Congressional Record* would more closely reflect what was said on the floor, Representative Robert Walker (R-PA) jumped to his feet and demanded that Meek's comments be stricken from the record. The Acting Speaker ruled that Meek was out of order. "In minutes, Representatives of both stripes cleared out of their offices and committee meetings like baseball players emptying the dugouts for a donnybrook over an umpire's call. Then they put the issue to a vote and, on strict party lines, 217–178, the Republican majority prevailed."[a] After the Acting Speaker's ruling was upheld, the story took on a life of its own. At least thirty news stories on the incident (according to a LEXIS-NEXIS database count) appeared within a week. C-SPAN clips of the exchange surfaced on CNN and the major commercial networks.

The day after President Bill Clinton's 1995 State of the Union message, Robert Dornan (R-CA) took to the floor objecting to the president's invitation to a Medal of Honor winner to sit with Mrs. Clinton. After complaining that this was improper because Clinton had given "aid and comfort to the enemy" during the war in Vietnam, his words were challenged.[b] Despite their ability to stop the punishment, the Republican majority showed their willingness to punish one of their own and chastised Dornan by having his words "taken down." For his initial comments and his unwillingness to apologize, Dornan was barred from speaking on the floor for a day. In his refusal to apologize, he repeated the charges against Clinton a number of times, again getting them in the *Record*. As the fulminating Dornan walked from the floor, freshman Representative Sonny Bono (R-CA) asked him, "Bob, *why* did you do that?" Recognizing that C-SPAN was quick to point out that this was not just another set of empty promises. He challenged the voters by saying that if the Republicans were elected, in January the viewers would "be able to tune into C-SPAN and say, 'These guys are for real.' "[58] The mainstream media treated the contract as a gimmick, and many commentators felt that its specificity would do more political harm than good. Even retiring Republican Representative Fred Grandy (R-IA) suggested that the Republicans were successful in nailing Clinton on "his propensity to overpromise and underdeliver," and Grandy noted, "Here we are doing the same . . . thing."[59] President Clinton derided it as a "Contract *on* America." Few publications or news programs dealt with its substance in any great detail. After *TV Guide* published the contract as a paid advertisement, Gingrich reasserted its seriousness by saying, "We told the American people to tear it out, put it on their refrigerator door and then if we get a majority, tune in on January 4th on C-SPAN and watch us keep our word."[60]

Republican Pollster Frank Luntz, who provided much of the public-opinion data on which the contract was based, argued that the public would be watching. Shortly after the 1994 elections he commented, "I bet C-SPAN's viewership goes up 20 or 30 percent in the first 100 days because people are going to be tuning in to see if these issues actually get debated."[61]

Changing Public Expectations of the Media

C-SPAN's unfiltered, gavel-to-gavel coverage of political events provides a new challenge for journalists. At one time, the public was dependent on journalists to describe and interpret events. If the media said a candidate "looked tired" or "responded with a bit of anger," the public had little choice but to take them at their word. With more sophisticated viewers having direct access to events, there is the potential for a "chasm" between the public and the journalists. As former presidential adviser and now CNN commentator John Sununu put it: "The public can see Congress in action on C-SPAN, can see press conferences covered live. . . . Therefore when they turn to CBS, NBC, and ABC, a lot more of them are saying, 'That's not what I saw today. Why are these guys saying this?' "[62]

Writer Tom Rosenstiel reflected on an experience as a C-SPAN guest when a viewer challenged his interpretation of a speech by Ross Perot, a speech that the viewer had also seen on C-SPAN. Going back over the tape, Rosenstiel found there was some validity in the caller's complaint. Rosenstiel said it was "a humbling" experience and concluded: "For years, we [the media] held the upper hand with the public. We were observers who saw events and reported them. Now with CNN, C-SPAN, and other cable outlets, anyone with the time and interest can see the same events." Interested members of the public "can make their own interpretations of the candidate's real motivations. They don't need the professional insiders."[63] Because of C-SPAN, "the public business is vaulted over the heads of journalists."[64] Voters now want "contact without the media middleman. They want to judge for themselves, to look at a candidate and determine whether he's principled and can be trusted. C-SPAN, and 'Larry King Live' and 'Donahue,' therefore, become the primary tools of political communication."[65]

As a result of this new awareness of the media's shortcomings in presenting the full story, it may be that "the press is seen as part of the problem now that people can watch so many events themselves on C-SPAN."[66] Increasingly, stories about the media include a backhanded compliment to C-SPAN by prefacing a litany of media shortcomings with the words, "Unlike C-SPAN . . ."

CHANGING NEWS-GATHERING TECHNIQUES

For some journalists, C-SPAN has changed their entire approach to news gathering. As stated above, it was once expected that a journalist had to "be

The Channel Changer

Journalists discovered how C-SPAN could change the entire way they gathered news. (Cartoon by Jeff MacNelly, originally published by Tribune Media Services, reprinted by permission.)

there" to cover an event. With many events now covered live on C-SPAN, "a journalist can cover many of the official events in Washington without leaving the office."[67] Whereas the VCR and remote control will never completely replace the reporter's pen and notebook, C-SPAN has expanded the reach of many reporters—plugging them into events that in the past would have been either bypassed or taken from the wire services.

To some degree, the existence of C-SPAN has taken the networks off the hook. NBC journalist Tim Russert believes that by covering many events, C-SPAN "liberated the networks" and allowed them "to focus more on reporting."[68] The networks used the existence of C-SPAN to lessen the sting of criticism when they stopped full coverage of the national political party conventions.[69]

CHANGING NEWS-MAKING TECHNIQUES

Politicians seek media attention to promote their careers and causes. The contemporary politician often grates at the selectivity of the mainstream media and the economic necessities that force most coverage to be in the form of short sound bites. During recent campaigns, major party candidates have been given less than ten seconds at a time to make their points on the commercial television networks (see preface). To some degree, C-SPAN serves as an antidote to sound-bite journalism and overinterpretation. With C-SPAN's long-form approach, commitment to balance, and a significantly larger "news hole" to fill, politicians are relieved from "presenting themselves through the filter of professional, and sometimes cynical, political press."[70]

Following the Audience

The 1992 presidential campaign was a marked departure from those of previous years. In preparation for the campaign, both Bill Clinton and Ross Perot used C-SPAN to become known. As the campaign progressed, both took the advice of "following the audience" and showed up on commercial and cable talk shows and call-in programs such as *Larry King Live* and *Donahue* and on more clearly entertainment venues such as *Arsenio Hall* and MTV. C-SPAN did a simulcast of the *Donahue* show and linked its public affairs "junkies" with an audience more interested in entertainment. President George Bush and his advisers initially viewed such exposure on "infotainment" programs as "unpresidential," but they soon bowed to the reality that his message was not getting through via the traditional networks. One of the clearest departures in this campaign was the number of times candidates used the talk-show format to interact directly with the public. In talk-show host Larry King's view, in 1992 "the public switched places with the campaign press. This time around, voters interviewed the candidates while the journalists watched. It happened on ABC's 'Good Morning America,' NBC's 'Today Show,' and 'CBS This Morning'; on C-SPAN and on 'Donahue.' "[71]

Using the Presence of the Cameras

Recognizing the commercial media's thirst for conflict, members of Congress use the presence of C-SPAN to reach a larger audience. Coverage of an event by C-SPAN can help set the political agenda by "seeding" other media with a story. During the 1988 presidential campaign, Representative Pat Williams (D-MT) took to the floor to criticize vice-presidential candidate Dan Quayle (R-IN) for praising a Job Corps program even though Quayle had voted to shut the center down. Williams commented: "There's a word for it, my colleagues. It's called hypocrisy." Republican Dan Lundgren (R-CA) objected that House rules did not allow personal innuendo, and the word "hypocrisy" was ruled out of order. Williams asked for unanimous consent to a suggestion: "The last sentence of my one-minute statement, the sentence in which I characterize Senator Quayle's actions as hypocrisy, [should] be stricken." Despite objections from the Republicans, Williams was able to repeat the word "hypocrisy" a number of times while withdrawing the initial charge. The withdrawal statement, including the word "hypocrisy," remained in the *Record*, and the C-SPAN coverage brought the incident extensive media attention from commercial networks.[72]

The "use" of C-SPAN is not limited to one party. In the 1992 campaign, Representative Bob Dornan (R-CA), well aware of the reach of C-SPAN, used the Special Orders period in the House to criticize presidential candidate Bill Clinton for going, while a student, to Moscow as a guest of the KGB. Dornan asserted that Clinton lacked patriotism because he had opposed the war in Vietnam and circumvented the draft. Without C-SPAN coverage, Dornan probably would not have even brought the issue to the floor, nor would it have been viewed worthy of news coverage. The fact that thousands of people already knew about the charges alerted the media and made it a story. The mainstream media reported Dornan's claims, often including C-SPAN footage. A few days later Larry King even asked President George Bush for his reaction when he appeared on *Larry King Live*. Bush tried to raise himself above the fray by being noncommittal "but did suggest that the story was at least worth looking into."[73]

The ultimate effect of Dornan's charges was not clear. Despite Dornan's hopes, it may be that his largely unsubstantiated claims "did more to damage the Republican Party than the Democrat running for President."[74] Attention was drawn away from the larger issues of the campaign, and much debate ensued about the appropriateness of both the format and the substance of Dornan's charges. Among undecided voters, Clinton may have received more sympathy than denunciation.

Given C-SPAN's commitment to covering events gavel-to-gavel with no editing, politicians can "use" C-SPAN more easily than the commercial networks. Thus political activists may sometimes abuse their access to the public. As we saw earlier, for example, this was not the last time Representative Dornan used C-SPAN to criticize Bill Clinton and to attempt to redefine the media agenda.

CHANGING THE STATUS AND EVALUATION OF OTHER MEDIA

To a large degree, the audience for public affairs information is fixed, and the capacity of public institutions to absorb new agenda items is limited. Gains in C-SPAN audiences and influence come at the expense of other media.

Those frustrated with one segment of the media often look to other segments for relief. After years of frustration with what they perceived as the liberal commercial media, conservative sources gloated after the 1994 elections. An editorial in *Investor's Business Daily* argued: "Once again, the media are focusing on power and politics rather than policy. And even then, the pundits are getting it wrong. . . . How long will it take the national media to catch on to the new agenda. . . . The Washington press corps has lost some of its influence to alternative sources like C-SPAN, Rush Limbaugh and the Internet. That trend will certainly continue if the liberal media and wrong-headed pundits, like a pack of frenzied termites, continue to gnaw at the conservative mandate because it's not what they predicted and still want."[75]

C-SPAN officials bristle at any suggestion that theirs is a conservative network. Their commitment to ideological and partisan balance is legion, but in a commercial media world, where the party in power is often defined as more newsworthy, balance seems like a real advantage. The conservatives' relative pleasure with C-SPAN cannot finally be evaluated until they have been in power and under commercial media scrutiny for a number of years. There is no inherent reason why C-SPAN should favor only the Republicans. In fact, under different conditions, with the Republicans now being held responsible for the behavior of Congress, this might be another example of a technology serving as a "double-edged sword." As one journalist put it: "A lot of voters have learned a bit about the way the system has worked, a factor attributable chiefly to C-SPAN. Keeping in mind the truism about the danger of a little knowledge, C-SPAN could represent a threat to the GOP future. What will these eager learners think of Jesse Helms, Strom Thurmond, Alfonse D'Amato and Bob Packwood?"[76]

It is yet to be determined whether the party in power can use C-SPAN to its own advantage or whether the medium is best suited to venting the frustrations of those out of power. Since the efforts of Newt Gingrich and the Conservative Opportunity Society in the mid-1980s (see chapter 2), C-SPAN has been viewed as a strategic tool for evening the playing field. Brian Lamb is proud of the fact that it is the "one place in American media where the elected official has a direct access to the audience. They never have to ask permission."[77] C-SPAN became the "network of the minority." After the Democrats' fall from power, senior Democrat David Obey (D-WI) looked back at the effective use of C-SPAN by the Republicans and concluded that the Democrats never should have opened up Congress to C-SPAN's prying cameras.[78]

Other "outsiders" also find C-SPAN's relatively open access and long-form format refreshing. Consumer advocate Ralph Nader commented: "People like us are now on C-SPAN every once in a while. We are shut out from the network news. Once in a while you get a four-second comment where they slice one sentence in half. . . . On C-SPAN you can actually talk in sentences."[79]

In a broader sense, C-SPAN and other nontraditional media have affected the power and influence of competing mainstream media and journalists. Media critic Barbara Matusow commented on the media after the 1992 election: "The media's pecking order had been scrambled, with Larry King at the top and less showy talk show hosts like C-SPAN's Brian Lamb growing in stature and clout. . . . [They] have relatively small audiences, but are watched by so many journalists that ultimately millions more hear."[80]

C-SPAN and the Cable Industry

C-SPAN and its supporting entity, the cable industry, are more than just observers of American politics. The cable industry is subject to numerous government regulations and has specific political interests about which of these regulations are best.

The decision to launch C-SPAN was less of an economic or philosophical decision than a political one. C-SPAN is unlike other channels: cable operators pay to include C-SPAN and get no revenue in return. Many members of the cable industry supported C-SPAN because they felt that it would help improve the public image of their industry.[81] Not far from many cable operators' minds was the assumption that by being good citizens and providing public exposure to those who would be writing the regulations, they could enhance their chances of affecting those regulations in a positive way. Tom Wheeler, head of government relations for the National Cable Television Association, saw the potential for cable to spruce up its image and counter the feeling that it "wasn't making a contribution" to public service and was in fact guilty of "piracy" of broadcast signals. Wheeler felt that by broadcasting Congress, the industry would gain attention and prestige among political decision makers. "It gives cable operators something to stick their chests out about and be proud of."[82] More cynically, journalist Phil Rosenthal concluded that C-SPAN is a "genuine bit of goodwill that can be cited when hitting customers with another rate hike."[83]

To the cable operators' credit, they recognized that without both the image and the reality of impartiality, C-SPAN's public relations advantage was nil. C-SPAN's connection with the cable industry can lead to embarrassment when issues involving the cable industry are involved. If C-SPAN covers or decides not to cover hearings or policy forums dealing with cable issues, viewers often assume that the network is doing the cable industry's bidding. Although C-SPAN credibly protests that the industry takes a hands-off approach to the content of its programming, maintaining the image of independence is a continuing problem.

At times the issues hit even closer to home. During a C-SPAN interview in early 1995, Speaker Newt Gingrich stated that public funding for the Public Broadcasting System (PBS) should be discontinued. He used C-SPAN as an example of what the private sector could do. Gingrich gushingly praised C-SPAN: "One of the reasons I have loved what you've done for America is you went out as an entrepreneur, you sold the cable industry into putting on its own channel . . . you've gotten people to watch because they like you. And we don't give you $100 million or $200 million a year."[84] PBS host Bill Moyers responded: "I find it intriguing that his . . . attack on public broadcasting came in an interview on C-SPAN, which I greatly admire, but C-SPAN is the creature of the cable industry, run by friends of the new speaker of the House. I think there is a correlation between this ideology of publicly supported politicians in the service of a commercial industry that, frankly, would like to see public television not exist."[85] Ervin Duggan, CEO of PBS, entered the fray: "C-SPAN, that wonderful service, might not exist were it not for the pioneering of public television in clearing its schedule to cover the Watergate hearings."[86] Although uncomfortable with being in the middle of a political battle, C-SPAN and the cable industry enjoy basking in the adulation of all sides and privately welcome being on the more popular side of private enterprise, as opposed to government subsidies.

Cable operators are proud of C-SPAN and use its success to promote their own brand of private entrepreneurship. Although there is no evidence that this philosophy finds its way inordinately into C-SPAN programming, it may feed perceptions such as those of Moyers. At the network's fifth anniversary, numerous members of the board added their praise:

C-SPAN is a prime example of how the cable industry, without any government regulation or edict, can use its combined resources to provide a unique and valuable service to the public (Robert Magness, Telecommunications Inc.).

C-SPAN represents one of the best applications of narrowcasting today (Doug Dittrick, Tribune Cable Communications).

C-SPAN is the best example of the fulfillment of our industry commitment to delivering new and constructive forms of communication (Robert Rosencrans, United Artists Cablesystems).

[C-SPAN] is a unique program, best of all, it's cable's own (James N. Whitson, Sammons Communications).

C-SPAN is the epitome of what is best about cable television: unique programming which is impossible in the environment of the ratings-conscious networks (Jack Frazee, Centel Cable).[87]

In its specific battles over government regulation, there is no evidence that the cable industry simply prevails because of the public service it provides with C-SPAN. In fact, regulations promulgated in recent years have been hard on cable. It is impossible to tell whether these regulations would have been even more unpalatable without the C-SPAN initiative.

C-SPAN, the Public, and the Emerging Cyberdemocratic Age

At the dawn of the information age, Marshall McLuhan argued that "instant information creates involvement."[88] C-SPAN is a practical application of that dictum. Its programming has been described as being "as addictive as crack cocaine: One whiff of C-SPAN and every viewer has the potential to become a raging citizen-warrior."[89] Although the emphasis must remain on the word "potential," recipients of the public's attention credit C-SPAN with activating many of their constituents. Representative Dick Cheney (R-WY) told of being in the middle of a House debate when he got a call from a constituent who had been watching and had an amendment to offer. The close attention impressed Cheney, although he chose not to offer the proposed amendment.[90] Other members of Congress reported similar episodes:

After the debate this morning, I received a call from a constituent in South Dakota who had seen it on C-SPAN (Senator Larry Pressler [D-ND]).

[During the debate, there] were a lot of phone calls in the Cloak Room. People who were watching everything on C-SPAN were calling and telling us their opinion (Representative Bill Richardson [D-NM]).

By the time I got back to my office [after giving a speech covered by C-SPAN], the phones had lit up, not just from back home in Ohio, but from all over the country (Senator John Glenn [D-OH]).[91]

C-SPAN has led to an increased number of "issue-oriented citizens who feel empowered by picking up the phone and chewing out a politician."[92] Early in the history of C-SPAN, field-workers for the National Republican Congressional Campaign Committee asserted that in congressional districts receiving C-SPAN, "the level of involvement was higher and the level of information was certainly higher."[93] Assertions of C-SPAN's contributions to democracy come from all sides. Speaker Tip O'Neill talked about C-SPAN "expanding democracy."[94] Representative Lionel Van Deerlin (D-CA) implied that C-SPAN is the fifth historical step in democratizing American politics, ranking right after the direct election of senators, women's suffrage, the Voting Rights Act, and the "one person, one vote" redistricting decision of the Supreme Court.[95]

Even though democracy and citizen access are often considered unmitigatedly good phenomena, a number of writers have begun questioning a system that is too "plugged in" to the people back home.[96] C-SPAN undoubtedly feeds a society tied to its elected officials by an umbilical chord of fax machines, electronic mail, and telephone calls. In the din of immediate public response, elected officials may not be able to make the educated judgments necessary for good policy. The Founders expected that elected representatives would "cool" popular passions and that those officials should be "buffered from the often-shifting winds of public opinion."[97] In analyzing the new information environment of which C-SPAN is a part, journalist Robert Wright sounds an alert on the way C-SPAN and other technologies keep Washington tuned in to the public:

From C-SPAN's studios just off Capitol Hill, lawmakers chat with callers live—including callers who have been monitoring their work via C-SPAN cameras on Capitol Hill. More messages pass through the Beltway than ever before. And contrary to popular belief, politicians pay attention. What we have today is more of a cyberdemocracy than the visionaries may realize.... Intensely felt public opinion leads to the impulsive passage of dubious laws

> ... the emerging cyberdemocracy amounts to a kind of "hyperdemocracy":
> a nation that, contrary to all Beltway-related stereotypes—is thoroughly
> plugged into Washington—too plugged in for its own good.[98]

Some members of the House and Senate are worried about how C-SPAN is used to activate the public during key congressional debates. It has become more common for members of Congress to alert their supporters to an upcoming debate and to address the C-SPAN audience by stating, "Contact your House or Senate member directly on this bill." Senator Simpson complained that this "is a very unfortunate way to do the Nation's business." He added, "I do not believe we should make policy based on C-SPANII coverage."[99] Minority Leader Robert Michel initially complained that electronic communication has "made a spectator sport out of the legislative process," but he later admitted, "Our problems lie not within our television cameras, but in ourselves."[100] Supporters of more direct democracy and those who oppose the "inside the Washington beltway" mentality, on the other hand, view public access and activation in a more positive light. They see technology as a way of reinvigorating democracy and making direct citizen involvement a reality.

Much of the debate revolves around one's trust in the wisdom of the public at large and one's judgment of the performance of the representatives in Congress. C-SPAN founder Brian Lamb regularly argues both on and off the air that the public "is much smarter than the politicians give them credit for." Journalist Jeff Greenfield explained, "C-SPAN relies on a blend of modern technology and Jeffersonian faith in people."[101] Futurists Alvin Toffler and Heidi Toffler included C-SPAN when they stated: "Today's spectacular advances in communications technology open, for the first time, a mind boggling array of possibilities for direct citizen participation in political decision-making ... and since our 'pseudo representatives' are so 'unresponsive,' we the people must begin to shift from depending on representatives to representing ourselves."[102]

Those observers less sanguine about the public draw on the words of James Madison, who feared popular "passion" and who sought to "refine and enlarge the public views by passing them through the medium of a chosen body of citizens, whose wisdom may best discern the true interest of their country and whose patriotism and love of justice will be least likely to sacrifice it to temporary or partial considerations."[103]

The potential for public activation through C-SPAN goes beyond our clearly representative institutions. Even the White House reports an impact, with the switchboards lighting up with an increasing number of calls from citizens wanting to express their views. President Bill Clinton attempted to capitalize on the popularity of his town-meeting approach to politics and encouraged voters to involve themselves in a more participatory democracy. C-SPAN has created "more informed debates and swifter, two-way reaction."[104]

C-SPAN audiences not only are willing to give their time to viewing public affairs but also seem willing to open their checkbooks. Political fund-raisers relying on telephone banks have seen the impact of "real-time" political debate. During the 1994 debate on the crime bill, one national Republican phone bank that typically averaged contributions of $150 per hour found its contributions shooting up to $400 per hour during C-SPAN coverage.[105]

Members of the media are often amazed at the willingness of the C-SPAN audience to take action. After appearing on C-SPAN, one journalist commented, "There are more of these [C-SPAN] aficionados than one might expect. . . . [My] appearance on C-SPAN . . . resulted in more calls from around the country than anything written for the *Financial Times* in the past 26 years."[106]

Conclusion

The evidence from a number of realms is clear. C-SPAN has become a significant factor in redefining the relationship between citizens and their government and between political players in different institutions. Politics is about power, and C-SPAN finds itself in the middle of numerous power games. Each player attempts to "use" C-SPAN for his or her own personal or organizational benefit. Politics is an honorable method of settling conflicts over policy preferences when there is no correct answer. It involves informing and activating both the public and the policymakers. C-SPAN's success will not be determined by whether it can "avoid politics"—a patently absurd goal for a network committed to covering the political process—but rather by whether it can maintain its position as an "honest broker," providing access to the public and the policymakers in a fair and honest way. Almost two decades of experience have given credence to media critic Tom Shales's early evaluation that C-SPAN is "the driving of the golden spike that unites the people with their government through television. It is a crucial, pivotal moment in the life of the newly emerging wired republic."[107]

CHAPTER NINE

The Future of C-SPAN in the Age of Cyberdemocracy

When yu' can't have what you choose, yu' just choose what you have.
>   Owen Wister, *The Virginian*

Politics ought to be the part-time profession of every citizen who would protect the rights and privileges of free people and who would preserve what is good and fruitful in our national heritage.
>   Dwight Eisenhower, 1954

Until recently, the choices of ways for good citizens to monitor Congress and other aspects of public affairs were limited. In the early years of the Republic, the print media focused on Congress. The electronic media were less fascinated by the slow pace and complex procedures of the policy process. By the 1950s, news coverage of Congress was being cut in favor of covering the more telegenic presidency. Reporting on Congress lacked timeliness, with stories that either projected upcoming decisions or, more likely, reported retrospectively on decisions that had already been made. People who wanted a say in the policy process seldom knew when or how to make their views known. Interest groups helped inform and activate some members of the public, but many interests lacked organization and leadership. Motivated by decisions that had already been made, most communications to members of Congress arrived too late to have much impact. The official behavior of most members of Congress was hidden from even the most diligent constituents. Voting records seldom showed up in newspapers, and television news time was so limited that only a few members ever became the subject of stories. The public was generally unaware of key stages in the political process, such as committee hearings. Members of Congress largely controlled what their constituents knew about them, nearly guaranteeing a distorted and self-serving picture—and feeding the tendency for almost uncontested victories by incumbents.

For many people the bleak scenario of an uninformed public still prevails, but unlike in earlier times, the blame today falls more on the public than on the public institutions and their members. A new era of expanded options for public affairs monitoring has arrived, and an increasing number of citizens are taking advantage of it. C-SPAN has become an important part of that new mix of public-access options. In many ways C-SPAN has been at the forefront of the new wave; in other ways, it has supplemented and complemented additional offerings.

Contemporary Media Environment

The more things change, the more things change—contrary to the old saying. When C-SPAN appeared on the political radar screen in 1979, the three commercial networks dominated national news, talk radio was rare and devoid of political content, and talk television was unheard of. The ill-defined concept of public affairs television consisted of minimal public affairs programming on commercial stations, which were forced to air the programs by the relicensing procedures of the Federal Communications Commission (FCC).

Almost two decades later, C-SPAN is part of a new and expanded mix of public affairs opportunities for viewers. The FCC has largely fallen victim to changes in technology and the philosophy of deregulation, working on the margins rather than at the heart of the regulatory process. The free market approach to media availability predominates, with only sporadic attempts by politicians to "perfect" the exigencies of the market. The implicit societal goal of mass audiences and nationalized political information and

Violent Language

C-SPAN's full-format "raw" footage may be too much for some viewers. (Cartoon by Pat Crowley, *The Hill*, reprinted by permission.)

outlooks, a goal common throughout the 1970s, has given way to the recognition that there may well be virtue as well as inevitability in fragmented niche markets. Cable television as created in the United States is clearly a market-driven phenomenon. In a variety of ways, C-SPAN showed the way. Without C-SPAN, television talk shows such as *Larry King Live* would probably exist, but without C-SPAN, the idea of an entire channel devoted to televised trials or talk television would be unlikely. Without C-SPAN, many foreign parliaments, state legislatures, and city councils would still be meeting behind closed doors on the assumption that televising their procedures would be devoid of an audience or would totally disrupt their procedures.

The media byword for the 1990s and beyond is "choice." In writing about the decline of the networks, Ken Auletta noted: "If you go back to 1976, let us say when Jimmy Carter was nominated, then nine out of ten viewers were watching just three networks. CNN didn't exist. C-SPAN didn't exist. Comedy Central didn't exist. And today, you know, only six out of ten viewers are watching those three networks. Inevitably, it's democracy of choice."[1]

Early on, people recognized that C-SPAN offered this choice. "More than any other network, C-SPAN makes good on cable's early promise to offer diverse channels that narrowcast for specific audience interests, and ignore the dictates of mass market economics."[2] The old image of the networks framing political issues and determining what Americans think—if such an image ever was completely true—clearly does not fit the new reality. As media critic Max Frankel stated, "Americans are [now] more likely to be defined by their *choice* of media than they are indoctrinated by the *content* of the media."[3]

The lessons for the other media are not entirely clear. Media critic Tom Rosenstiel clearly laid down the challenge: "Journalism is in the end about providing people with information that will help them live their lives. It is about lifting people, offering signposts to what is important, binding the culture, and appealing to our better nature. If the traditional press does not provide that, a new press will arise in its place, probably through a new television delivery system."[4]

To some degree the commercial television networks have imitated C-SPAN in the talk-show format. For example, in 1992 the *Today* show invited the presidential candidates to accept calls during an extraordinarily long segment. In other realms, the networks have launched pseudo–public affairs programs, such as *Hard Copy*, which blur the line between news and entertainment. Newspapers, fearing the domination of television, have made their graphics look more like television. *USA Today* even distributes its papers in a street-corner box that looks like a television set. The commercial television and print media, in their constant striving for audiences or readers and, ultimately, advertising dollars, fail to recognize the public desire for and the societal advantages of a broad set of choices. While C-SPAN officials can sound magnanimous about recognizing that C-SPAN is only part of the information mix for the informed citizen and that no one can—or probably should—watch C-SPAN twelve hours a day, the other media outlets attempt to create a product that will attract the exclusive attention of their target audience. Brian Lamb's vision of the future clearly includes traditional media such as newspapers. In the increasingly dominant television era, newspapers will play an important role, helping citizens "make sense out of the coming deluge of cathode ray infotainment." If nothing else, in a five-hundred-channel environment, viewers will need newspapers "more than ever before to sort all this out."[5]

The existence of C-SPAN indirectly hurts other media by serving as a benchmark for what good public affairs coverage can or should be. The phrase "C-SPAN-like coverage" is often used to place the superficial, sound-bite-driven public affairs efforts of the commercial networks in a negative light. For much of the public, the commercial media have become part of the problem. In 1995, less than 60 percent of the public gave the national network news an "A" or "B" rating, criticizing it particularly for being too cynical and too adversarial.[6] Distrust of the media has led to a situation in which many "voters now want contact [with politicians] without the media middleman. They want to judge for themselves, to look at a candidate and determine whether he's principled and can be trusted. C-SPAN and 'Larry King Live' and 'Donahue,' therefore, become the primary tools of political communication."[7]

As C-SPAN attempts to compete in this new environment, it is to some degree a victim of its own success. C-SPAN must now run faster just to stay in place. It must compete with other public affairs programming for an audience. The Court Channel, the arrival of NBC's America's Talking Channel, and a number of proposed state versions of C-SPAN (see later discussion) threaten to take away viewers.[8] More partisan venues, such as the conservative National Empowerment Television (NET)—which prides itself on being called "C-SPAN with an attitude"—and the Republican National Committee's GOPTV, have the potential for capturing part of the public affairs audience as well as making the line between balanced public affairs reporting and partisan propaganda less clear. Inside C-SPAN, officials are concerned about "cream skimming," the potential for new cable (or even broadcast channels) to build audiences by covering the most popular programming for which C-SPAN developed an audience.

C-SPAN also helped create a public affairs audience whose members had more sophisticated demands about production values and higher expectations about their level of involvement in the media. On the production side, the audience is unwilling to accept low-budget home-video-style presentations and graphics. The contemporary public affairs audience is also more demanding in its populist expectations about contributing to the national dialogue. The public affairs talk-radio and talk-television phenomenon—in which C-SPAN was an early driving force—changes the chemistry of American politics, increasing the speed with which ideas pervade the political system. But not everyone is enamored with the rise of the talk media. Moderates and liberals bemoan the conservative dominance in talk-show hosts and callers. Rush Limbaugh, with his syndication to over 600 stations and an estimated audience of over twenty million, far outdistances his closest liberal competitor, Jim Hightower, who is carried on only 180 stations.[9] In addition, many conservatives, moderates, and liberals alike argue that the talk media form a venue that tends to thrive on controversy and criticism more often than it creates positive solutions to problems.[10]

The subjects of C-SPAN coverage have also changed. Once hesitant about any form of television, most members of Congress now recognize its importance for politics. As veteran Capitol Hill reporter Cokie Roberts described the contemporary Congress: "You walk around Congress, with a microphone. You use it as a weapon. Down boy, down boy. Access isn't a problem here." She went on to repeat the joke about the current expert at using the media: "What has two legs and is attracted to [television] lights? Newt Gingrich."[11]

Today, more time in Congress is spent on either using the current electronic media or establishing a new legal environment in which televised coverage will help Congress and its members. Blocking television access is largely a thing of the past. But congressional members' collective accommodation of television came slowly, in small steps. One clear step away from an emphasis on interpreting themselves through printed formats was the move of the document room from its prime location just off the Senate chamber. As one group of media observers put it: "Think of the demise of the document room as a metaphor for the changing universe of Congress. Today's members have become almost totally preoccupied with achieving a state of grace known as high visibility. They came to Washington to produce laws; they stayed to produce quotes."[12]

C-SPAN has become one of the favorite places for members of both chambers to produce quotes—either on the floor or in interviews—which will then permeate the political stratum in their original form and through the rest of the media via sound-bite summaries. C-SPAN has been "discovered" by political activists of all types. Whereas it was once hard to get guests for call-in shows and other programs, candidates, elected officials, appointees, interest group representatives, and even members of the other media now often seek out the exposure that C-SPAN provides. In the 1992 presidential campaign, candidate Bill Clinton recognized the advantages of using C-SPAN and other alternative media early, but George Bush hung back, fearing that such exposure would be "unpresidential." The newfound popularity puts a burden on C-SPAN to increase its vigilance to ensure that it will not be used to favor certain political interests inordinately.

On the horizon, C-SPAN must keep its eye on emerging technologies. Just as C-SPAN is a creature of the cable revolution, the emergence of new methods of distribution through phone lines or household direct-broadcast dishes could undermine C-SPAN's political and financial support. To some degree, C-SPAN's success shows the power of a delivery mechanism (cable television) as much as the appeal of program content.

Continuing Issues: Congress and Camera Control

As Brian Lamb stated: "This is an issue of control. Politicians want to control the picture that people have of them."[13] It is appropriate in the congressional setting to remember that the decision to allow C-SPAN to disseminate gavel-to-gavel coverage of legislative proceedings was the result of a compromise. Compromises arise when participants in the decision have a commitment to action but differ on their ultimate goals. The key issue in the late 1970s was control. The House had three basic options: allowing the networks to take complete control over program creation and dissemination; taking complete control of program creation and dissemination itself; or splitting control. The House recognized that it had to find a way to communicate its message unadulterated by the typical news-making rules dominated by the commercial networks. The commercial networks were willing to install cameras in the chamber, but only under the condition that the networks would control the criteria of newsworthiness. Members of Congress foresaw—probably rightly—that the coverage of Congress under network control would involve short sound bites of conflicts, which would create an even more negative image of their institution. The option of creating a government channel completely under the control of Congress was both expensive and potentially counterproductive. Such coverage would do Congress little good if it was seen as simply a public relations gimmick. C-SPAN's willingness to accept floor feeds from cameras controlled by the House reduced potential costs for and criticism of Congress. The partnership worked relatively well for both parties. When the Senate decided to begin broadcasts in 1986, it opted for the same arrangement.

Starting from a relatively weak position, C-SPAN accepted congressional control of the cameras with a set of production guidelines far removed from typical definitions of newsworthiness. Although a symbiotic arrangement existed, C-SPAN clearly needed the cooperation of Congress more than Congress needed the cooperation of C-SPAN.

Like all viable organizations, Congress reacted to the environment and learned. During the 1980s it became clear that C-SPAN's broadcasts were not going to dramatically change Congress or its environment. As C-SPAN began to expand coverage under the control of its own cameras, C-SPAN's philosophy of gavel-to-gavel coverage of speeches, news conferences, and committee hearings led to programming that was significantly different from typical network news. Trust in the network grew among those being covered. Although there are examples of C-SPAN helping to make or break political careers, influencing the nature of policy decisions, and changing the level of public information, none of the implications were clear enough or one-sided enough to raise the ire of those not benefiting from the change. The common reaction of "killing the messenger" rather than showing concern for the message has not occurred. Those people and causes damaged by C-SPAN tended to respond by criticizing themselves, saying, "We should have done a better job of getting our message out," or "We will just have to try harder next time." Criticizing C-SPAN has become politically incorrect. Critics of the kind of public access that is provided by C-SPAN are generally written off as people who are trying to hide something.

CHANGING THE RULES: ROUND 1

In most of the institutions and processes covered by C-SPAN, little attention has been paid to the cameras. The C-SPAN camera was just one more at a press conference or a speech—its only claim to fame was that it kept rolling through the whole event and the video was broadcast in its entirety.

On the other hand, Congress—especially the House—was more self-conscious, seeing the C-SPAN distribution and the chamber's control over the cameras as a political tool. The initial rules for both House and Senate camera operators were relatively straightforward: keep the camera on the person who was speaking. No cutaway or reaction shots were allowed.

As we saw in chapter 2, Representative Newt Gingrich (R-GA) and his Conservative Opportunity Society colleagues used C-SPAN's coverage of Special Orders in the House to expand their political base. Frustrated with minority members who were garnering too much publicity, Speaker Tip O'Neill (D-MA) attempted to embarrass them by panning the chamber during Special Orders speeches. The Democratic majority never really saw the need for using Special Orders as a coordinated political strategy, and Speaker Tom

Foley (D-WA) maintained the practice of panning. The panning continued in the House until 1994.

CHANGING THE RULES: ROUND 2

Acceptance of the practice of panning during Special Orders in the House became an issue in 1994 when a poll conducted by Mark Mellman for the Democratic leadership found that regular C-SPAN viewers regarded Congress more negatively than did nonviewers. Part of the disenchantment seems to have come from the public's misunderstanding of how Congress works. Images of an empty chamber as the camera pans during Special Orders were identified as one of the sources of negative feelings.[14] The panning of empty seats gave the impression that members were shirking their duty.[15] When members of both parties recognized that pictures of empty seats provided a negative image of Congress, proposals were floated to ban Special Orders, limit them, or change the nature of coverage.

The route to reform was fraught with political potholes. The Democrats seemed ready to grasp at any straw to reverse the negative image of the chamber, an image for which they were held responsible. The Republicans were reluctant to give up the access to the national audience provided by Special Orders. Any attempt by the Democrats to ban such after-hours speeches would open the proponents to charges of "gagging" the opposition.

After almost a year of bipartisan negotiation, compromise was reached. A maximum of four hours would be allowed for Special Orders, with no session going beyond midnight. The two parties would alternate at being the first to speak, and the cameras would not pan the chamber. To provide more opportunities for debate, members would be allowed to make five-minute speeches on Monday and Tuesday mornings, with formal, Oxford-style debates scheduled for prime time every three weeks.[16] The experimental Oxford-style debates were covered by C-SPAN and came off without any problem, but they also seemed to have little effect either in Congress or among the public. They were eventually shelved as an idea that sounded good but was of little importance.

CHANGING THE RULES: ROUND 3

The 1994 elections, and the corresponding Republican takeover of Congress, ushered in a new team of leaders and a new outlook. Speaker Gingrich, frustrated with what he saw as Democratic constraints on information in the emerging information age, pushed for a new openness. He promoted his goal of a "new dialogue" and argued, "That's why we're going to change C-SPAN's access to Congress . . . we're going to do a lot of things in public that the old order did in private." Changed chamber rules and leadership openness were part of a larger "third wave"—a move toward greater public access that included computerized availability of congressional records and procedures.[17]

Gingrich's thinking was furthered by C-SPAN's initiative. Recognizing a new environment, Brian Lamb sent a letter to the leaders in both chambers and asked for more control and access. He urged Speaker Gingrich and Senate Majority Leader Robert Dole (R-KS) to "consider opening the 104th Congress *fully* to television cameras." Lamb proposed the following changes:

- Allow C-SPAN cameras to cover House floor debates (current rules prevented House cameras from panning the chamber or taking reaction shots)
- Open the Speaker's press conferences to television (cameras were kept out of these daily on-the-record briefings between the Speaker and reporters)
- Install a permanent camera just off the floor (allowing interviews with members during votes)
- Open the House Rules Committee and all other legislative committee hearings (coverage had been sporadic)
- Open all House-Senate conference committees to cameras (cameras were often shut out of this final and important step in the legislative process)[18]

Gingrich immediately opened the Speaker's press conferences to cameras, and C-SPAN began covering them in their entirety. Confident in his verbal abilities, Speaker Gingrich felt that the ability to get his ideas directly to the public would "undermine the credibility" of the reporters who attacked him and would allow him "to win his fight with the press."[19] Senate Majority Leader Dole expressed a willingness to open his pre-session "dugout" meetings to television cameras, but he ran up against Senate rules. Because the press would question Dole on the Senate floor, Dole said the Senate would have to pass a resolution allowing C-SPAN to operate while the Senate was technically not in session.[20] Senator Robert Byrd (D-WV) decried the breaking of Senate tradition and railed against "too much politics."[21] Much of his opposition undoubtedly stemmed from an attempt to limit the Republican leader's public exposure. Not surprisingly, the rules change was put on the back burner, since the Senate's right of filibuster would allow one member to block the change. Dole has held occasional dugout sessions outside the chamber.

The question of giving C-SPAN complete control over the cameras was sent to a bipartisan study commission. This was not a new idea. When the House had first allowed television cameras, Representative Al Gore had predicted, "When the House becomes comfortable with these changes, the news media will be allowed to bring their own cameras in."[22]

Peter Hoekstra (R-MI) chaired the House task force. The optimism concerning openness was tempered when Hoekstra was asked, on a C-SPAN call-in program, whether the task force would hold open meetings. Hoekstra hesitated a bit, not seeming to realize the implications. He quickly recovered and answering: "You know I guess we could do a conference. We could do it on C-SPAN. And we have nothing to hide. . . . It would be ironic to hold this meeting [on] opening up the process behind closed doors."[23] The demands of other business overwhelmed the new Republican majority, however; almost a year after Hoekstra's comments, public hearings had yet to be convened.

In the meantime, Speaker Gingrich ordered House cameras to provide a fuller picture by using broad shots of the entire chamber and by using cutaway shots during speeches to show the reactions of other members. House directors were allowed to use an 80/20 rule: 20 percent of the time they could use reaction shots.[24] Although Gingrich might honestly have thought that panning would open up the system and allow television viewers to see the House more as a visitor in the gallery might, some observes saw a more political motivation. Since he had made his name on C-SPAN in the prepanning era, when official shots added aura to the Special Orders speeches given to an empty chamber (see chapter 8), Gingrich might also have thought that adding a broader picture would dull the luster of these sessions for his opponents. In journalist Charlotte Grimes's words: "As a self proclaimed revolutionary, he also knows that it is not a good idea to leave your own weapons lying around for the opposition. The whole picture is unlikely to create new stars on an empty stage."[25]

After a few months, some of the experiment with openness confronted the grim reality of political self-interest. Speaker Gingrich's televised press conferences had become raucous and confrontational. The Speaker's press secretary, Tony Blankley, explained: "Some of the questioning was a tad flamboyant and got in the way of serious discussion of the news. It provided an opportunity for obscure journalists to come in and harangue him on their pet projects."[26] Privately, a number of Republican members were concerned that Gingrich was taking attention away from other House leaders and that Gingrich's off-the-cuff comments often created stories that later had to be modified or repudiated. Political commentators Jack Germond and Jules Witcover pointed out: "The new Speaker has found that the cameras are not always benign and sometimes can be merciless. And he has found that it is possible to be too much in the face of the voters." Gingrich discovered that daily exposure on C-SPAN led to media overexposure. Network news officials noted, "We have him [on film], so we end up using him a lot . . . and we do it sometimes even when he doesn't make a lot of news."[27] But once the doors had been opened to the cameras, they could not be selectively shut. Rather than simply banning the cameras, Gingrich canceled the daily press conferences, which had been a Capitol Hill fixture for over thirty years.

The livelier coverage of floor proceedings quickly ran into opposition from members. Within a week, thirty Republicans asked the Speaker to reverse the policy. Opposition ran the gamut from disapproval of the lack of attention for the member who was speaking to intrusiveness into members' behavior. Some members feared that their deal-making would be captured by the cameras. Others worried that they would be shown in their sneakers after arriving from the House gym or that they would be caught picking their noses. The cameras showed Representative Eva Clayton (D-NC) cleaning her nails and Representative Bill Orton (D-Utah) studiously scrutinizing his tie. Reflecting C-SPAN's faith in its audience, Vice-President Susan Swain commented, "If members are caught occasionally being human, our viewers are sophisticated enough to understand that."[28] Despite overwhelming editorial support from newspapers around the country, the experiment in increased openness was dropped after a few weeks, and the issue was sent to a joint task force on the media.[29]

Other initiatives have been more permanent in redefining the visual coverage of Congress. House rules adopted for the 104th Congress allow cameras into virtually all committee meetings. To create a more valid record of congressional proceedings, Gingrich and the Republican leaders accepted Lamb's argument: "The exclusion of cameras from these kinds of sessions only contributes to the public's present skepticism about the political process."[30]

In an attempt to reduce the embarrassment of a video record that failed to match the printed *Congressional Record*, the Republican-controlled 104th House changed the rules for "revising and extending" remarks. The old rules allowed members to add, delete, or change what was said on the floor. Critics had long pointed out that the *Congressional Record* was a "smoke screen behind which politics is conducted"[31] and that the right to revise and extend turns "absentees into prime movers and verbal clubfoots into gazelles."[32] By watching C-SPAN, careful viewers and the media had picked up harsh verbal attacks that later appeared in the official printed record as mild differences of opinion. Under a little-noticed but important reform pushed through by the new Republican majority, members of the House were no longer able to edit history. The lawmakers probably realized that, since their debate was televised live on C-SPAN, trying to sanitize the written record only made them look foolish. The new rule states that the *Congressional Record* must be "a substantially verbatim account" of House proceedings, subject only to technical, grammatical, and typographical corrections.[33] Words not actually spoken on the House floor but submitted for publication now appear in a different typeface.

Finally, both chambers and their members began to cooperate with C-SPAN's desire to provide real-time context. C-SPAN cameras were allowed to be set up outside the chambers to capture the comments of members while votes were in progress.

Congressional rules that determine what, when, and how C-SPAN can cover the legislative branch change regularly. Despite philosophical commitments to openness or exhortations about tradition, the motive for change is almost always political. The party in control will attempt to fashion coverage that places its institution or cause in the most favorable light. As members of what many thought would be a permanent minority, the Republicans had little stake in making Congress look good until after the 1994 elections. One of the positive results of the change in partisan control may well be that members of both parties now see the advantage of fashioning rules that will make their chamber look competent.[34]

The battle over congressional control of the visual image that C-SPAN presents will be increasingly exacerbated as different political groupings recognize its strategic advantage. The House Republicans' effective use of C-SPAN beginning in the mid-1980s for Special Orders speeches may have given the impression that C-SPAN was inherently a Republican vehicle. It is perhaps more valid to characterize C-SPAN's gavel-to-gavel approach and its commitment to balance in programming as a vehicle particularly suited to a political minority denied adequate coverage by the commercial networks. Early in the 104th Congress, the Democrats rediscovered C-SPAN and Special Orders. As Dan Turton, the floor assistant to Minority Leader Richard Gephardt (D-MO), put it: "When we were in the majority, we couldn't get people to do it. Now Special Order requests have become much more popular. Almost every single day, we fill our time."[35] As the Democrats began to coordinate Special Orders speeches, Republicans found themselves looking at this vehicle, which had helped propel them into power, in a new perspective.[36]

Technologies seldom wear a political label. They are tools to be used by clever practitioners. The ongoing battle over coverage rules both within Congress and in the broader political arena will be driven by perceptions of political advantage. As a "guest" in most political forums, C-SPAN will continually be faced with attempts to adjust the rules governing its presence. In most cases, C-SPAN's role will be to react to short-term opportunities and challenges.

A Longer-Term View: Emerging Issues

In other realms, societal conditions and technological changes will provide C-SPAN with another set of opportunities and challenges. The network will be able to control some of these, but it will be able only to react to others.

THE NEW MEDIA MIX

To understand C-SPAN's future, we need to recognize how different today's media mix in the United States is from that of the days when C-SPAN began. The trends that started in the 1970s are likely to continue—even accelerate—as we enter the twenty-first century.

In the 1970s the three major networks dominated television, garnering over 90 percent of the prime-time audience. College courses, book titles, and popular conversation used the term "mass media" both as a description of the current reality and as an indication of a desirable goal for information dissemination. The vehicle for dissemination was clearly defined as *broad*casting, and the audience was seen as a mass of inert consumers that accepted whatever the mass media decided to broadcast. Public affairs and news programming, viewed as an economic loser, was driven by federal regulations requiring public service.

By the mid-1990s both the reality and the goals were undergoing significant transformation. The television networks were struggling to hold on to 50 percent of the prime-time audience. The initiatives of Ronald Reagan's presidency had largely removed regulation of program content. The commercial media increasingly attempted to recapture the fleeing audience with glitzy graphics, brief sound bites, and an emphasis on personal-interest stories. As the line between information and entertainment became blurred, the word "infotainment" entered our vocabulary.

C-SPAN proved to be a prime example of media specialization. The operative term has become "*narrow*casting" rather than "broadcasting," and C-SPAN has captured a niche market. Rather than aiming its signal at an undifferentiated public, C-SPAN—and the growing array of niche channels—assume that the potential audience is in fact a wide variety of distinct publics, differentiated by their interests. As media critic Douglas Davis put it: "No large, multicultural body of information-hungry minds poured into hundreds of channels of digitized data can be thought of as a monolithic consumer anymore. We are increasingly contradictory, our minds and our tastes honed by an unprecedented variety of ideas, products and experiences."[37] C-SPAN never intended to capture a mass audience but rather planned to focus on those thoughtful individuals with a deep interest in public affairs. The network took advantage of the desire for personal choice.

C-SPAN pioneered long-form public affairs television—the antithesis of sound-bite journalism. In the process it discovered a market and encouraged imitation. As Brian Lamb noted: "Our common television culture really rose out of television tyranny. Americans' viewing choices were limited by a government system that unnecessarily restricted broadcast licenses and put this vast resource in the hands of just a few companies. Instead of dying on the vine, C-SPAN now finds itself with lots of competition in long-form hearing coverage."[38] CNN, Court TV, National Empowerment Television, and a host of other channels now overlap C-SPAN programming to some degree and compete for its audience.

C-SPAN also was a trailblazer—perhaps unintentionally—in encouraging active political viewing. The couch potato of the 1980s has been partially replaced by the channel surfer and talk-show participant of the 1990s. C-SPAN gave birth to national public affairs talk shows, particularly on television, and, by example, encouraged their expansion. Today, talk politics has become a major force in the political process. Nearly half of all Americans report listening to talk radio or talk television "sometimes."[39] Larry King— who first got the idea for talk television from C-SPAN's coverage of his radio program—has seen *Larry King Live* become a place to announce for the presidency (Ross Perot in 1992) and to hold major policy debates (Ross Perot and Vice-President Al Gore on NAFTA in 1993). By the 1992 presidential campaign, even the major networks had experimented with talk-show formats. Talk shows seem to have a special appeal to the baby-boomer population, which had tuned out on politics but is now tuning back in.[40]

C-SPAN also tapped—and perhaps contributed to—the growing public unwillingness to passively accept the selected, predigested, and homogenized public affairs information that the mainline media presented as "news." Distrust of and disenchantment with the traditional media have grown (see chapter 6). "People believe increasingly that they can make their own judgments, based on direct access to the primary sources of information. They don't need or want others pre-chewing their political food."[41] As one of the new Republican leaders in the House put it, there has been a "real sea change that's occurred in how people receive information. For a long time, political and government information in Washington came through traditional media sources. For many Americans, it doesn't any longer. It comes from talk radio shows. . . . It comes from C-SPAN, where they're able to turn on the television every night and find out directly . . . what's going on in the Congress."[42]

As journalist Jonathan Alter noted, the old media are "far from dead."[43] They still capture a large audience. Few viewers of niche networks such as C-SPAN ignore the commercial networks completely. Yet C-SPAN and other niche-market programming channels have become a factor that needs to be taken into account. Although C-SPAN is not concerned with television ratings—at least not publicly—an increasing number of media-analysis firms include C-SPAN in their audience estimates. As we will see later, C-SPAN would be wise not to rest on its laurels. A new wave of change has already begun to beat on the shores of the media environment, and many observers would argue that the ebbing away of the traditional media audiences will be nothing compared with the tidal wave that will arrive with the digitized multimedia storm.

NEW PROGRAMMING OPTIONS

The most obvious changes for C-SPAN involve expanding its programming by offering more depth and breadth. The driving concept of covering "public affairs" provides C-SPAN with a rather broad definition of potential programming. Several new or expanded venues look promising.

Congress

Although C-SPAN has been called a "window on Congress," providing a unique view of the legislative process, it is a window with a limited view. All windows place some limitations on their users, of course. Shifting the analogy a bit, one media observer commented, "C-SPAN now functions like a video knothole in the congressional fence."[44] One factor limiting the C-SPAN view is the set of congressional rules regarding the nature of floor coverage. Another limitation stems from C-SPAN's philosophy and capacity for covering events over which it has full control, such as committee meetings and news conferences. Although thirty to forty committee and subcommittee meetings occur every day, C-SPAN chooses to cover only a few. The major constraint is the number of camera crews available. Secondarily, camera crews are seldom dispatched to congressional events unless they can be covered live. The overlapping schedules of meetings and press conferences generally mean choosing between them.

Technology is ripe for solving the logistics problem. C-SPAN is willing to wire all congressional hearing rooms with permanent robotic cameras that could capture all meetings either for live or for delayed broadcast. Since virtually all of these sessions are currently open to the public—and to C-SPAN cameras—there should be little congressional opposition. And since press conferences are designed for coverage, an expanded C-SPAN presence at such events should be welcomed.

A range of other meetings, such as party caucuses, state delegation meetings, and special-interest caucuses, could also be covered.[45] Since most of these meetings are currently closed to the public, members of Congress would have to determine whether the advantages of openness outweigh those of secrecy. Lest anyone think that opening formal meetings of any type will place the entire legislative process out in the open, the experience of committees must be reviewed. After committees opened their doors to the public, much of the key decision making occurred in the hallways, in the anterooms, and at premeeting meetings. There is a limit to "sunshine" procedures. Certain issues must be discussed in private. The legislative process would grind to a halt if everything had to be done in public. Just as there is a legitimate "private space" in each of our lives, public officials must be allowed to have some private space to develop policy agreements. Despite this recognition, a great deal more of the legislative process could be shown.

The emphasis on an expanded role for C-SPAN in covering Congress is natural, since House and Senate sessions make up the network's core programming. But too much emphasis on Congress as the only realm for C-SPAN expansion belies the fact that currently only 15 percent of broadcasting time is allocated to congressional sessions. C-SPAN has already discovered other realms.

A White House Channel

C-SPAN coverage of the White House has run into two problems. The commitment to covering House and Senate sessions live and gavel-to-gavel, no matter what, reduces C-SPAN's flexibility in covering White House press conferences and speeches, which often occur while Congress is in session. Second, C-SPAN is committed to serving as an alternate source of programming of events that the other media fail to cover. The other networks tend to cover the major White House events in their entirety. C-SPAN does provide viewers access to these events on a delayed basis.

But there are many White House events, such as bill-signing ceremonies and press briefings, that are not currently well covered. The idea of allowing cameras into cabinet and staff meetings is tantalizing. The Clinton White House is particularly interested in providing more television access. It sees the political advantage of getting to the public directly and repositioning the way that C-SPAN "tilts our perception of what government does toward Congress."[46] The downside of expanded White House coverage is largely political. Since the White House has decades of experience in mastering public relations, it would be difficult to distinguish between legitimate public affairs forums and purely self-serving political events.[47] Additionally, if the C-SPAN focus is on programming *from* the White House, the incumbent president's increasing use of the network could jeopardize C-SPAN's commitment to balance. A channel *about* the presidency has more potential for offering a balanced view. Coverage could include academic conferences, historical footage, and critics of the current president. C-SPAN's ability to engage its audience by offering call-in programs after presidential events would provide a useful service.

Opening the Federal Bureaucracy

Some of the most critical government decisions occur in the bureaucracy. Decisions as to how laws are implemented are often based on hearings presided over by bureaucrats. The move toward openness in government means that many of these meetings have already been opened to the public. Anyone attending one of these meetings would have a rather distorted view of the "public," however, since both the audience and those asking to testify are usually paid lobbyists or a very narrow representation of those people uniquely aware of the policy process or deeply involved with it. Television cameras seldom record such events as the Federal Elections Commission establishing rules for political fund-raising on the Internet, the Federal Trade Commission outlining regulations for the use of used parts in new equipment, or the FCC deciding on the rights of foreigners to own U.S. media outlets. The information the public receives usually emanates from print journalists well after the key decisions have been made.

While he chaired the House Banking Committee, Henry Gonzales (D-TX) went so far as to suggest that the Federal Reserve Board should videotape its closed-door meetings in which money supply decisions are made. These meetings could be made available for broadcast by C-SPAN after a delay that would preclude market manipulation. Gonzales argued, "Those responsible for the nation's money supply must be fully accountable for their actions to the American public."[48] A similar argument could be made for public access to most bureaucratic decision-making meetings.

C-SPAN has only sporadically penetrated the federal bureaucracy. Following legislation to the implementation stage would provide a broader picture of the policy process.

Federal Courts: Forbidden Ground

One of C-SPAN's continuing frustrations is its inability to cover public sessions of the Supreme Court. From a logical perspective, the analogy between the public galleries of Congress and the public gallery of the Supreme Court seems unassailable. Both the public and the press are allowed to hear all the oral arguments in front of the Court. Yet despite continued requests, C-SPAN has been unable to get a majority of the Court to allow cameras. Opponents on the Court are more amenable to gavel-to-gavel coverage but fear what the networks would do using sound bites. Justice Antonin Scalia worried the cameras would make the "oral argument show biz."[49]

Much like the experience in Congress, opposition in the Supreme Court tends to be generational. Two of the most recent justices are supporters. During her confirmation hearings, Justice Ruth Bader Ginsburg stated: "I don't see any problem with having proceedings televised. I think it would be good for the public."[50] Responding to a similar question during his confirmation, Justice Clarence Thomas voiced qualified support: "I have no objection beyond my concern that the cameras be as unobtrusive as possible . . . it's good for the American public to see what's going on."[51]

C-SPAN and the rest of the media keep hoping that the Supreme Court will follow the lead of the forty-seven state court systems, which allow some form of television coverage. Between 1991 and 1994 a three-year experiment in the federal court system revealed no significant negative results. Yet despite the fact that surveys of trial participants indicated that the presence of the cameras "did not disrupt court proceedings, affect participants in the proceedings or interfere with the administration of justice," the experiment was abandoned by a two-thirds vote of the U.S. Judicial Conference.[52] The hope was that positive results would put pressure on the Supreme Court, but none seemed to be forthcoming. The perceived media circus associated with coverage of high-visibility criminal trials, such as the Menendez brothers and the O. J. Simpson trials, undermined any immediate move toward expanded C-SPAN access to the courts.

State and Local Government

State and local government offers an immense range of potential programming options for C-SPAN. Many state legislatures already allow cameras to cover some or all of their proceedings. More than a half dozen states have established systems for gavel-to-gavel distribution of the proceedings of at least one chamber (California, Kentucky, Massachusetts, Minnesota, Nebraska, New York, Oregon, and Rhode Island).[53] A number of the state systems were directly inspired by or modeled after C-SPAN. There is an active debate in other states about installing cameras.

Required "open meeting" laws in many states create a large number of possible venues for C-SPAN cameras. So far, C-SPAN has limited itself to coverage of governors' "State of the State" speeches, inaugurations, and selected state legislative sessions. The task of selecting which meetings to cover and the cost of remote camera crews temper some of C-SPAN's enthusiasm for fuller coverage of state and local public affairs.

International: Capturing the World

While the jet airplane has physically reminded many Americans that the world is getting smaller, worldwide satellite transmissions that electronically take viewers from place to place in real time are a more viable vehicle for representing the shrinking world to most citizens. The potential for expanding state and local government programming pales in comparison with the potential for expanding international coverage. Over fifty national legislatures allow cameras into their chambers. In many instances, opening legislative proceedings to the electronic media preceded the existence of C-SPAN, but in a number of key cases, such as Great Britain and Japan, the experience of C-SPAN served as a model to be adapted to local conditions and laws. In Britain, for example, the limited spread of cable television mitigated against using cable as the primary distribution vehicle.

C-SPAN currently includes sporadic coverage from a limited number of foreign legislatures. It has broadcast the legislatures of Argentina, Australia, Canada, France, Germany, Great Britain, Hungary, Ireland, Japan, Mexico, Poland, and the Soviet Union. C-SPAN's Sunday evening rebroadcast of the British prime minister's appearance before Parliament during "Question Time" has developed something akin to a cult following. C-SPAN viewers have also been treated to sessions of the European parliament, speeches by foreign leaders, press conferences by U.S. and foreign leaders in foreign settings, and programs interpreting foreign political systems. Coverage of such international events requires special agreements with the institutions being covered, the purchase of satellite time, and in many cases, expenditures for camera crews and translation.

Current international programming is largely sporadic and lacking context. Expanded international programming could be designed to allow viewers to follow a particular political system or to follow public policy themes across a number of systems.

DELIVERING THE NEW PROGRAMMING

Disseminating expanded programming presents C-SPAN with a "chicken versus the egg" problem. It is not clear whether the network should create the programming and then wait for the delivery system or wait for the delivery system and then create the programming. Although limited experimentation continues under the existing two-channel cable delivery system, there is little room in current scheduling for a major expansion of programming. Cost is a factor, but C-SPAN is ready to make the investment in staff and equipment for additional programming.

The major choke point at this time is not the availability of new programming but rather the lack of enthusiasm by cable systems to carry another C-SPAN channel. The combination of "must carry" legislation (see chapter 2) and limited channel capacity leaves little room for new C-SPAN channels. In addition, the slower-than-anticipated growth of C-SPANII has tempered the board's enthusiasm for additional channels at this time. The predicted arrival of five-hundred-channel systems in the near future, however, has encouraged C-SPAN to keep itself ready for the expansion. C-SPAN has three new channels on the drawing boards. One would cover a wider ranger of congressional committees, the second would focus on international programming, and the third would emphasize business and finance. Just as it did two decades ago, C-SPAN assumes that with expanded channel capacity, the need for additional programming will offer it an opening not only for the three planned new channels but also, possibly, for a host more.

NEW DELIVERY OPTIONS

C-SPAN was created as a delivery vehicle to disseminate the House television signal via cable television. The fledgling cable industry was willing to jettison its journalistic load and accept the House feed on the House's terms. Only slowly did the industry get into the business of venturing beyond Congress and producing its own programming. As technology changes, C-SPAN is faced with a dilemma: does it blindly repay its loyalty to cable, or does it look out for its own interests first?

Funded, Founded, or Forgotten by Cable?

To some degree the cable industry is a victim of its own success. In less than two decades it penetrated the population to an extent that most commercial entities would envy. For most Americans, cable is no longer an option but is the same as any other utility to be paid for each month. C-SPAN has benefited the cable industry well, acting as its public affairs flagship. C-SPAN programming and publications prominently displayed the phrase, "Privately funded to serve the public by America's cable television companies." The cable industry used this demonstrable commitment to the public interest in its battles over licensing and regulation. C-SPAN also helped build a subscriber base as a growing and vocal set of viewers opted for cable because of C-SPAN.

One longtime C-SPAN board member described C-SPAN as the "low maintenance child of the cable industry . . . the good kid . . . the one to be proud of." He also pointed out that like the good kid in the family, C-SPAN could be overlooked: there is the temptation to forget about its growth and development while expending effort on the "problem children."[54] C-SPAN could be in jeopardy if the cable industry takes it for granted.

The contemporary cable industry is experiencing a dramatic change. The age of small, mom-and-pop cable operations has given way to the almost daily announcements of acquisitions and the creation of massive Multiple System Operators (MSOs). As cable has become more businesslike, financial bottom lines increase in importance. Cable's loyalty to C-SPAN—cable's "public affairs flagship"—was once based on the personal experiences of board members or on an idealistic commitment to better informing the public. Today, loyalty must satisfy the demands of a cost-benefit analysis. Some cable executives, while praising C-SPAN, point out that their "sacrifice" in providing the network did not do them much good in the battle over the 1992 cable bill.

The uncertainty about future technologies has created new lines of cleavage. Although the cable industry was never "one big happy family," the areas of disagreement over regulation were relatively limited. In the old days, cable operators fought over the rights to cable particular geographic areas. Today, some cable companies maintain a complete commitment to cable delivery, while others are actively exploring and backing new technologies such as wireless cable and direct-broadcast satellite systems (see later discussion). One cable industry executive and C-SPAN board member, who did not want to be identified, blurted out: "In the future it will be ridiculous to talk about the cable *industry*. With the growing diversity of delivery options, there is little we will have in common."[55]

In the process of building the demand for cable, the industry has created an audience that expects expanded choices and options. The five-hundred-channel cable system option based on compressed video seems to be within sight, although most analysts predict that the more likely figure is closer to two hundred channels. The availability of more channels offers C-SPAN the opportunity to expand its programming but also raises the potential that it will get lost in the glut of new options. On the positive side, C-SPAN's marginal cost in creating new channels is considerably lower than the cost of creating a network from scratch, and its low direct cost to cable companies makes it an unlikely target for cost cutting. Market analysts have pointed out that even with expanded channels, the difficulties of launching a new network are significant in this age of "picky viewers." Those independent networks most likely to succeed will need large infusions of money, existing libraries of video materials, and/or sister networks to help feed them programming.[56] C-SPAN has the latter two and could expand its channels relatively inexpensively.

From the audience perspective, the program-rich new environment will require more work to receive maximum benefit. The time it will take to read through the traditional printed television guide will eat up most of one's discretionary television time. Channel-surfing through hundreds of channels will no longer be a realistic option for most viewers. More sophisticated programming guidance will be necessary. Either the cable industry will come to grips with this problem or the situation will spawn a new set of computerized products designed to help one navigate through the options. As cable moves toward on-demand programming (see later discussion) the problem will be exacerbated.

Direct-Broadcast Systems

The Achilles' heel of the cable industry is the cable itself. Installation and maintenance costs saddle local cable companies with a continuing upkeep burden. Local managers look with dread at approaching storm clouds and major construction projects, knowing that disruptions in service are the likely results—causing the phones to ring off the hook as dissatisfied customers complain that their cable is out.

Direct-broadcast systems reduce the number of things that can go wrong, and they put the burden on the consumer. Receiving cable programming via a satellite dish has long been possible, but the costs and inconvenience of a three- to nine-foot dish have dampened the demand. Although C-SPAN has never been scrambled, access to much of the programming required special equipment. The arrival of small and less expensive dishes (seven hundred dollars) for home use has ushered in the direct-broadcast era. Direct broadcast, or direct to home (DTH), reached about a half million homes in its first year. Although the figures look small compared with the cable potential of over sixty million homes, such figures would have looked significant in the early years of cable. DTH established a foothold in the distribution realm and created a mechanism for generating ongoing revenue. Like cable operators, DTH systems charge customers a fee and in turn compensate program providers such as C-SPAN.

To some degree, the cable industry in the 1990s acted much like network television in the 1970s, attempting to first ignore and then thwart an emerging technology. Many of the once radical pioneers of the cable industry saw the proponents of the new technology as unrealistic upstarts. Other cable operators decided that it was "better to join them than to fight them" and became partners in PrimeStar's DTH initiative.

C-SPAN initially hesitated to sell its programming to DTH systems but eventually accepted the 1992 cable law requirement that it must sell to all distribution venues. The decision belied C-SPAN's logo of being "*funded by cable television*," so this was changed with a stroke of the pen to "*created by*."

C-SPAN has not faced its last technology-driven distribution challenge. Considerable work is being done on distribution of television signals over phone lines. For the foreseeable future, the well-entrenched cable system will continue to serve as the backbone of C-SPAN distribution, but C-SPAN now has a programming base that makes it appealing to alternative distribution systems that can pay its modest fees.

Other Distribution Mechanisms

Various alternative mechanisms for delivering television signals, such as wireless cable and the use of telephone lines to transmit digitized video, are in the developing stage. The entrance of telephone companies into the delivery of video has prompted numerous questions concerning access to programming and regulation. For example, if video delivery by telephone companies is treated the same way as telephone service, the "common carrier" rules would deny telephone companies the right to control the programming they deliver.

C-SPAN is approaching these new technologies with an open mind. On the one hand it has a commitment to—and a board of directors dominated by—the cable industry; on the other hand, it cannot afford to let old loyalties preclude new opportunities.

The Digitized Future: C-SPAN On-Demand

To a degree, television-viewing patterns are dictated by technology. Until the 1980s, virtually all home viewing was in real time. Viewers chose programs from among those offered by the networks during particular time slots. The decision to watch one program meant a de facto decision to bypass another. C-SPAN was one of the few networks that repeated its programming, providing alternative viewing times.

The arrival of the VCR allowed viewers to expand their options—watching one program while taping another, or taping a program to watch later. For those able to master the recording function of their VCR—or for those with children who had done so—time-shifting the programs turned some viewers into their own schedulers. For many viewers, however, mastering the technology and finding the necessary programming information were too intimidating.

The impending arrival of digitized video-on-demand offers a tremendous potential for allowing viewers to construct an individualized set of programs. In the future, C-SPAN viewers will be able to order only the programs they want and view those programs in any schedule they choose.

C-SPAN LOOKS TO THE FUTURE

The buzzword for the future in much of the information industry is "positioning" or, in the more negative terminology, "hedging." Recognizing the rapid changes of the last few decades and the unclear future, wise organizations are attempting to prepare for alternative visions of the future. Everyone agrees that organizations ignore technological change at their own peril. Sitting out a round of technological change may well deny one the opportunity to play in the next round.

On the one hand, the C-SPAN programming philosophy of providing long-form coverage of major public affairs events is unlikely to change. Coverage of Congress will remain a driving force. C-SPAN staff constantly point out, "We want to do what we have done well in the past, only do it better." Over the years, C-SPAN has encouraged both individual staff members and formalized study groups to think about the future. The C-SPAN 2000 group (discussed in chapter 4) is a good example. Whereas much of the planning involves improving current programming, two initiatives—one technological and one organizational—portend more extensive changes.

Maintaining a C-SPAN Lane on the Information Superhighway

The impending merging of technologies presents C-SPAN with a challenge. C-SPAN's bread and butter has always been video. For viewers to understand much of the video, textual and graphical information is useful. Until recently C-SPAN yielded the distribution of such information and the active facilitation of political involvement to others. C-SPAN felt that information was the basis for activation and that the network had done its part by providing the information. Commonly, C-SPAN ended much of its programming with an address or telephone number that viewers could use to get further information and options for participation.

The emerging information revolution blurs the line between various information formats and provides for more active involvement in information gathering and use. The mixing and matching of audio, video, graphics, and text is the wave of the future, and C-SPAN is attempting to ride it. The very name of the network's high-tech operation—the "convergence room"—reveals its view of the future. Beginning with a Gopher site on the Internet, C-SPAN allowed people to freely search for public affairs information. Textual information—such as C-SPAN programming, committee assignments for members of Congress, and the text of legislation—was readily available to people with the equipment and skill to use it. Next C-SPAN established a site on America Online (and later Prodigy), with three thousand to ten thousand people per day downloading material. With the arrival of the more graphical World Wide Web, C-SPAN developed a "home page" that would lead users to both C-SPAN information and other public affairs information. This serves as C-SPAN's first attempt to disseminate some of its extensive collection of video in digitized form.

In 1995, C-SPAN began sponsoring interactive computer forums on America Online. Anyone with a computer, modem, and access to America Online could settle into a "seat" in the C-SPAN "auditorium." Participants could either sit back and monitor the discussion on their computer screen or pose questions to the participating members of Congress.[57]

The specific applications are less important than the fact that C-SPAN is continuing to take advantage of new technology to better inform the public and provide a connection between citizens and elected officials.

Preparing a New Generation of Leaders: The Post-Lamb Period

Just as the untimely obituary of Mark Twain inspired the author to respond, "The reports of my death are greatly exaggerated," the end of the Brian Lamb era at C-SPAN is not clearly in sight. Lamb continues to play an active and creative role as chief executive. He wears his role as "founding father" well, inspiring employees to look to the future and somewhat self-consciously recounting the "good old days" when pressed.

Like all other good leaders, though, he has attempted to create an organiza-

C-SPAN's Home Page

Author Stephen Frantzich logs on to the C-SPAN home page on the World Wide Web, which gives him access to C-SPAN programming and general political information. (Photo courtesy of the U.S. Naval Academy.)

tion that will continue to carry out his dream of facilitating public awareness of public affairs when he does depart. Brian Lamb is a realist. He understands that with the proper care, C-SPAN will outlive him. Unlike many other innovators, he is trying to prepare the organization for his eventual departure. He would like to be remembered as someone who had a good idea, worked to accomplish it, and had the sense to leave the project in good hands. He has clearly attempted to "wean" the organization from dependence on him while at the same time he still takes an active role in its current activities. He is very willing to delegate responsibility and share credit. When the C-SPAN bus won the Cable Golden Ace Award in 1995, Lamb insisted that two vice-presidents, Susan Swain and Rob Kennedy, officially accept the award on air. The gesture was more than the typical C-SPAN "no stars" philosophy; it was a clear attempt to share glory and imbue his probable successors with demonstrable evidence of his confidence in them.

The New C-SPAN Leadership

The new generation of C-SPAN leadership, Rob Kennedy (left), Su-
san Swain, (second from right), and Bruce Collins (right), join Brian
Lamb and board of directors member June Travis in front of the
C-SPAN bus. (Photo courtesy of C-SPAN.)

One longtime board member commented that he used to think of C-SPAN
as "a long running play with an irreplaceable director that would some day
close. On closing day everyone would comment, wasn't that a great run."
Now the board member believes that Lamb was wise enough to "build a
great theater" that has enough intrinsic value and a solid enough foundation
to exist well beyond the time when the original "director" departs.[58]

A Brief Reprise: Back to the Future

As C-SPAN approaches its twentieth anniversary, it has come a long way
from the day it first began broadcasting in March 1979—and even further
from Brian Lamb's vague idea of a channel to distribute the House television signal. Lamb's success in turning the idea into reality is a testimony to
the fact that idealism is not always unrealistic. Lamb was clever enough to
hitch his wagon to the star of the cable industry, whose motives were based

more on regulatory tolerance and economic success.

Slow and steady growth capitalizing on new opportunities has served C-SPAN well. C-SPAN is an established phenomenon. It would be hard to turn the clock back now. Both the public and the political practitioners have come to depend on it. Both within the network and among outside observers, C-SPAN has largely fulfilled—or superseded—initial expectations. As the wheel of technological change is given another—an even faster—spin, C-SPAN will have to move quicker just to stay in place. There is every indication that it is in position to do so.

C-SPAN brings to the American public the great thoughts—and many of the more base expositions—of the politicians of our day. Future generations will pick and choose from among these ideas to guide public affairs in their time. If C-SPAN had been around in previous centuries and decades, it would have captured basic truths that either presage or undergird the modern conception of C-SPAN. Although the great public affairs thinkers of the past would no doubt be appalled at many aspects of modern society, they would probably applaud C-SPAN and its efforts. The great Greek thinker Thucydides laid down the challenge for us all to maintain an interest in public affairs when he argued that a citizen "does not neglect the state because he takes care of his own household." He added: "Even those of us who are engaged in business have a very fair idea of politics. We alone regard a man who takes no interest in public affairs, not as a harmless, but as a useless character, and if few of us are originators, we are all sound judges of a policy."[59]

Thomas Jefferson saw the new form of government created in the United States as an experience based on the assumption that "man may be governed by reason and truth." For this to be accomplished, he felt that citizens must have available to them "all avenues of truth," which in his day meant freedom of the press.[60] Jefferson could not see beyond the newspaper as the vehicle for informing the public. Were he alive today, he would undoubtedly accept a modern translation of his famous dictum: "The basis of our government being the opinion of the people, the very first object should be to keep that right; and were it left to me to decide whether we should have a government without C-SPAN (newspapers), or C-SPAN (newspapers) without government, I should not hesitate a moment to prefer the latter."[61]

A few decades later, commenting on American democracy, the French politician Alexis de Tocqueville—another captive of the print media—stated: "In countries where the doctrine of sovereignty of the people ostensibly prevails, the censorship of the press is not only dangerous, but absurd. When

the right of every citizen to a share in the government of society is acknowl-
edged, everyone must be presumed to be able to choose between various
opinions of his contemporaries and to appreciate the different facts from
which inferences may be drawn."[62]

Were Tocqueville to reappear today to evaluate the health of American
democracy, he would undoubtedly stop by the C-SPAN offices and heartily
approve of the network's efforts—at least for America. Even though he had
some doubts about the universal value of both democracy and freedom of the
press, at worst he would probably highlight C-SPAN as an interesting and
uniquely American phenomenon consistent with many of our other governmental choices. As an author with a keen commercial sense, he would also
be likely to accept an invitation to appear on *Booknotes*.

Imagine what a call-in show with Thucydides, Jefferson, and Tocqueville
would have been like. Consider how their contemporary societies might have
changed if their words had been spread to the mass of the citizenry in real
time. Much of the valuable public rhetoric of the past was heard by only a
few people, and the bulk of it has been lost to history. When Brian Lamb was
asked what questions he would ask Tocqueville if the French writer were a
guest of *Booknotes*, he laughed and then with relish laid out a set of typical
C-SPAN questions he thought the audience would want answered:

"Mr. Tocqueville, where are you from?"
"How was it that you happened to come to America?"
"Who paid for your trip?"
"What aspects of your background helped you to be so insightful at such
a young age?"
"Is there democracy in America?"[63]

If Tocqueville, Jefferson, or Thucydides were asked to testify before a congressional committee or speak to the National Press Club, C-SPAN would
be there. The events would be covered gavel-to-gavel, with no sound bites
or editing. It would almost be like being there. The town hall of earlier years
would be re-created through technology.

C-SPAN now provides the mass public with access to the arguments that
guide our public policy and records those arguments for posterity. For anyone who is truly committed to a democracy and who has faith in the public,
it is hard to argue that we as a society are not better off because of C-SPAN.
Let the cameras roll.

Notes

Prologue

1. Lance Bennett, *News: The Politics of Illusion* (New York: Longman, 1988), p. xii.
2. In 1961, 57 percent of American adults classified newspaper as their most important source of news, 52 percent reported television, 34 percent indicated radio, and 9 percent reported magazines (multiple responses permitted). By the 1970s, television became the primary source for two-thirds of Americans, newspapers dropped to less than 50 percent, and radio decreased to 20 percent. *American's Watching: Public Attitudes toward Television 1993* (New York: Network Television Association and National Association of Broadcasters, 1993) pp. 29, 31. (Reported in Harold Stanley and Richard Niemi, *Vital Statistics on American Politics*, 4th ed. [Washington, D.C.: Congressional Quarterly Press, 1994] p. 74.)
3. Ibid.
4. See Ken Auletta, *Three Blind Mice: How the TV Networks Lost Their Way* (New York: Vantage Books, 1992).
5. Quoted in Joe S. Foote, *Television Access and Political Power* (New York: Praeger, 1990), p. 61.

6. Tom Rosenstiel, *Strange Bedfellows: How Television and the Presidential Candidates Changed American Politics, 1992* (New York: Hyperion, 1993), pp. 346–47.

7. Quoted in Merrill Brown, "C-SPAN, Almost Ten Years Old, Continues to Broaden Its Vision," *Channels of Communication* 8, no. 10 (November 1988): 16.

8. Calculated from National Election Studies, reported in Stanley and Niemi, *Vital Statistics*, p. 73.

9. Walter Goodman, "What's Bad for Politics Is Great for Television," *New York Times*, November 27, 1994, p. H33.

10. Don Kowet, "TV Coverage Confronts the Soaring Snooze-Factor," *Washington Times*, July 21, 1988, p. A4.

11. See Denis Rutkus, "Political Broadcasting as a Political Forum," *Congressional Research Service Report*, 92-108-GOV, January 2, 1992, pp. 201–2.

12. Robert Cirino, *Don't Blame the People* (New York: Vintage Books, 1971), p. 2.

13. See Thomas Patterson and Robert McClure, *The Unseeing Eye: The Myth of Television Power in National Elections* (New York: Putnam, 1976). In *Out of Order* (New York: Alfred A. Knopf, 1993), Thomas Patterson points out that the pattern has not changed, only become more cynical.

14. Tim Crouse quoted in Lewis Wolfson, *The Untapped Power of the Press* (New York: Praeger, 1985), p. 106.

15. Quoted in ibid., p. 35.

16. Ibid., p. 44.

17. Thomas Jefferson quoted in Charles J. Sanders, "Reason to Keep Cameras Rolling in Courtrooms," *New York Law Journal*, February 24, 1995, p. 5.

18. Quoted in Suzy Platt, *Respectfully Quoted* (Washington, D.C.: Government Printing Office, 1989), p. 295.

19. Norman Ornstein, Thomas Mann, and Michael Malbin, *Vital Statistics on Congress, 1993–1994* (Washington, D.C.: Congressional Quarterly Press, 1994), pp. 115, 120.

20. Linda Ellerbee, *And So It Goes: Adventures in Television* (New York: G. P. Putnam's Sons, 1986), pp. 64–65.

21. Representative Barber Conable (R-NY) quoted in Stephen Bates, *The Media and Congress* (Columbus, Ohio: Publishing Horizons,, 1987), p. 23.

22. David Obey (D-WI) quoted in ibid., p. 22.

23. Quoted in *New York Times*, February 23, 1958, p. 46.

24. John F. Kennedy, *Public Papers of the Presidents of the United States* Washington, D.C.: Government Printing Office, 1963), p. 376 (presidential press conference, May 9, 1962).

25. Quoted in Richard Armstrong, "Gutter Politics in the Global Village," *National Review*, April 20, 1984, p. 32.

26. Alan Balutis, "Congress and the President and the Press," *Journalism Quarterly*,

Fall 1976, pp. 509–15.

27. Jacqueline Adams quoted in Stephen Bates, *The Media and Congress* (Columbus, Ohio: Publishing Horizons, 1987), p. 43.

28. See Timothy Cook, *Making Laws and Making News* (Washington, D.C.: Brookings Institution, 1989), p. 60, and Stephen Hess, *Live from Capitol Hill* (Washington, D.C.: Brookings Institution, 1991), p. 105.

29. Thomas Whiteside, "Onward and Upward with the Arts: Cable–1," *New Yorker*, May 20, 1985, p. 45.

30. Matt Stump and Harry Jessell, "Cable: The First Forty Years," *Broadcasting*, November 21, 1988, p. 35.

31. Whiteside, "Onward and Upward with the Arts: Cable–1," p. 45.

32. Thomas Southwick, "Passages," *Cable World*, March 27, 1995.

33. Ibid.

34. Stump and Jessell, "Cable: The First Forty Years," p. 35.

35. Ibid.

36. Thomas Whiteside, "Onward and Upward with the Arts: Cable–2," *New Yorker*, May 27, 1985, p. 43.

37. Hank Whittemore, *CNN: The Inside Story* (New York: Little Brown, 1990), pp. 20–22; Caroline Ely, "Three for C-SPAN," *C-SPAN Update*, December 2, 1985, p. 9.

38. Whiteside, "Onward and Upward with the Arts: Cable–1," p. 51.

39. Whittemore, *CNN: The Inside Story*, p. 31.

40. Stump and Jessell, "Cable: The First Forty Years," p. 35.

41. Stanley and Niemi, *Vital Statistics*, p. 63.

42. Auletta, *Three Blind Mice*, p. 560.

43. Stanley and Niemi, *Vital Statistics*, p. 63.

44. National Election Study, 1976 (database online), Center for Political Studies, University of Michigan.

Chapter 1

1. *Congressional Record*, March 19, 1979, p. H-5411.

2. Jana Fay, interview with the authors, tape recording, C-SPAN offices, Washington, D.C., January 13, 1995.

3. Caroline Ely, "Timing Is Everything," *C-SPAN Update*, July 15, 1985, p. 2.

4. Fay interview, January 13, 1995.

5. Brian Lamb, interview with the authors, tape recording, C-SPAN offices, Washington, D.C., March 22, 1995.

6. Colman McCarthy, interview with James Lardner, 1994, from Lardner interview notes, C-SPAN files, Washington, D.C., 1994.

7. Henry Goldberg, interview with James Lardner, 1994, from Lardner interview

notes.

8. Tom Wheeler, interview with James Lardner, 1994, from Lardner interview notes.

9. Quoted in Michael Leahy, "Brian Lamb and C-SPAN," xerox draft case study, 1994, Vanderbilt University, Nashville, Tennessee.

10. Lamb quoted in Kathleen Hillenmeyer, "From Congress to Your Living Room," *St. Anthony Messenger*, April 1995, p. 37.

11. Lamb quoted in James Lardner, "The Anti-Network," *New Yorker*, March 14, 1994, p. 51.

12. Brian Lamb, interview with the authors, C-SPAN offices, Washington, D.C., October 15, 1994.

13. Lamb quoted in Thomas J. Meyer, "No Sound Bites Here," *New York Times Magazine*, March 15, 1992, p. 46.

14. Lamb quoted in Valerie Haddon, "C-SPAN: The Unblinking Eye of America," *Washington Times*, September 21, 1982, p. 1B.

15. Lamb quoted in Meyer, "No Sound Bites Here," p. 46.

16. Paul H. Weaver, "Is Television News Biased?," *Public Interest* 27 (1972): 69.

17. Quoted in Brooke Gladstone, "Cable's Own Public Television Network," *Channels*, September/October 1984, p. 34.

18. Quoted in Connaught Marshner, "Brian Lamb on the C-SPAN TV," *Conservative Digest*, January 1987, p. 82.

19. Lamb interview, October 15, 1994.

20. Ibid.

21. Quoted in Cox Wire Service, March 16, 1994.

22. Quoted in Marilyn Duff, "C-SPAN: The Way TV News Should Be," *Human Events*, November 1994, p. 5.

23. Sharon Geltner, "Brian Lamb: The Man Behind C-SPAN II," *Saturday Evening Post* 258 (January-February 1986): 46.

24. Michael Leahy, "Brian Lamb and C-SPAN," Vanderbilt University, unpublished case study, p. 5.

25. See Caroline Ely, "The Meeting of Congress and Cable," *C-SPAN Update*, December 9, 1985, p. 3.

26. Evans quoted in Caroline Ely, "Strength of Belief," *C-SPAN Update*, September 9, 1985, p. 2.

27. John Evans, interview with the authors, C-SPAN offices, Washington, D.C., June 27, 1995.

28. Since Ronald Garay's book-length discussion of allowing cameras into the House covers the process in great detail, our discussion will cover simply the high points, with an emphasis on issues related to C-SPAN. For more detail on the deliberations in the House, see Ronald Garay, *Congressional Television: A Legislative History* (Westport, Conn.: Greenwood Press, 1984).

29. Orrin E. Dunlop Jr., "Reading Some Radio Mail: Radio Gets Taste of Congress,"

New York Times, December 11, 1932, p. 6.

30. Edward W. Chester, *Radio, Television, and American Politics* (New York: Sheed and Ward, 1969), p. 62.

31. Casey Peters, "The Cable Satellite Public Affairs Network: C-SPAN in the First Decade of Congressional Television" (master's thesis, UCLA, 1992), p. 17.

32. U.S. Congress, Joint Committee on Congressional Operations, *Congress and Mass Communications: Hearings*, 93rd Cong., 2d sess., 1974, p. 2.

33. Caroline Ely, "The Odd Couple," *C-SPAN Update*, June 24, 1985, p. 2.

34. See Terry Hall, *Congress on the Tube: Who's Watching?* (Norman, Okla.: Bureau of Government Research, 1980), p. 4, and Garay, *Congressional Television*.

35. Michael Robinson, "Three Faces of Congressional Media," in Thomas Mann and Norman Ornstein, *The New Congress* (Washington, D.C.: American Enterprise Institute, 1981), p. 67.

36. Tony Coehlo, interview with James Lardner, 1994, from Lardner interview notes.

37. Ibid.

38. Gillis Long, "Television in the House of Representatives," *Capitol Studies* 6 (Spring 1978): 8.

39. Van Deerlin quoted in Richard Krolik, "Everything You Wanted to Know about C-SPAN," *Television Quarterly* 25, no. 4 (1992): 93.

40. *Congressional Record*, October 27, 1977, p. H-35433.

41. Leahy, "Brian Lamb and C-SPAN," p. 8.

42. Lardner, "The Anti-Network," p. 53.

43. Ely, "The Meeting of Congress and Cable," p. 3.

44. Quoted in ibid.

45. Quoted in Ely, "The Odd Couple," p. 3.

46. Ely, "The Meeting of Congress and Cable," p. 2.

47. Quoted in Mary Holley, "Two Early Proponents Recall the Beginning of House TV," *C-SPAN Update*, March 27, 1989, p. 6.

48. Quoted in Lardner, "The Anti-Network," pp. 53–54.

49. Quoted in Ely, "The Meeting of Congress and Cable," p. 2.

50. Holley, "Two Early Proponents," p. 6.

51. See Krolik, "Everything You Wanted to Know about C-SPAN," p. 93.

52. "Cable Keeps a Promise," *Cablevision*, September 20, 1982.

53. Evans interview, June 27, 1995.

54. Pat Gushman, "C-SPAN's Historic Undertaking," *Cablevision*, February 12, 1979, p. 40.

55. Lamb interview, March 22, 1995.

56. Thomas Southwick, "Hundt Follies," *Cable World*, November 28, 1994.

57. See Caroline Ely, "C-SPAN's General Counsel," *C-SPAN Update*, October 7, 1986, p. 2.

58. James Perry, "Beginning of House Television Marked with Skepticism," *Wall*

Street Journal, March 26, 1979, p. 16.

59. Newt Gingrich, interview with James Lardner, 1994, from Lardner interview notes.

60. Geltner, "Brian Lamb," p. 46.

61. C-SPAN board members, interviews with the authors, C-SPAN offices, Washington, D.C., June 27, 1995.

62. Irwin Arieff, "House TV Gets Mixed Reviews but Cancellation Isn't Likely," *Congressional Quarterly Weekly Report*, March 15, 1980, p. 735.

63. Laurence Zuckerman, "The Raw Material News Network Hits Its Stride," *Columbia Journalism Review* 24 (March-April 1980): 44.

64. Rose quoted in Gladstone, "Cable's Own Public Television Network," p. 35.

65. Quoted in Geltner, "Brian Lamb," p. 46.

66. Lamb quoted in Rich Burhardt, "Can House Copyright TV?," *Roll Call*, November 19, 1981, p. 11.

67. Both quoted in Geltner, "Brian Lamb," p. 46.

68. Zuckerman, "The Raw Material News Network," p. 44.

69. Quoted in Kenneth Adelman, "Real People: On C-SPAN Substance Can Be More Interesting Than Style," *Washingtonian* 28 (December 1992): 32.

70. Gail Rubin, "Then and Now," *C-SPAN Update*, April 1, 1985, p. 6.

71. Fay interview, January 13, 1995.

72. Krolik, "Everything You Wanted to Know about C-SPAN," pp. 94–95.

73. Fay interview, January 13, 1995.

74. Ibid.

75. Brian Lamb, *C-SPAN: America's Town Hall* (Washington, D.C.: Acropolis Books, 1988), p. 147.

76. Dan Feldstein, "Washington's Link to C-SPAN," *C-SPAN Update*, July 8, 1985, p. 3.

77. The first person to run afoul of the chamber rules was an incumbent member. Raphael Musto (D-PA) had just won a special House election and was already campaigning heavily for the 1980 election for a full term in office. Musto, presumably innocent of wrongdoing, used footage from his swearing-in ceremony in a television ad. His opponent in the Democratic primary complained. The negative publicity did not cost him the nomination, and the House did not discipline him ("Musto Campaign Breaks House TV Rule," *Congressional Quarterly Weekly Report*, April 26, 1990, p. 1106).

78. Lamb, *C-SPAN*, p. 259.

79. Ibid.

80. Quoted in Sarah Pekkanen, "C-SPAN Accuses Roth Opponent," *The Hill*, October 19, 1994, p. 1.

81. J. Leonard Reinsch, *Getting Elected* (New York: Hippocrene Books, 1988), p. 324.

82. John Carmody, "The TV Column," *Washington Post*, March 31, 1992, p. B6.

83. John Carmody, "The TV Column," *Washington Post*, April 1, 1992, p. C8.

Chapter 2

1. Jana Fay, interview with the authors, C-SPAN offices, Washington, D.C., March 13, 1995.

2. Caroline Ely, "C-SPAN Made the Map in 1984," *C-SPAN Update*, March 31, 1986, p. 3.

3. Gingrich quoted in Rosemary Harold, "The Twentieth Century Evolution of 'Special Orders' Speeches," *C-SPAN Update*, May 9, 1988, p. 5.

4. O'Neill quoted in Bruce Collins, "Television Is Here to Stay," *C-SPAN-Update*, March 19, 1984, p. 2.

5. Steven V. Roberts, "House Democrats Seeking Better TV Image," *New York Times*, May 10, 1984, p. A29.

6. Mark Starr and Gloria Borger, "Not Ready for Prime Time?," *Newsweek*, May 28, 1984, p. 24.

7. Thomas P. (Tip) O'Neill, *Man of the House* (New York: St. Martin's, 1987), p. 424. O'Neill quoted in *Congressional Record*, May 15, 1984, p. H-12201.

8. *Philadelphia Daily News*, January 15, 1994.

9. Ely, "C-SPAN Made the Map," p. 3.

10. Neil Hickey, "Eye on Democracy," *TV Guide*, March 30, 1985.

11. Robert S. Byrd, *The Senate, 1789–1989: Addresses on the History of the Senate* (Washington, D.C.: Government Printing Office, 1989), p. 27.

12. Ibid.

13. Roy S. Swanstrom, *The United States Senate: 1787–1801*, S. Doc 99-19 (Washington, D.C.: Government Printing Office, 1985), pp. 38–40.

14. Byrd, *The Senate*, p. 28.

15. Like our discussion of the decision to televise House proceedings, our account here will focus on the high points rather than attempt to be comprehensive. Since C-SPAN played a more active role in opening the Senate to television cameras, our discussion will be more detailed than it was for the House. For a more comprehensive discussion of the political process by which the Senate opted to televise its proceedings, see Richard F. Fenno Jr., "The Senate through the Looking Glass: The Debate over Television," *Legislative Studies Quarterly* 14, no. 3 (August 1989): 313–48, and Christopher Maloney, "Lights, Camera, Quorum Call: A Legislative History of Senate Television" (master's thesis, College of William and Mary, 1990).

16. Fenno, "The Senate through the Looking Glass," p. 327.

17. American Enterprise Institute Symposium, "C-SPAN Tenth Anniversary Retrospective," telecast, April 3, 1989.

18. *Congressional Record*, April 14, 1982, p. S-3474.
19. Baker quoted in Kurt Sayenga, "Senate Approves Permanent TV," *C-SPAN Update*, August 11, 1986, p. 1.
20. Steven V. Roberts, "Senators Ponder Value of Letting TV in the Door," *New York Times*, September 16, 1985, p. B6.
21. Baker quoted in Tom Shales, "The Floor Show," *Washington Post*, December 1, 1981, p. B1.
22. See Fenno, "The Senate through the Looking Glass," pp. 320–321.
23. U.S. Congress, Senate, Committee on Rules and Administration, *Television and Radio Coverage of Proceedings in the Senate Chamber: Hearings on S. Res. 20*, 97th Cong., 1st sess. (Washington, D.C.: Government Printing Office, 1981), pp. 4–5.
24. Howard Baker, "We're Losing Political and Historical Treasures," *TV Guide*, July 21, 1984, p. 33.
25. Byrd, *The Senate*.
26. Byrd quoted in Roberts, "Senators Ponder Value," p. B6.
27. Robert Byrd, "The British-American Parliamentary Meetings," *Congressional Record*, June 7, 1990, p. S-7578.
28. *Broadcasting*, August 11, 1992, p. 42.
29. Both quoted in ibid.
30. Mike Michaelson, interview with the authors, tape recording, C-SPAN offices, Washington, D.C., March 22, 1995.
31. Simpson quoted in Linda Greenhouse, "Justices Back Cable Regulation," *New York Times*, June 28, 1987, p. D1.
32. Fenno, "The Senate through the Looking Glass," pp. 325, 334.
33. Russell Long, "Keep the Cameras out of the Senate," *USA Today*, September 18, 1985, p. 10A.
34. U.S. Congress, Senate, Committee on Rules and Administration, *Television and Radio Coverage*, pp. 82–83.
35. C-SPAN transcript, August 1, 1988, reported in *C-SPAN Update*, August 22, 1988.
36. Both quoted in Tom Shales, "As the Hill Turns: C-SPAN's Riveting Mini-Series," *Washington Post*, May 17, 1984, p. E1.
37. Ibid.
38. Both quoted in Roberts, "Senators Ponder Value," p. B6.
39. Quoted in *Broadcasting*, March 3, 1986, p. 36.
40. Fenno, "The Senate through the Looking Glass," p. 319.
41. *Congressional Record*, February 20, 1986, p. S-1460–61.
42. Jacqueline Calmes, "Senate Agrees to Test of Radio, TV Coverage," *Congressional Quarterly Weekly Report*, March 1, 1986, p. 520.
43. *Congressional Record*, June 2, 1986, p. S-12044.

44. Ibid., pp. S-12042–43.

45. Howard Rosenberg, "Senators Answer TV Role Call," *Los Angeles Times*, June 4, 1986, pt. 6, p. 6.

46. *Congressional Record*, June 2, 1986, p. S-12052.

47. Fenno, "The Senate through the Looking Glass," p. 335.

48. *Broadcasting*, July 21, 1986, p. 35.

49. Ibid., p. 36.

50. See W. Mark Crain and Brian L. Goff, *Televised Legislatures: Political Information Technology and Public Choice* (Boston: Kluwer Academic Publishers, 1988).

51. All quoted in *C-SPAN Update*, May 25, 1987.

52. Maloney, "Lights, Camera, Quorum Call," p. 42.

53. Byrd, *The Senate*, p. 719.

54. Susan Swain, interview with Lisa Wechsler, tape recording, April 21, 1995. Notes made available to the authors.

55. Timothy J. Burger, "GOP Rally Sept. 27 Could Attract 300+," *Roll Call*, September 19, 1994.

56. Quoted in the *Congressional Record*, September 24, 1993, p. S-12484.

57. For similar comments, see Elizabeth Newlin Carney, "Squeeze on for C-SPAN," *National Journal* 26, no. 24 (June 11, 1994): 1356.

58. *Congressional Record*, September 22, 1992, p. S-14607.

59. "Hill Enacts Cable TV Law over Vote," *1992 Congressional Quarterly Almanac* (Washington, D.C.: Congressional Quarterly Press, 1993), pp. 171–83.

60. Joe Flint, "Net's Cause an fX," *Variety*, June 1, 1994, p. 16.

61. Jane Hall, "fX Affects C-SPAN's Viewership," *Los Angeles Times*, May 31, 1994, p. B9.

62. Flint, "Net's Cause an fX," p. 1.

63. Lamb quoted in Dusty Saunders, "Angry C-SPAN Viewers Up in Arms over TCI's Scheduling Squeeze Play," *Rocky Mountain News*, June 8, 1994, p. 17d.

64. Lamb interview, October 15, 1994.

65. Lamb quoted in Carney, "Squeeze on for C-SPAN," p. 1356.

66. Ibid.

67. John Carman, "Big Cutback of C-SPAN on Viacom," *San Francisco Chronicle*, May 26, 1993, p. E1.

68. Robert Snyder, "Excuses for Dropping C-SPAN Are Weak," *Waterloo Courier*, May 2, 1994.

69. Elizabeth Kolbert, "Some Cable Systems Are Cutting C-SPAN for Other Channels," *New York Times*, June 20, 1994, pp. A1, D5.

70. Richard Katz, "fX Celebrates While C-SPAN Licks Wounds," *Multichannel News*, June 6, 1994, p. 3.

71. Kolbert, "Some Cable Systems Are Cutting," p. D5.

72. Ibid.

73. Collins quoted in Vincente Pasdeloup, "It's Back to Square One for Cable after Must-Carry Ruling," *Cable World*, July 4, 1994, p. 23.

74. Lardner, "The Anti-Network," p. 50.

75. Lamb quoted in ibid.

76. Greenhouse, "Justices Back Cable Regulation," p. D1.

77. Collins quoted in Joan Biskupic, "Supreme Court Connects Cable to Free Speech Protections of Press," *Washington Post*, June 28, 1994, p. A1.

Chapter 3

1. Lamb interview, March 22, 1995.

2. Jack Frazee, interview with the authors, tape recording, C-SPAN offices, June 27, 1995.

3. Caroline Ely profiled Saeman in *C-SPAN Update*, December 2, 1985, pp. 8, 9. His comment comes from her interview.

4. Ibid. From the beginning, Lamb selected the chairs.

5. Fay interview, January 13, 1995.

6. Lamb quoted in "C-SPAN, Public Affairs Network Goes 24 Hours," *Cablevision*, September 20, 1982, p. 50.

7. The awards were given at the C-SPAN board meeting. "Network Recognizes Super Citizens," *C-SPAN Update*, December 1982, p. 4.

8. Channel capacity continues to plague C-SPAN's efforts to reach viewers. At the outset, Saeman and other board members envisioned a thirty-five-channel world. " '35 Channels to Be the Norm' Say Industry Leaders," *C-SPAN Update*, December 1982, pp. 1, 4.

9. Bob Brewin, "Pulling the Plug on Congress," *Soho News*, March 9, 1982, p. 21.

10. Lamb interview, March 22, 1995.

11. *C-SPAN Update*, December 2, 1985, p. 9.

12. The pamphlet bears the date February 1982, C-SPAN files.

13. Lamb interview, March 22, 1995.

14. Memo from Jana Fay to the authors on budget and finance requests, February 8, 1995. These figures are based on operating expenses.

15. Fay interview, January 13, 1995.

16. Mike Michaelson, "PotPouri," October 10, 1981, pp. 6, 7 (staff minutes). The bulk of the material in this collection of notes and press clippings covers the earliest years of Michaelson's tenure at C-SPAN.

17. Fay interview, January 13, 1995.

18. *C-SPAN Update*, December 2, 1985, p. 9.

19. Lamb interview, March 22, 1995.

20. See "Uniting the Industry," *Cable Television Magazine*, January 1, 1985, p. 21.

Throughout its history, C-SPAN has had to advertise itself to the industry.

21. *C-SPAN Update*, December 2, 1985, p. 8.

22. *C-SPAN Update*, September 10, 1986, p. 3.

23. *C-SPAN Update*, December 2, 1985, pp. 1, 4, 8.

24. *Weekly Report*, September 24, 1984. *Weekly Report* is C-SPAN's in-house newsletter.

25. Frazee interview. Frazee would later head the board and would go on to the chairmanship of Centel. Kennedy, who returned to C-SPAN full-time in 1987, is now one of two executive vice-presidents.

26. "Executive Summary," Five Year Plan, C-SPAN files.

27. Fay budget memo, February 8, 1995.

28. "Projected Cash Flows," Five Year Plan.

29. In addition to providing updated schedule information, the *C-SPAN Update* was also to play a role in educational community marketing. Much of the marketing effort, however, would be directed to increasing an awareness within the cable industry.

30. Brian Lamb, interview with the authors, tape recording, C-SPAN offices, Washington, D.C., June 29, 1995.

31. The six telephone companies had an uneasy relationship with cable leaders. Recent legislation may make them major players in the industry.

32. Summary of Executive Committee Minutes, New York, September 27, 1989, C-SPAN files.

33. Steve Knoll, "C-SPAN Strives to Expand," *New York Times*, April 8, 1984, sec. 2, p. 28.

34. *Cablevision*, October 8, 1984, p. 28; Richard Fleeson, interview with John Sullivan, tape recording, C-SPAN offices, March 11, 1995.

35. Frazee interview.

36. The first five-year plan would be constantly updated and revised throughout the next several years in response both to crises and to growth.

37. Lamb interviews, March 22 and June 29, 1995. Almost from the beginning, the board had urged Lamb to put C-SPAN on a business footing.

38. Fay interview, January 13, 1995.

39. Jana Fay, Pam Fleming, and Barry Katz, group interview with the authors, tape recording, C-SPAN offices, March 13, 1995.

40. Paul FitzPatrick to Cable Operators, Washington, D.C., September 13, 1985, C-SPAN files.

41. Fay interview, January 13, 1995.

42. Ibid.

43. Ibid.; Bruce Collins, interview with the authors, C-SPAN offices, December 13, 1994.

44. Frazee interview.

45. A management "core group" consisting of Brian Lockman, Bruce Collins, Mike Michaelson, Brian Lamb, and Paul FitzPatrick was announced in the *Weekly Report*, September 16, 1985. Paul FitzPatrick to Brian Lamb, September 10, 1985, C-SPAN files.

46. Quoted in Mary Holley, "C-SPAN President Leaves Network to Embark on New Venture," *C-SPAN Update*, June 29, 1987, pp. 11, 12.

47. The list of items that needed attention is a further suggestion of just how rapidly the organization had grown.

48. Collins interview.

49. Wheeler quoted in Sharon Geltner, "Matching Set: Brian Lamb and Cable's C-SPAN," *Washington Journalism Review*, September 1984, p. 30; Lamb interview, June 29, 1995.

50. Geltner, "Matching Set," p. 30.

51. Geltner, "Matching Set," seems to be the only publication that mentions this idea. We have been unable to locate such a story elsewhere.

52. "Cable Castings," *Broadcasting*, in *Weekly Report*, April 29, 1985.

53. Collins interview.

54. "Guidelines for C-SPAN Supervisors," May 14, 1985, C-SPAN files.

55. Frazee interview. Looking back on the incident, Frazee believes that C-SPAN's successes may have motivated some employees to ask for better benefits.

56. Collins interview.

57. Brian Lamb to "Dear C-SPAN Colleague," May 14, 1985, C-SPAN files. Based on informal talks with current employees, the "Lamb lunches" are still important for morale.

58. "Marketing Strategy," Five Year Plan.

59. Ibid.

60. Frazee interview.

61. Lamb quoted in Michael O'Daniel, "Cable Has a Secret," *Emmy Magazine*, November/December 1983, p. 67.

62. Memo to John Evans from Brian Gruber, Director of Marketing, March 28, 1985, C-SPAN files.

63. Bill Brown to Kevin Rorke, CableVision of Central Florida, May 24, 1985, C-SPAN files.

64. It would be one thing to have full coverage and another to have viewer awareness. Officials at C-SPAN worked also to get local affiliates to use the presence of C-SPAN on their systems as a marketing device as ATC had done, often with little success.

65. Pam Fleming, interview with the authors, tape recording, C-SPAN offices, January 9, 1995.

66. Letter to the editor, *C-SPAN Update*, December 2, 1985, p. 10.

67. *Weekly Report*, December 15, 1986.

68. Brian Lamb to board members, May 16, 1986, C-SPAN files.
69. *Weekly Report*, May 18, 1986, brought the news of FitzPatrick's impending departure. FitzPatrick is now president of the News Talk Television channel.
70. Lamb interview, March 22, 1995.
71. Since this negotiation, C-SPAN has continued to use protected transponder sites.
72. Despite this proviso, C-SPAN has found some systems splitting time between the two channels. The latest occurrence came about as a response to "must carry" legislation in the 1990s.
73. John Wolfe, "C-SPAN's Persuasive Powers Face a Key Test," *Cablevision*, May 19, 1986, p. 30.
74. Quoted in ibid.
75. In 1995, there are approximately twice as many homes with C-SPAN as there are with C-SPAN II.
76. Wolfe, "C-SPAN's Persuasive Powers," p. 30.
77. *Weekly Report*, June 29, 1987, announced Kennedy's return.
78. Introduction to Lamb, *C-SPAN*, p. xv. Among other C-SPAN watchers whom the writers attempted to reach were Ivan Lendl and former President Richard Nixon. The writers lost an interview with actor Peter Falk because of technical problems.
79. Other reporters find C-SPAN every bit as useful. Jack Nelson, Washington bureau chief of the *Los Angeles Times*, was also profiled. He appears occasionally on the *Washington Journal*.
80. "C-SPAN Tenth Anniversary Retrospective," April 3, 1989, C-SPAN telecast, Purdue University Public Affairs Video Archives; memo to Brian Lamb from the "10th anniversary group," September 14, 1988, C-SPAN files.
81. Frazee interview.
82. At its meeting in June 1995, the board discussed reducing the size because of the spate of mergers and acquisitions in the industry.
83. Lamb interviews, March 22 and June 29, 1995.
84. Hostetter is the only board chair ever to be featured on *An American Profile*.
85. Testimony of Brian Lamb before the Subcommittee on Telecommunications and Finance, Committee on Energy and Commerce, June 27, 1991, C-SPAN files.
86. "Olympics Pre-emption Idea Earns Divided Reaction within Industry," *Cable World*, March 25, 1991, p. 4.
87. Marty Lafferty to Amos Hostetter, New York, New York, May 1, 1991, C-SPAN files.
88. "Closed Circuit," *Broadcasting*, May 20, 1991, p. 6; Peggy Ziegler, "C-SPAN Puts Foot Down on Olympics Pre-emption," *Cable World*, May 20, 1991, p. 7.
89. *Cable World*, May 20, 1991, p. 7.
90. Fay interview, January 13, 1995.
91. Geltner, "Matching Set," p. 30.
92. Brett Betsill, interview with John Sullivan, tape recording, C-SPAN offices,

March 16, 1995.

93. Collins interview.

94. Fleming interview, January 9, 1995, and Fay interview, January 13, 1995.

95. Collins interview.

96. Fleming interview, January 9, 1995.

97. Rubin's account appears in *C-SPAN Update*, April 1, 1985, p. 6.

98. Fleming interview, January 9, 1995.

99. Collins interview.

100. "New Employees," *C-SPAN Update*, April 15, 1985, p. 2.

101. Fay interview, January 13, 1995, and Fleming interview, January 9, 1995.

102. Ibid.

103. The authors have been unable to locate an earlier document. Judging from the comments of early employees, however, many of the elements were there from the beginning.

104. The following quotations are from Lamb's letter to Speaker Tip O'Neill, March 5, 1982, C-SPAN files.

105. This concern for balance would lead to record keeping in all departments, although Craig Brownstein, who now tracks such issues for the assignment desk, claims the political categories can get fuzzy. He tracks both philosophies and issues covered and is now working on a way to factor in prime time versus non–prime time.

106. Bruce Collins to All Personnel, n.d., C-SPAN files. The formal C-SPAN Mission Statement appears in the preface to this book.

107. Like most other C-SPAN activities, the manual was written by committee.

108. This document was not finalized until the early 1990s. The employee evaluation system for which this document serves as a guide was put into place in 1993.

109. The incident occurred during Lamb's talk with new interns on January 11, 1995 (authors' notes).

110. Rich Fahle, phone interview with John Sullivan, May 3, 1995.

111. Lew Ketcham, who produced the C-SPAN special on the debates, was one of the few who did not attend. He recalled one interesting incident arising from the clips he selected to use on the program. At the end of the telecast, Brian Lamb praised Ketcham on the air, but he said, "They're some things we'll have to talk about." Lamb was upset because Ketcham had shown Lamb emerging from a bus at one of the debate sites, to the applause of the crowd. C-SPAN is a "no stars" operation, and Lamb wanted to keep it that way. Lew Ketcham, interview with John Sullivan, tape recording, C-SPAN offices, February 7, 1995.

112. Ibid.

113. Such warnings began in the *Weekly Report* in the early 1980s and are repeated during each campaign season.

114. Data is available by age, race, gender, and job category. "C-SPAN EEO

Employment Statistics," April 21, 1995, C-SPAN files.

115. *C-SPAN, The First Ten Years: A Pictorial History of America's Network, 1979–1989* (Washington, D.C.: National Cable Satellite Corporation, 1989).

116. Lamb spoke before a group of University of Virginia students on April 21, 1995, at the C-SPAN offices.

117. Swain interview, April 21, 1995.

118. Lamb quoted in Rex Polier, "See Congress Run," *Western's World*, September 1983, pp. 72, 77.

119. Lamb interview, March 22, 1995.

120. Fleming interview, January 9, 1995.

121. Susan Swain, interview with the authors, tape recording, C-SPAN offices, Washington, D.C., May 25, 1995.

122. Fay interview, January 13, 1995.

123. Jobs are always posted in-house first. Because experience with the mission is so crucial, opportunities to move into vacancies are good.

124. Betsill interview.

125. Ketcham interview.

126. Steve Scully, interview with John Sullivan, tape recording, C-SPAN offices, February 8, 1995.

127. Connie Brod, interview with John Sullivan, tape recording, C-SPAN offices, March 15, 1995.

128. Fay budget memo, February 8, 1995. Bus maintenance for FY '94 was $192,069.

129. "Industrial-Strength Television," *Los Angeles Times*, December 5, 1993, sec. M, p. 4.

130. Terry Murphy, interview with John Sullivan, tape recording, C-SPAN offices, April 20, 1995.

131. Ibid.

Chapter 4

1. Marcia Gelbart, "Dems Learning to Use House Special Orders, but GOP Master Genre," *The Hill*, March 8, 1995, p. 22.

2. "C-SPAN Programming Practices," Focus Group Report, June 28–July 4, 1989, C-SPAN files.

3. Speech by Brian Lamb to the C-SPAN Seminar for Professors, January 9, 1995, Washington, D.C.

4. Focus Group Report.

5. The final report of the focus group included a summary of findings. The reports of each team were attached. Lamb's report was the only one not produced by a group ("BPL" Focus Group).

6. Quoted in Adelman, "Real People," p. 36.

7. Letters from Brian Lamb to Senator Robert Dole, Washington, D.C., November 21, 1994, and Brian Lamb to Speaker Newt Gingrich, Washington, D.C., November 21, 1994, C-SPAN files.

8. Howard Kurtz, "Under Speaker's Order, C-SPAN Pans the House," *Washington Post*, March 29, 1995, sec. A, p. 21; "An Open Letter from C-SPAN to the 104th Congress" ran as an advertisement in various trade publications.

9. John Splaine to Susan Swain, memo on "Directing Hearings," November 2, 1994, C-SPAN files.

10. Daniel Schorr, statement from the floor, forum on "Media Coverage of the Clinton Administration," sponsored by the Twentieth Century Fund at the National Press Club, June 28, 1994, C-SPAN telecast, Purdue University Public Affairs Video Archives.

11. *C-SPAN, The First Ten Years*, pp. 4, 12.

12. Michaelson interview, March 22, 1995; Lou Prato, "Politics in the Raw," *Washington Journalism Review*, September 1992, 36.

13. Garney Gary, interview with John Sullivan, tape recording, C-SPAN offices, March 17, 1995.

14. Ellen Schweiger, interview with John Sullivan, tape recording, C-SPAN offices, March 17, 1995.

15. "House Committees Bar C-SPAN," *News Media and the Law*, Summer 1992, p. 9.

16. C-SPAN in-house documents, "Directing Hearings" and "Covering Two Camera Events," n.d., C-SPAN files.

17. "Top of the Week," *Broadcasting*, October 21, 1991, pp. 23, 24.

18. Fay budget memo, February 8, 1995. C-SPAN's operating expense budget for that year was $15,495,240.

19. Peter Ross Range, "C-SPAN: The Little Network That Could," *TV Guide*, July 18–24, 1992, p. 15.

20. C-SPAN live viewer call-in program, "House Television Coverage with Representative Peter Hoekstra," December 15, 1994, C-SPAN telecast, Purdue University Public Affairs Video Archives.

21. Jack Kemp, on C-SPAN's *An American Profile*, December 25, 1988, C-SPAN telecast, Purdue University Public Affairs Video Archives.

22. Ketcham interview.

23. Benjamin Bradlee, on C-SPAN's *An American Profile*, August 26, 1991, C-SPAN telecast, Purdue University Public Affairs Video Archives.

24. Lamb quoted in Diane Wertz, "C-SPAN: The Comfort Zone," *Newsday*, March 20, 1992, p. 73.

25. Focus Group Report.

26. Core Product Committee, "Mid-Course Report for C-SPAN 2000," n.d., C-SPAN files.

27. Mike Michaelson, "Program Ideas for C-SPAN's Presentation of Congress:

C-SPAN 2000, Strategy Planning for the Future," March 24, 1992, C-SPAN files.

28. Core Product Committee, "Mid-Course Report."

29. Lee Winfrey, "The Cable Watch," *Philadephia Inquirer*, May 26, 1989, sec. C, p. 8.

30. Lamb speech, January 9, 1995.

31. January 1st Groups: "The House of Representatives," November 8, 1994, and "Senate Group Recommendations," n.d., C-SPAN files. The third group, which studied coverage of congressional events, was led by Connie Brod, who now heads the *Washington Journal* team.

32. This and the following quotations are from "Senate Group Recommendations."

33. By June 1995, a Senate coverage team was in place and had begun to implement some of the less radical suggestions.

34. Lamb quoted in Wertz, "C-SPAN," p. 73.

35. The following quotations all come from a taped interview with Jim Mills conducted on Capitol Hill by John Sullivan, March 14, 1995.

36. Fay interview, January 13, 1995.

37. The C-SPAN version of this story appears in Lamb, *C-SPAN*, pp. 343–46.

38. Caroline Ely, "The Biggest Role of the Dice," *C-SPAN Update*, February 24, 1986, p. 8.

39. "Calls Create Humorous Memories," *C-SPAN Update*, March 27, 1989, p. 3; Betsill interview.

40. Ibid.; Brod interview.

41. "Dutch Couple Spends Sleepless Night Answering Misdialed Calls," *C-SPAN Update*, August 1, 1988, p. 1.

42. C-SPAN Mission Statement.

43. C-SPAN Press Release; Brian Lamb, interview with John Sullivan, tape recording, C-SPAN offices, February 3, 1995.

44. "Philosophy, Guidelines and Procedures" p. 7, C-SPAN files.

45. Theodore Parker, "The Political Destination of America and the Signs of the Times," excerpted in *Theodore Parker: An Anthology*, ed. Henry Steele Commager (Boston: Beacon Press, 1960), p. 176.

46. Lamb interview, February 3, 1995.

47. "Ten Features That Make the C-SPAN Call-In Different," n.d., C-SPAN files. This document is usually distributed at C-SPAN Seminars for Professors. Given its thrust, it may well have been prepared especially for that purpose.

48. Splaine's connection with C-SPAN is an interesting one. He first came to Lamb's notice because he would call Lamb directly, always off the air, to point out what he saw that betrayed the C-SPAN mission. He was particularly concerned about the questioning style and about the potential bias of camera angles. He now conducts in-house seminars on both aspects and has become a trusted adviser. He played an important part in C-SPAN's coverage of the Lincoln-Douglas debates.

Just as he prepared a monograph for teachers and viewers on the debates, he also prepared one to accompany the network's coverage of "Campaign '96."

49. Lamb interview, February 3, 1995; Marshner, "Brian Lamb on the C-SPAN TV," p. 87.

50. Lamb interview, February 3, 1995.

51. Tim Russert, "Round Table on Sunday Morning Talk Shows," October 6, 1992, C-SPAN telecast, Purdue University Public Affairs Video Archives.

52. Lamb interview, February 3, 1995.

53. Collins interview.

54. Fay interview, January 13, 1995.

55. Caller comment, *Joel Spivak Show*, Michael Harrison, guest host, C-SPAN simulcast, December 22, 1994, Purdue University Public Affairs Video Archives.

56. Collins interview.

57. Comments such as these are compiled in daily and weekly reports by the C-SPAN viewer services department. These examples come from October 19, 1994, and January 19, 25, 1995. For Lamb's appreciation of the work of that department, see *Washington Journal*, March 10, 1995, C-SPAN telecast, Purdue University Public Affairs Video Archives.

58. Michael Kinsley, "The Intellectual Free Lunch," *New Yorker*, February 6, 1995, pp. 4, 5.

59. Lamb interview, February 3, 1995.

60. C-SPAN viewer services department, "Year-end Report for 1994," C-SPAN files.

61. Accounts in *C-SPAN Update* and *Weekly Report* differ slightly in detail.

62. These words were originally applied to talk radio in 1987 by Murray Levin, *Talk Radio and the American Dream* (Lexington, Mass.: Lexington Books, D. C. Heath and Co., 1987), p. 19. They are still applicable today.

63. Howard Kurtz, "Radio Daze," *Washington Post*, October 24, 1994, sec. B, pp. 3, 4.

64. "C-SPAN 2000: Core Product Committee," n.d. This one-sheet document was probably used for a luncheon meeting with board members. The board supplied guidance for the development of C-SPAN 2000 and supplemented the efforts of in-house staff members, who did much of the planning with the help of outside experts.

65. Swain and Murphy's announcement was contained in a memo to all employees, October 31, 1994, C-SPAN files. In a taped interview, on April 20, 1995, Terry Murphy indicated that the idea to revamp the call-in format came before, and was separate from, the January 1st Groups, which evaluated congressional coverage.

66. Swain and Murphy memo.

67. On rare occasions, outside focus groups have been used, but most programming evaluation is done by in-house teams selected from across the company and by those in charge of a particular show. For example, during the developmental phase of the *Washington Journal*, producers and staff held brief discussions after each

program.

68. Connie Brod, "To: All Hosts," April 26, 1995, C-SPAN files.

69. Talk by Terry Murphy, "C-SPAN Week," May 23, 1995, C-SPAN offices, Washington, D.C.; Brod interview. C-SPAN Week is an employee-orientation program.

70. Viewer services department comment line, January 23, 1995. Lamb takes any hint of bias seriously. On June 29, 1995, he mentioned several such complaints to the authors, noting that he was having difficulty finding a good bust of Andrew Jackson to use on the set.

71. Brod, "To: All Hosts." The following quotations are taken from this memo unless otherwise noted.

72. Brod interview. Calls of this sort seem most often to be directed to Brian Lamb, who has become a lightning rod for viewers.

73. By the summer of 1995, clips from the C-SPAN school bus had also become part of the *Journal*. Initially they were used both as promos and as bridges between segments of the show. The decision to add a second bus with microwave capabilities should provide an out-of-the-beltway look for the show.

74. *Washington Journal*, March 27, 1995, C-SPAN telecast, Purdue University Public Affairs Video Archives.

75. Swain and Murphy memo.

76. Betsill interview.

77. See chapter 6, where such surveys are discussed in detail. On June 29, 1995, Lamb told the authors that the surveys were conducted primarily for the board. Each one, of course, is followed by a news release.

Chapter 5

1. C-SPAN Mission Statement.

2. Kenneth Boulding, *The Image* (Ann Arbor: University of Michigan Press, 1956), p. 134.

3. For an account of Agnew's attacks on the press, see Joseph C. Spear, *Presidents and the Press: The Nixon Legacy* (Cambridge, Mass.: MIT Press, 1984), pp. 113–21. Agnew's Des Moines speech, November 13, 1969, can be found in the *New York Times* of the following day, p. 24. David Stone's *Nixon and the Politics of Public Television* (New York: Garland Publishing, 1985) provides an interesting insight into the Nixon administration's role in shaping telecommunications policy.

4. *Washington Journal*, May 15, 1995, C-SPAN telecast, Purdue University Public Affairs Video Archives.

5. Industry sources estimate that a thirty-second commercial on a popular show like *Home Improvement* can cost $350,000 per spot. See *Washington Post*, April

19, 1995, sec. A, p. 20, for a discussion of preempting such programming for political broadcasts.

6. Summary of board minutes, December 13, 1979, C-SPAN files.

7. Summary of executive committee minutes, September 10, 1980, C-SPAN files.

8. Krolik, "Everything You Wanted to Know about C-SPAN," pp. 94–95; "C-SPAN: Carving Out a New Programming Niche," *Broadcasting*, November 3, 1980, pp. 48, 49, 52.

9. *Broadcasting*, November 3, 1980, pp. 48, 49.

10. Ibid., p. 49.

11. Summary of board minutes, December 11, 1980, C-SPAN files.

12. *Broadcasting*, November 3, 1980, 52.

13. Bill Carter, "C-SPAN Fills Schedule with Public Service TV," *Baltimore Sun*, August 20, 1984, sec. C, p. 4.

14. "Fine Tuning: C-SPAN Plans Alternative Coverage of 1984 Election," *Cable-marketing*, January 1984, p. 56.

15. *Weekly Report*, December 5, 1983; *Cable Marketing*, January 1984, p. 56.

16. *Weekly Report*, February 27, 1984.

17. David Crook, "The Thinking Man's Channel, *Los Angeles Times*, March 22, 1984, sec. 6, pp. 1, 10.

18. *Weekly Report*, February 27, 1984.

19. Various dollar amounts floated around in the press during the campaign season.

20. Whittemore, *CNN: The Inside Story*, p. 263.

21. John Corry, "Why Convention Coverage Should Be Gavel-to-Gavel," *New York Times*, July 8, 1984, sec. 2, p. 1.

22. *Weekly Report*, April 4, 1988.

23. Caroline Ely, "C-SPAN Insider: Directing and Shooting a Convention Live, Gavel-to-Gavel," *C-SPAN Update*, August 13, 1984, p. 3.

24. Even as late as the 1992 conventions, C-SPAN staff still had similar concerns. Most events covered in Washington, D.C., require no more than two crews. In 1996, C-SPAN had only seven permanent crews.

25. Carolyn Lamson, "C-SPAN Insider: Post Conventional Wisdom," *C-SPAN Update*, July 30, 1984, p. 3.

26. Brod interview.

27. Caroline Ely, "Thanks: C-SPAN Had Crucial Support in San Francisco," *C-SPAN Update*, August 6, 1984, pp. 1, 3; Caroline Ely, "GOP Rides into Dallas," *C-SPAN Update*, June 11, 1984, p. 1.

28. Lamson, "C-SPAN Insider," p. 3.

29. Ely, "C-SPAN Insider," p. 3.

30. Ibid.

31. Leslie Gransto, "Following the Voters Instead of the Candidates," *New York Times*, September 23, 1984, sec. 2, p. 28.

32. David Bianculli, "Convention: Cuomo, Carter, C-SPAN Star," *Philadelphia Inquirer*, July 17, 1984, sec. C, p. 6.

33. Michael Dougan, "Television," *San Francisco Examiner*, July 22, 1984, sec. H, p. 6.

34. Jim Finnegan, "C-SPAN in San Francisco," *Manchester (N.H.) Union Leader*, July 16, 1984, p. 21.

35. All three letters appeared in "The Mail" section of *C-SPAN Update*, September 10, 1984, p. 7. It appears that, over time, the *Update* editors tried to balance the mail they published. Another letter in the same issue complained of "complicity" on the part of hosts. One reader said bluntly: "Canceling—it's too much. Can't afford cable bias—will read fiction."

36. Raymond W. Smock to Mike Michaelson, U.S. House of Representatives Office for the Bicentennial, July 23, 1984, C-SPAN files.

37. Carter, "C-SPAN Fills Schedule," p. 4.

38. Bob Brewin, "Democratic Convention: Public TV Blew It," *Village Voice*, August 7, 1984, pp. 28, 29.

39. Tom Shales, "The Networks' Hit and Run Play," *Washington Post*, July 16, 1984, sec. B, p. 3.

40. Brewin, "Democratic Convention," pp. 28, 29.

41. Susan Swain gave this figure to the *Baltimore Sun*, August 20, 1984, sec. C, p. 4. In June 1984, C-SPAN was carried by fifteen hundred cable systems, reaching a potential eighteen million households.

42. Fact sheet, *Grassroots '84*, n.d., C-SPAN files. *Grassroots '84* was as much a marketing campaign for cable as it was a way to cover the election.

43. The cities visited and the sponsoring cable companies were the following: Mission Viejo, Calif. (Times-Mirror); Santa Ana, Calif. (Group W); Monterey, Calif. (Western); Seattle, Wash. (Group W); Denver, Colo. (ATC); Tulsa, Okla. (United); New Orleans, La. (Cox); Jacksonville, Fla. (Continental); Traverse City, Mich. (Centel); South Bend, Ind. (Heritage); Harrisburg, Pa. (Sammons); New Haven, Conn. (Storer); Westchester County, N.Y. (United Artists); Cleveland, Ohio (Viacom).

44. *Weekly Report*, August 13, 20, 1984; Lurette Arrowsmith, "A Road Trip across the United States: *Grassroot '84* Starts August 31," *C-SPAN Update*, July 30, 1984, pp. 1, 7.

45. Teri Sorensen, "C-SPAN Insider: The Making of a Household Word," *C-SPAN Update*, October 29, 1984, p. 3.

46. Betsill interview.

47. Gary interview.

48. "Grassroots Producers on the Air," *C-SPAN Update*, November 19, 1984, p. 2.

49. Frazee's speech appeared in *C-SPAN Update*, April 1, 1985, pp. 1, 6. The speech was given before the Washington Metropolitan Cable Club on March 27, 1995.

50. Lardner interview notes, C-SPAN files.

51. "C-SPAN Tenth Anniversary Retrospective," April 3, 1989.

52. Fay budget memo, February 8, 1995; Auletta, *Three Blind Mice*, p. 439; Whittemore, *CNN: The Inside Story*, p. 301. See also Richard Salant, "Network News: Prospects for Its Future—If Any," *Los Angeles Times*, May 20, 1988, sec. VI, pp. 1, 20.

53. Mike Michaelson, "Planning Strategy for Future Negotiations with Broadcast Networks re: Pool Charges," n.d., C-SPAN files.

54. Ibid.

55. Neil Postman and Jay Rosen, "When Television News Has a Breakdown: Now That's Entertainment," *Los Angeles Times*, January 31, 1988, sec. V, p. 5.

56. Carl Rutan, "C-SPAN Tenth Anniversary Retrospective," April 3, 1989, C-SPAN telecast, Purdue University Public Affairs Video Archives.

57. Quotations in Andrew Rostenthal, "Re-writing the Rules of the Presidential Campaign," *New York Times*, reprinted in *C-SPAN Update*, November 2, 1987, pp. 1, 3. As had been done in 1984, the *Update* would reprint favorable articles from major newspapers for the benefit of its readers.

58. Mary Holley, "Mapping Out C-SPAN's Presidential Campaign Coverage," *C-SPAN Update*, November 9, 1987, p. 4.

59. Rutan, "C-SPAN Tenth Anniversary Retrospective," April 3, 1989.

60. Holley, "Mapping Out C-SPAN's Presidential Campaign Coverage," p. 4.

61. "C-SPAN Tenth Anniversary Retrospective," April 3, 1989.

62. Gibson quoted in Rosemary Harold, "The Great C-SPAN Fly Around: Previewing Coverage of the February Iowa Caucuses," *C-SPAN Update*, November 23, 1987, pp. 6, 7.

63. Steve Daley, "The Media Farms the Message—but They're Digging in the Wrong Field," *Chicago Tribune*, February 2, 1988, sec. 5, p. 7.

64. Rutan quoted in Dan Hunter, "Cable's New Clout," *Columbia Journalism Review*, January-February 1988, p. 36.

65. The episode, of course, made the C-SPAN list of hits. It was included in "C-SPAN Tenth Anniversary Retrospective," April 3, 1989, from which the quotation is taken.

66. Jack W. Germond and Jules Witcover, *Whose Broad Stripes and Bright Stars? The Trivial Pursuit of the Presidency, 1988* (New York: Warner Books, 1989), p. 234.

67. "C-SPAN Tenth Anniversary Retrospective," April 3, 1989.

68. *Weekly Report*, October 10, 1988. Grandmaison, the Democratic Party state chair, presented C-SPAN an award for its coverage of his party's primary.

69. *C-SPAN Update*, November 9, 1987, p. 5, contains a description of Katz's role and a brief history of his service at C-SPAN. "On the Road with Kathleen Brown," *Weekly Report*, May 2, 1988, describes the Pennsylvania and New York trip.

70. The following quotations, unless otherwise noted, come from two documents

prepared by Mike Michaelson to assist C-SPAN in planning its 1988 convention coverage. "Conventions 1984 Step By Step Planning" contains no date, but "A Working 'How to' Report on the 1984 Conventions" is dated May 1, 1987. Internal evidence suggests that the former memo was written at about the same time as the latter. Both are in the C-SPAN files.

71. Jay Smith to David Andersen, Atlanta, Ga., July 22, 1988. A copy of this letter was distributed to C-SPAN employees by means of the *Weekly Report*, September 12, 1988.

72. Smith's reaction is typical of editors who allow C-SPAN to cover editorial meetings. A number have reported receiving more mail and calls than they expected. The same is true for those talk-show hosts whom C-SPAN simulcasts regularly. Almost every host who appeared over the Christmas holiday season in 1994 seemed surprised at the caller response generated by C-SPAN.

73. Tom Shales, "On the Air: Live from New Orleans—It's RNC," *Washington Post*, August 15, 1988, sec. C, p. 8.

74. Jeff Gralnick in ibid.

75. Richard Harwood, "Perserve the Political Convention," *Washington Post*, July 24, 1988, sec. C, p. 6.

76. Tom Shales, "Network Uncoverage: Peeking in Periodically," *Washington Post*, July 18, 1988, sec. B, p. 3.

77. The year 1988 appears to have been the high-water mark for local television attendance. The number dropped in 1992.

78. Shales, "On the Air."

79. Tom Rosenstiel, "Democrats Learn to Control Show," *Los Angeles Times*, July 21, 1988, sec. 1, p. 5.

80. Fay budget memo, February 8, 1995.

81. A copy of the advertisement, indicating that it appeared in *USA Today*, August 12, 1988, is in the C-SPAN files.

82. *Weekly Report*, October 10, 1988.

83. Ibid.

84. Michael Barone, "In Defense of Sound Bite Politics," *Washington Post*, September 21, 1988, sec. A, p. 19.

85. Chuck Crouse, "C-SPAN's Growing Role in a Tumultuous Presidential Year," *Communicator*, August 1992, p. 12.

86. Lamb interview, June 29, 1995.

87. This image seems to be a common one at C-SPAN. Both Brian Lamb and Connie Brod also used it in their interviews.

88. Crouse, "C-SPAN's Growing Role," p. 12.

89. Richard Katz, "The New Political Scene," *Cablevision*, August 10, 1992, p. 45.

90. Rosenstiel, *Strange Bedfellows*, p. 245.

91. Reuven Frank, *Out of Thin Air* (New York: Simon and Schuster, 1991), p. 403.

92. Marc Gunther, *The House That Roone Built: The Inside Story of ABC News* (Boston: Little Brown and Company, 1994), pp. 335, 342, 349.

93. Howard Kurtz, *Media Circus: The Trouble with America's Newspapers* (New York: Times Books, Random House, 1993), p. 143.

94. Memos to Clinton from his advisers, April 27, 1992, in Peter Goldman, Thomas M. DeFrank, Mark Miller, Andrew Murr, and Tom Matthews, *Quest for the Presidency, 1992* (College Station: Texas A&M University Press, 1994), pp. 657–66.

95. Memo from Fred Streeper, April 28, 1992, in ibid., pp. 674–75.

96. Memo from Celinda Lake, October 10, 1992, in ibid., p 726.

97. Steve Scully, interview with the authors, tape recording, C-SPAN offices, Washington, D.C., April 21, 1995.

98. Lamb quoted in Ed Bark, "C-SPAN's Candid Camera Catches On," *Dallas Morning News*, May 6, 1992, sec. A, p. 11.

99. Murphy quoted in Edwin Diamond, "Primary Source," *New York*, March 9, 1992, p. 19.

100. Ibid., pp. 16–19.

101. Swain quoted in Patrick Sheridan, "C-SPAN Plans 1000 Hours of Election Coverage," *Broadcasting*, November 11, 1991, p. 38.

102. Scully interviews, February 8 and April 21, 1995.

103. Ibid.

104. Quoted in Patrick Sheridan, "C-SPAN Holds Firm: No Tapes of Kerrey Joke," *Broadcasting*, December 2, 1991, pp. 38, 42.

105. Quoted in Bob Minzesheimer, "Enjoying Politics at the C-SPAN Pace," *USA Today*, February 19, 1992, sec. A, p. 9.

106. Scully interviews.

107. David Kissinger, "C-SPAN Picks Up Network Slack," *Variety*, February 17, 1992, p. 91.

108. Bark, "C-SPAN Candid Cameras," sec. A, p. 11.

109. Susan Baer, "On C-SPAN, It's Politics, Politics, Politics, Politics," *Baltimore Sun*, April 26, 1992, sec. A, p. 15.

110. Scully interviews.

111. Ibid.

112. Susan Swain appeared on a panel on "TV and the Presidential Debates" sponsored by the Annenberg Washington Program. It aired on C-SPAN on August 15, 1993, Purdue University Public Affairs Video Archives.

113. Ibid.

114. "Road to the White House Aggregate Times for All Candidates," April 92–July 92, C-SPAN files.

115. Cumulative Transcript Rankings, November 17, 1990, to November 30, 1994, C-SPAN files.

116. Scully interview, April 21, 1995; Meyer, "No Sound Bites Here," p. 46.

117. Jane Hall, "Is the Party Over?," *Los Angeles Times*, August 16, 1992, sec. Calendar, p. 4.

118. Fay budget memo, February 8, 1995.

119. Tom Shales, "The Dems on TV: Going Full Bore," *Washington Post*, July 15, 1992, sec. F, p. 4.

120. Lee Winfrey, "TV and the Campaign Trail," *Philadelphia Inquirer*, July 15, 1992, sec. D, p. 1.

121. Lamb quoted in Jane Hall, "C-SPAN: The Network That Dares to Be Dull," *Los Angeles Times*, July 13, 1992, sec. F, p. 1.

122. Memo from John Splaine to Brian Lamb, Susan Swain, and Rob Kennedy concerning Houston coverage, July 23, 1992, C-SPAN files.

123. Winfrey, "TV and the Campaign Trail," sec. D, pp. 1, 5.

124. Deborah Snyder, *Concord Monitor*, July 13, 1992, sec. A, pp. 1, 8.

125. Peggy Noonan, "Red Meat and Astroturf: Decoding the Convention," *Washington Post*, August 23, 1992, sec. C, p. 1.

126. Steve Scully, phone interview with John Sullivan, July 11, 1995; John Carmody, "The TV Column," *Washington Post*, August 20, 1995, sec. C, p. 8.

127. Noonan, "Red Meat and Astroturf," sec. C, p. 1.

128. Murphy quoted in Hal Boedeker, "C-SPAN Goes beyond Game-Style Coverage, Debates: Tonight It's Final Jeopardy!," *Miami Herald*, October 19, 1992, sec. C, pp. 1, 5.

129. Swain and Lamb quoted in Rod Granger, "C-SPAN Election Night Coverage a Matter of 'Reflection,'" *Multichannel News*, November 9, 1992, p. 16.

130. The *C-SPAN Digest* chronicled cable industry support for its readers throughout the campaign.

131. Fay budget memo, February 8, 1995.

132. Rosenstiel, *Strange Bedfellows*, pp. 5, 260–61.

133. John Splaine, memo to Terry Murphy on election-night coverage, November 25, 1992, C-SPAN files.

134. Elizabeth Kolbert, "The Longest Campaign Has Started," *New York Times*, September 19, 1993, sec. 4, p. 6.

135. The following quotations come from Scully interviews—February 8, April 21, and July 11, 1995—cited earlier.

136. Speech by Brian Lamb to C-SPAN Seminar for Professors, January 9, 1995, Washington, D.C.

137. Victoria Gits, "Grassroots '84: A Day on the Road with C-SPAN's Brian Lamb," *Cablevision*, October 8, 1984, p. 28.

138. Frazier Moore, " 'Booknotes' Covers Writers the Right Way," *Washington Times*, April 10, 1994, sec. D, p. 2.

139. Media backgrounder for *Booknotes* fifth-anniversary telecast, C-SPAN files.

Videotapes of *Booknotes* are available from the Purdue University Public Affairs Video Archives.

140. *C-SPAN Update*, April 3, 1989, p. 1.

141. Lamb interview, March 22, 1995.

142. Media backgrounder.

143. Sarah Trahern, interviews with John Sullivan, tape recording, C-SPAN offices, April 21, 24, 1994.

144. John Splaine, interview with the authors, tape recording, C-SPAN offices, Washington, D.C., February 8, 1994.

145. Howard Rosenberg, "Can't Put Down Booknotes," *Los Angeles Times*, April 8, 1994, sec. B, pp. 8, 11.

146. Lamb quoted in Moore, " 'Booknotes' Covers Writers," sec. D, p. 2.

147. The quotation comes from a list of *Booknotes* programs identifying the producers and containing occasional comments about particular shows. It has neither an author's name nor a date but appears to have been written no later than March 1993.

148. Rosenberg, "Can't Put Down Booknotes," sec. B, p. 8.

149. The conversation first aired on July 31, 1994, but re-aired during the holiday season.

150. Lamb interview, March 22, 1995.

Chapter 6

1. Robin Warshaw, "The Real Drama of All-Natural TV," *Philadelphia Inquirer Magazine*, August 14, 1988.

2. Cathleen Schine, "Like, Politics, Wow," *Vogue* 176 (November 1986).

3. Mary Holley, "I Got Pretty Downhearted at Times," *C-SPAN Update*, September 18, 1987, p. 4.

4. Lamb quoted in Frazier Moore, "A Good Look at Government," AP press report, January 24, 1994.

5. Diane Werts, "C-SPAN: The Comfort Zone," *Newsday*, March 10, 1992, p. 47.

6. Margo Hammond, "Authors on the Air," *St. Petersberg Times*, May 9, 1993, p. 7D.

7. Peter Ainslie, "Confronting a Nation of Grazers," *Channels*, September 1988, pp. 54–62.

8. Douglas Davis, *The Five Myths of Television Power; or, Why the Medium Is Not the Message* (New York: Simon and Schuster, 1993), p. 223.

9. "The Battle (Zap! Click!) of the Sexes," *New York Times*, July 7, 1991, p. D1.

10. Seinfeld quoted in ibid.

11. Times Mirror Center for the People and the Press, "The Role of Technology in American Life," May 1994, p. 15 (press release, Washington, D.C.).

12. Walter Kirn, "Welcome to the Network of the Nineties," *Mirabella*, June 1991.

13. This was a repeated phrase in numerous interviews with C-SPAN staff members.
14. Robert Dornan, *Congressional Record*, March 18, 1994, p. H-1549. This is just one of many congressional references to the size of the C-SPAN audience. Most estimates range from 1–1.3 million.
15. *Congressional Record*, February 15, 1995, p. H-1780.
16. Lamb interview, February 3, 1995. See also Stephen Hess, "The Decline and Fall of Congressional News," in Thomas E. Mann and Norman J. Ornstein, *Congress, the Press, and the Public* (Washington, D.C.: American Enterprise Institute and Brookings Institution, 1994), p. 155, for additional audience estimates.
17. Paul Kanjorski, *Congressional Record*, August 4, 1993, p. H-6089.
18. Richard Ray, *Congressional Record*, June 10, 1992, p. H-4453.
19. "Databank," *Cable World*, May 15, 1995, p. 56.
20. Adelman, "Real People," p. 32.
21. Davis, *The Five Myths of Television Power*, p. 134.
22. Unless otherwise noted, the data discussed in this chapter comes from the following polls. The descriptive name in parentheses is used in citations and figures.

1980 (HALL, 1980). Terry Hall, *Congress on the Tube: Who's Watching?* (Norman, Okla.: Bureau of Government Research, 1980).

1982 (BOAKES, 1982). Janet Carol Boakes, "A Study of Television Viewership of Congressional Proceedings Carried on C-SPAN, the Cable Satellite Public Affairs Network" (master's thesis, University of Pennsylvania, Annenberg School of Communications, 1983).

1983 (ARBITRON, 1983). Arbitron Ratings Company, "C-SPAN Special Cable Study." Based on 781 interviews with cable subscribers in ten geographically dispersed cable systems.

1985 (ROBINSON, 1985). Michael Robinson and Maura Clancey, "The C-SPAN Audience after Five Years," Media Analysis Project, George Washington University. Based on a random national sample of 959 cable homes in forty-one states.

1986 (AP, 1986). Associated Press and Media General, "Opinion Poll." Based on a random national sample of 1,473 adults; questions limited to viewing Congress on television; no specific questions on C-SPAN.

1987 (MARYLAND, 1987). University of Maryland Survey Research Center, "The C-SPAN Audience." Based on a national random sample of 3,944 households in the continental United States.

1988 (MARYLAND, 1988). University of Maryland Survey Research Center, "Survey of C-SPAN Audience." Based on a national random sample of 2,379 households in the continental United States.

1991 (SRI, 1991). Statistical Research Inc., "C-SPAN Awareness/Viewing

Study." Based on a national random sample of 1,053 adults in the continental United States.

1992 (SRI, 1992). Statistical Research Inc., "C-SPAN Election Study." Based on a national random sample of 1,417 adults in the continental United States.

1992–95 (TIMES MIRROR, 1992, 1993, 1994, 1995). Times Mirror Center for the People and the Press. Based on large, national, random adult samples (often over 3,000). LEXIS-NEXIS database, RPOLL file.

1994 (LUNTZ, 1994). Luntz Research Associates/Progress and Freedom Foundation, "The New American Agenda and the 1994 Elections." Based on a national random sample of 1,300 adults.

1994 (MELLMAN, 1994). Lazarus-Mellman-Lake. National poll for the House Democratic leadership.

1995 (SRI, 1995). Statistical Research Inc., "C-SPAN Pricing Study." Based on a national random sample of 500 adults in the continental United States.

23. ARBITRON, 1983.
24. MARYLAND, 1988. Measuring awareness can be tricky. The reported figures are for "aided" awareness, in which respondents are asked if they had "heard of" or were "aware" of C-SPAN. Over time, data from national surveys of cable subscribers by Beta Corporation indicates the following figures for aided recall:

    1989   64%
    1990   68%
    1991   75%
    1992   81%
    1993   82%

As would be expected, cable subscribers have higher awareness than the general population. "Unaided" recall is much lower. When asked to name the various cable channels in 1993, only about 10 percent of cable subscribers identified C-SPAN. Beta Research Corporation, "Evaluation of Cable TV Services," November 1993 (press release, Syossset, New York).

25. SRI, 1992.
26. SRI, 1995.
27. Lamb interview, February 3, 1995.
28. SRI, 1995.
29. Todd Gitlin, "Glib, Tawdry, Savvy, and Standardized: Television and American Culture, and Other Oxymorons," remarks at the John F. Kennedy School of Government/Home Box Office Conference on the Future of Television, Cambridge,

Massachusetts, February 8, 1993.

30. Most of the above surveys include some demographic data. A key methodological question involves defining a "viewer." Most of the surveys asked whether a person had viewed C-SPAN during the last year or last six months. Some asked whether the person had "ever" watched C-SPAN. Depending on how the question is asked, different demographic percentages emerge. In general, the more constrained the definition of a viewer, the more the general patterns are heightened. For the sake of timeliness, the data from the 1991 and 1992 SRI surveys will generally be reported.

31. SRI, 1992.

32. Maura Clancey, "The Political Knowledge, Participation, and Opinions of C-SPAN Viewers: An Exploration and Assessment of Mass Media Impact" (Ph.D. diss., University of Maryland, 1990). See also SRI. 1991.

33. MARYLAND, 1988.

34. Clancey, "Political Knowledge, Participation, and Opinions," p. 137. As the following table indicates, C-SPAN viewers are a bit more Republican and Independent than is the population as a whole, but they are also more likely to back the winners in a presidential contest.

Political Proclivities of the C-SPAN Audience: Partisan Identification

C-SPAN Viewers versus All Viewers (in parentheses)

Year	1987[a]	1988[b]	1991[c]	1992[d]	1992[e] ("regular" viewers)
Democratic	38%(41%)	32%(42%)	34%(39%)	28%(39%)	38%(39%)
Republican	32%(30%)	33%(30%)	33%(30%)	29%(28%)	33%(28%)
Independent/ Other	20%(29%)	35%(37%)	33%(31%)	43%(33%)	29%(33%)

Presidental Voting

1984	C-SPAN Viewers[f]	Cable Viewers[f]	National[g]
Reagan	63%	55%	59%
Mondale	37%	45%	41%

1992	C-SPAN Viewers (postelection)[f]	"Regular" C-SPAN Viewers (summer)[e]	National Vote Total
Clinton	46%	32%	43%
Bush	36%	25%	39%
Perot	18%	43%	19%

[a]C-SPAN Viewers: MARYLAND, 1987; All Viewers: *Gallup Poll Monthly*, no. 346

(July 1994), p. 57 (all years).

[b]C-SPAN Viewers: MARYLAND, 1988.

[c]C-SPAN Viewers: SRI, 1991.

[d]C-SPAN Viewers: SRI, 1992.

[e]TIMES MIRROR, 1992. Based on a national survey during May and June 1992 of 3,517 adults. Voting figures indicate voting intentions during the early summer of 1992. Figures are based on an analysis of 90 self-identified "regular" C-SPAN viewers (14 percent of the 43 percent of the population reporting some viewing).

[f]C-SPAN Viewers, 1984 Cable Viewers: ROBINSON, 1985; 1992 C-SPAN Viewers: SRI, 1992. For 1992, the 19 percent of respondents who refused to answer were eliminated.

[g]Stanley and Niemi, *Vital Statistics*, p. 106.

35. Polls cited in Tom Brazaitis, "Voters Seek Protection from Big Government: Clinton's Agenda, Vision Triggered Landslide," *Plain Dealer*, November 11, 1994, p. 19A.

36. See SRI, 1991, and SRI, 1992.

37. See Sheila E. Quinlin, "Campaigns and Cable," *C-SPAN Quarterly*, Spring 1992, p. 7, for a more general discussion of the political importance of the cable audience.

38. Clancey, "Political Knowledge, Participation, and Opinions," p. 166.

39. Lamb quoted in Robert Stewart, "Lawmakers Perfect Art of One-Minute Zinger," *Los Angeles Times*, October 4, 1992, p. A13.

40. Paul Huston and Robert Shogan, "Washington Insight," *Los Angeles Times*, February 7, 1994, p. A5.

41. TIMES MIRROR, 1992. Based on a national survey during May and June 1992 of 3,517 adults. Figures are based on an analysis of 90 self-identified "regular" C-SPAN viewers (14 percent of the 43 percent of the population reporting some viewing).

42. Ibid.

43. Erik Mankin, "Tales From Viewers Who Watch Every Day," *Los Angeles Herald Examiner*, July 3, 1984.

44. TV writer Douglas Davis, quoted in "The Art of Good Television," *C-SPAN Update*, July 25, 1988, p. 11.

45. Reed quoted in Alicia Mundy, "The God Squad," *Adweek*, February 13, 1985.

46. Lamb interview, February 3, 1995.

47. "C-SPAN to Sit in on Sentinel Forum," *Orlando Sentinel*, July 31, 1986.

48. "Mr. Reagan's Phone Calls," *C-SPAN Update*, May 1983, p. 1.

49. Betty Cuniberti, "Reagan: 'I'm Right,'" *Los Angeles Times*, June 26, 1984.

50. "Reagan: I'm a Fan of C-SPAN's Call-in," *C-SPAN Update*, December 10, 1984, p. 6.

51. See Lamb, *C-SPAN*, pp. 290–91.

52. Gergen quoted in Patrick Watson, "How America's 'First Viewer' Keeps Tabs on

Congress and Follows Public Opinion," *C-SPAN Update*, December 10, 1984, p. 6.

53. Quoted in *C-SPAN Update*, April 1, 1985, p. 9.

54. Howell Raines, "President Backs Spending on Arms before Governors," *New York Times*, March 1, 1983, p. A1.

55. Richard Berke, "Presidential Routine Takes on New Tone," *New York Times*, April 1, 1993, p. 41A.

56. Margaret Carlson, ". . . And Then Came the Carrot Cake," *Time*, March 1, 1993, p. 20.

57. C-SPAN simulcast transcript, December 18, 1994.

58. Bond quoted in Mary Holley, "The Next Best Thing to Being There," *C-SPAN Update*, March 19, 1987, p. 12.

59. As reported in internal memos, 48 percent of the 75 retiring members and 40 percent of the 110 new members completed survey forms.

60. Notes, C-SPAN focus group meeting with congressional staff, January 23, 1992, C-SPAN files.

61. Ibid.

62. Ford West, interview with the authors, American Fertilizer Institute, Fall 1994.

63. Thomas Southwick, "Breathing New Life into Election Coverage, *Multichannel News*, August 15, 1987.

64. Jeff Greenfield, "C-SPAN Comes Close to National Treasure," *Oregonian*, October 6, 1986.

65. Rosemary Harold, "Survey Questioned Iowans' News Habits," *C-SPAN Update*, March 21, 1988, p. 1.

66. Maura Clancey, "University of Maryland Survey Shows GOP Delegates to be Heavy Viewers of C-SPAN," *C-SPAN Update*, August 15, 1988, p. 1.

67. Based on a 1988 mail survey of an 1,100-person sample of designated elites carried out by C-SPAN.

68. Lamb, *C-SPAN*, pp. 375–77.

69. Maureen Dowd, "Clinton's Best Friends Find It a Tough Role," *New York Times*, May 27, 1993, p. A1.

70. Letter quoted in Mona Charen, "C-SPAN: Where TV Has Never Gone Before," *Reader's Digest*, May 1992, p. 115.

71. McCain quoted in ibid., p. 112.

72. "AT&T Counts 219,000 Attempts to Reach C-SPAN Open Phones Program," *C-SPAN Update*, October 24, 1988, p. 12.

73. Mary Holley, "Viewer Commits 30-Day Policy Breaker to Tape," *C-SPAN Update*, May 16, 1988, p. 1.

74. Bierbauer quoted in Lou Prato, "Enjoying the Heat," *Communicator*, March 1995, p. 3.

75. Morton M. Kondracke, "Why Are Voters 'Mad as Hell'? If You Ask, Duck!,"

Roll Call, October 20, 1994, p. 6.

76. Mundy, "The God Squad," p. A1.

77. Mary Holley, "A Lesson in Network Loyalty," *C-SPAN Update*, April 25, 1988, p. 1.

78. TIMES MIRROR, 1992. This national poll of 3,517 adults identified regular viewers of networks and programs such as C-SPAN (90 regular viewers), CNN (546 regular viewers), MacNeil/Lehrer (108 regular viewers), and NPR (143 regular listeners).

79. Clancey, "Political Knowledge, Participation, and Opinions," p. 70.

80. See Roger H. Davidson and Glenn R. Parker, "Positive Support for Political Institutions: The Case of Congress," *Western Political Quarterly* 25 (1971): 605.

81. National telephone poll of adults conducted by Mellman-Lazarus-Lake. Of all respondents, 8 percent reported watching C-SPAN every day, 15 percent several times a week, 13 percent once a week, and 10 percent less than once a week. These viewers were asked, "Has what you have seen on C-SPAN made you more favorable toward Congress, had no effect, or made you less favorable toward Congress?" See also Janet Hook, "An After-Hours Image Adjustment," *Congressional Quarterly Weekly Report*, February 12, 1994, p. 317.

82. Mellman quoted in Katherine Q. Seelye, "Gingrich, Now King of the Hill, Used Skill with Media to Climb to the Top," *New York Times*, December 14, p. 3.

83. Charles E. Cook, "Public Has No Idea What's Good about Congress, Poll Shows," *Roll Call*, February 3, 1994.

84. Quoted in Julia Malone, "Public Exposure Backfires," *Atlanta Journal and Constitution*, October 21, 1994, p. B4.

85. Ibid.

86. *CBS Evening News*, June 1, 1987.

87. Norman J. Ornstein, "If Congress Played a Better Host to Guests, Maybe It Would Improve a Crummy Image," *Roll Call*, March 31, 1994.

88. TIMES MIRROR, 1992.

89. James Madison, Letter to W. T. Barry, August 4, 1822, in *The Writings of James Madison*, ed. Gaillard Hunt (New York: G. P. Putnam's Sons, 1910), vol. 9, p. 103.

Chapter 7

1. Joel Achenbach, "The Land of the Desk Potatoes," *Washington Post*, January 17, 1993, p. F1.

2. Kenneth Lauden, *Communications Technology and Democratic Participation* (New York: Praeger, 1974), p. 30.

3. See Susan Douglas, *Inventing American Broadcasting* (Baltimore: Johns Hopkins University Press, 1987), p. 34, and Nathan Rosenberg, "Uncertainty and Technological Change" (paper prepared for Conference on Growth and Development:

The Economics of the Twenty-First Century, Stanford University, 1994).

4. See Stephen E. Frantzich, *Computers in Congress: The Politics of Information* (Beverly Hills, Calif.: Sage Publications, 1982).

5. Phyllis Watt, "Television Cameras in Congress" (Freedom of Information Report #483, School of Journalism, University of Missouri at Columbia, 1993). See also media critic Larry Sabato's comments in John Schachter, "Congress Begins Second Decade under TV's Watchful Glare," *Congressional Quarterly Weekly Report*, March 11, 1989, p. 507.

6. Quoted in "TV in the Senate; One Year Later," *Broadcasting*, June 1, 1987, p. 37.

7. Jesse Helms, *Congressional Record*, August 2, 1993, p. S-10040.

8. Robert Michel (R-IL), National Press Club Speech, reprinted in *C-SPAN Update*, December, 19, 1989, p. 8.

9. Elaine Povich, "Journalists' Roundtable," transcript, reprinted in *C-SPAN Update*, November 9, 1987, p. 3.

10. Thomas Foley, radio interview on Station KXLY, Spokane Washington, hosted by Alex Wood, August 26, 1993, LEGI-SLATE database.

11. Collins quoted in Hickey, "Eye on Democracy."

12. *Congressional Record*, February 2, 1982.

13. Quoted in Bill Roeder, "The Stage Struck Solons," *Newsweek*, May 21, 1979, p. 23.

14. Watt, "Television Cameras in Congress," p. 3.

15. Paul S. Rundquist and Ilona Nickels, "Senate Television: Its Impact on Senate Floor Proceedings," *Congressional Research Service Report*, July 21, 1986.

16. *Congressional Record*, February 2, 1982, p. S-627.

17. U.S. Congress, Senate, Committee on Rules and Administration, *Television and Radio Coverage*, pp. 90–101.

18. "How the House Fares On-Camera," *Newsweek*, June 15, 1981, p. 16.

19. Ornstein quoted in Bob Dart, "C-SPAN Celebrates 15th Anniversary of Showing Congress to America," Cox Wire Service, March 16, 1994.

20. Ornstein, "Yes, Television," p. 8.

21. C-SPAN transcript, August 1, 1988. For similar comments, see Senator J. Bennett Johnston (D-LA), quoted in Briant Nutting, "After One Year of Television, Senate Is Basically Unchanged, " *Congressional Quarterly Weekly Report*, May 30, 1987, p. 1140.

22. Timothy Clark, "Dry Spell Continues for Water Projects as Financing Agreement Eludes Congress," *National Journal* 16, no. 44 (November 3, 1984): 2079.

23. Kevin Fedarko, "Newt's Battle-Ready Armey," *Time*, November 28, 1994, p. 30.

24. Linda M. Harrington, "C-SPAN: TV's Political Insider in Washington," *Chicago Tribune*, October 31, 1993, p. C5.

25. *Congressional Record*, November 11, 1993, p. S-1607.

26. Bruce Collins, "When Words Can't Express It," *C-SPAN Update*, December 5,

1983, p. 6.

27. Mark West, "The Slide Show," *C-SPAN Update*, August 3, 1987, p. 5.

28. Collins, "When Words Can't Express It," p. 6.

29. Matejowsky quoted in Peter Kiley, "Press Clips," *C-SPAN Update* July 14, 1989, p. 8.

30. Quoted in "Senate Says Yes to TV," *Broadcasting*, August 4, 1986, p. 40.

31. Quoted in Ornstein, "Yes, Television," p. 6.

32. *Congressional Record*, February 2, 1982, p. S-627.

33. U.S. Congress, Senate, Committee on Rules and Administration, *Television and Radio Coverage*.

34. Ibid., p. 190.

35. Ornstein, "Yes, Television," p. 8.

36. Ralph Nader, interview with James Lardner, 1994, from Lardner interview notes.

37. Richard Cohen, "The Senate Show," *National Journal* 18, no. 10 (March 8, 1986): 610. See also Alan Ehrenhalt, "Media, Power Shifts Dominate O'Neill's House," *Congressional Quarterly Weekly Report*, September 13, 1986, p. 2135.

38. O'Neill quoted in Bob Franklin, "Televising Legislatures: The British and American Experience," *Parliamentary Affairs* 42 (October 1989): 501.

39. Dornan quoted in Stewart, "Lawmakers Perfect Art," p. A13.

40. Norman Ornstein and Michael Robinson, "The Case of Our Disappearing Congress: Where's All the Coverage?," *Congressional Record* (daily ed.), February 3, 1986, pp. S828–S830, originally in *TV Guide*, January 11, 1986, p. 4.

41. Mudd quoted in ibid., p. 5. See also Mann and Ornstein, *Congress, the Press, and the Public*, p. 4.

42. S. Robert Lichter and Daniel Amundson, "Less News Is Worse News: Television News Coverage of Congress, 1972–1992," in Mann and Ornstein, *Congress, the Press, and the Public*, p. 134.

43. Cook, *Making Laws and Making News*, p. 63.

44. Richard Harwood, "Sources Who Supply 'The News,'" *Washington Post*, January 7, 1995, p. A21.

45. Wofford quoted in Jack Anderson and Michael Binstein, "Rep. Hall Describes Genocide in Rwanda," *Washington Post*, June 16, 1994, p. B25.

46. Newt Gingrich, remarks at Republican Governors' Association Annual Conference, November 22, 1994, Federal News Service transcript, LEXIS-NEXIS database.

47. "Camera-shy," *Cincinnati Enquirer*, May 3, 1994, p. A6.

48. Newt Gingrich, "Tribute to the 200th Anniversary of the Attainment of the 1st Quorum in the 1st Congress," *Congressional Record*, April 4, 1989, pp. H-915, 921.

49. Robert Michel, "Tenth Anniversary of C-SPAN," *Congressional Record*, April 3, 1989, p. H-901.

50. Ralph Baruch (chairman of the National Academy of Cable Programming), speech inserted in the *Congressional Record*, December 11, 1987, p. E-4774.

51. Ronald Reagan, speech transcript, Moscow State University, Associated Press, June 1, 1988, LEXIS-NEXIS database.

52. Charles Cook, "Political Surveyor," *Roll Call*, February 3, 1994.

53. Mann and Ornstein, *Congress, the Press, and the Public*, p. 7, and Ronald Elving, "Brighter Lights, Wider Windows: Presenting Congress in the 1990s, in ibid., p. 192.

54. Richard Morin and David S. Broder, "Familiarity Is Breeding Contempt of Congress," *Washington Post*, July 3, 1994, p. A1.

55. Elizabeth Theiss-Morse and John R. Hibbing, "Familiarity Breeds Contempt: Education, Political Knowledge, and Disapproval of Congress," paper presented at the 1994 meeting of the Midwest Political Science Association, Chicago, Illinois.

56. Quoted in Adelman, "Real People," p. 33.

57. American Enterprise Institute, "Ten Years of Televising Congress," symposium, recorded on April 5, 1989, and broadcast on C-SPAN on April 9, 1989.

58. Gingrich quoted in Bob Dart, "C-SPAN's All-Seeing Eye Changes Congress," *Atlanta Journal and Constitution*, March 19, 1994, p. A10.

59. C-SPAN call-in transcript, April 3, 1989.

60. Media analyst Timothy Cook quoted in Stewart, "Lawmakers Perfect Art," p. A13.

61. Howard Fineman, "For the Son of C-SPAN, Exposure=Power," *Newsweek*, April 3, 1989, p. 23.

62. Quoted in ibid.

63. Quoted in American Enterprise Institute, "Ten Years of Televising Congress."

64. Ibid.

65. David Maraniss, "Armey Arsenal: Plain Talk and Dramatic Tales," *Washington Post*, February 21, 1995, p. A1.

66. Louis Romano, "Armey Muting Rhetoric," *Washington Post*, January 13, 1995, p. B3.

67. Fedarko, "Newt's Battle-Ready Armey," p. 30.

68. Ornstein, "Yes, Television," pp. 5–8.

69. T. R. Reid, "Congress: Best Little Soap Opera on Cable," *Washington Post*, April 29, 1984, p. B1.

70. Alan Ehrenhalt, *Politics in America* (Washington, D.C.: Congressional Quarterly, 1987), p. 203.

71. George Will, "Hurricane Bob: Dornan's Candidacy Is a Product of Term Limits and C-SPAN," *Newsweek*, May 1, 1995, p. 78.

72. Ibid.

73. Ibid.

74. Robert Stewart, "House Divided over Late-Night Speeches on TV," *Los Angeles*

Times, November 30, 1992, p. A14.

75. *Congressional Record*, March 19, 1979, p. H-5411.

76. *Congressional Record*, June 2, 1986, p. S-12044.

77. Ely, "Strength of Belief," p. 2.

78. Helen Dewar, "Senator Mitchell Elected Majority Leader," *Washington Post*, November 30, 1988, p. A1.

79. Steven V. Roberts, "Man in the News: George John Mitchell," *New York Times*, November 30, 1988, p. A1.

80. Ronald D. Elving, "Politics of Congress in the Age of TV," *Congressional Quarterly Weekly Report*, April 1, 1989, p. 722.

81. Byrd quoted in American Enterprise Institute, "Ten Years of Televising Congress."

82. Michael Wines, "Leadership Race in Senate Attracting Many Spectators," *New York Times*, June 23, 1994, p. A1.

83. John Zaller, "Unconventional Media Boosted Perot, Mainstream Brought Him Down," *Public Affairs Report*, March 1994, p. 14.

84. Larry King and Mark Stencel, *On the Line: The New Road to the White House* (New York: Harcourt Brace and Company, 1993).

85. Zaller, "Unconventional Media Boosted Perot," p. 1.

86. Lamb quoted in King and Stencel, *On the Line*, p. 20.

87. Bill Turque, "The Voters of Perot County," *Newsweek*, July 13, 1992, p. 22. See also Range, "C-SPAN," p. 13.

88. Turque, "The Voters of Perot County."

89. Zaller, "Unconventional Media Boosted Perot," p. 7.

90. Paul Tsongas, National Press Club luncheon, October 22, 1992, Federal News Service transcript, LEXIS-NEXIS database.

91. Ibid.

92. Paul Tsongas, interview with James Lardner, 1994, from Lardner interview notes.

93. Davis, *The Five Myths of Television Power*, p. 216.

94. Mike Gauldin, interview with James Lardner, 1994, from Lardner interview notes.

95. Ibid.

96. James Warren, "Taking the Long View," *Chicago Tribune*, March 15, 1992, p. 1.

97. Frank Greer, interview with James Lardner, 1994, from Lardner interview notes.

98. Susan King, "Q & A: Brian Lamb, the Election's Real Winner," *Los Angeles Times*, January 17, 1993, p. 83.

99. Quoted in Lardner, "The Anti-Network," p. 49.

100. Michael Kranish, "Observers See Real Clinton Talkathon," *Boston Globe*, December 16, 1992, p. 37.

101. Frazier Moore, "TV Made 35-Word Oath High Drama," *St. Paul Pioneer Press*, January 21, 1993.

102. Jack Germond, "The McLaughlin Group," December 16, 1994, LEXIS-NEXIS database transcript.

103. American Enterprise Institute, "Ten Years of Televising Congress."

104. Rosenstiel, *Strange Bedfellows*, p. 262.

105. Ibid.

106. King quoted in *C-SPAN Update*, January 23, 1994, p. 2.

107. O'Neill, *Man of the House*, p. 346.

108. Jim Wright, *Worth It All* (Washington, D.C.: Brasseys, 1993), p. 235.

109. "Inside Talk," *Minneapolis Star and Tribune*, January 10, 1994, p. 3B.

110. Elayne Rapping, "Cable's Silver Lining," *Progressive* 58, no. 9 (September 1994): 36.

111. Robert Dole, "Clinton-Gore Attempts to Rewrite History," *Congressional Record*, October 1, 1992, p. S-16014.

112. Murray Kempton, "Out of the Closet or Under the Rug?," *Newsday*, February 12, 1993, p. 13.

113. Diane Granat, "Televised Partisan Skirmishes Erupt in House," *Congressional Quarterly Weekly Report*, February 11, 1984, p. 246.

114. As reported in authors' interviews with C-SPAN staff.

115. Wright quoted in Granat, "Televised Partisan Skirmishes Erupt," p. 246.

116. Patrick Watson and Carolyn Lamson, "Home Is Where the House Is," *C-SPAN Update*, May 20, 1984, p. 2.

117. Gingrich quoted in Granat, "Televised Partisan Skirmishes Erupt," p. 247.

118. *Congressional Record*, March 10, 1987, p. H-1166.

119. *ABC Evening News*, February 9, 1992, LEXIS-NEXIS database.

120. Scully and Gingrich quoted in Steven Roberts, "Open Arms for On-Line Democracy," *U.S. News and World Report*, January 16, 1995, p. 10.

121. Quoted in Jack Anderson, "Democrats Organizing New Ways of Communicating," *Springfield (Ill.) State Journal-Register*, February 10, 1995, p. 8.

Chapter 8

1. Max Frankel, "Full-Text TV," *New York Times Magazine*, February 5, 1995, p. 32.

2. Carol Richards, "Establishing a New Standard," Gannett News Services, 1985.

3. Quoted in Geltner, "Brian Lamb," p. 46.

4. Quoted in U.S. Congress, Senate, Committee on Rules and Administration, *Television and Radio Coverage*, p. 90.

5. Newton Minow, *Equal Time* (New York: Atheneum, 1964), p. 52.

6. Speech before the Radio and Television News Directors' Association, New York, 1961.

7. Quoted in Range, "C-SPAN," p. 15.

8. Caroline Ely, "1984—Who's Watching," *C-SPAN Update*, December 26, 1983, p. 6.

9. Shales, "As the Hill Turns," p. E1.

10. Greg Dawson, "Don't Miss These Shows If You Like the Bizarre," *Orlando Sentinel Tribune*, January 13, 1983, p. E4.

11. *Congressional Record*, September 20, 1993, p. S-11953; *Congressional Record*, September 22, 1993, p. S-13161; "A Progress Report on Renewing American Civilization," *Congressional Record*, September 9, 1993, p. H-6586; "National Voter Registration Act of 1993: Conference Report," *Congressional Record*, May 7, 1993, p. S-5677; Bruce Collins, "Television Is Here to Stay," *C-SPAN Update*, March 18, 1984, p. 1.

12. Evans quoted in Ely, "Strength of Belief," p. 2.

13. James Crosswaite, quoted in *C-SPAN Update*, June 11, 1984, p. 1.

14. Mark Pazniokas, "Hearings Left Many People Angry," *Hartford Courant*, June 24, 1993, p. A1.

15. Ibid.

16. Ibid

17. Ibid.

18. Ann Devroy and David S. Broder, "Missteps Mired Clinton Package," *Washington Post*, April 11, 1993, p. A1.

19. Tim Curran and Susan Glasser, "Welcome Freshman of 104th," *Roll Call*, September 12, 1994.

20. King and Stencel, *On the Line*, p. 7.

21. Ibid.

22. See Irving Janis, *Victims of Groupthink* (Boston: Houghton-Mifflin, 1972).

23. Molly Ivins, "Attention News Junkies: You Don't Have to Rely on Brokaw/Jennings/Rather Versions," *TV Guide*, December 3, 1988, p. 49.

24. CBS Entertainment President Jeff Sagansky, quoted in Ed Bark, "How Does TV Fit into Your Future? Will the Tube Grow Up, or Just Grow?," *Dallas Morning News*, February 21, 1993, p. 1C. See also Rutkus, "Political Broadcasting as a Political Forum," p. 200.

25. Frank Rich, "Journal: The Gridiron Follies," *New York Times*. March 24, 1994, p. 23.

26. Ned Zeman, "Uproar over a 'Filthy Mouth,' " *Newsweek*, February 5, 1990, p. 4.

27. Paula Schwed, Gannett News Service, January 26, 1990.

28. Mary McGrory, "Let the Good Times Plod," *Washington Post*, February 1, 1990, p. A2.

29. Hall-Jamieson quoted in Jeanne Cummings, "Election '92 TV's Starring Role: Cable Allows Candidates to Target Their Messages," *Atlanta Constitution*, November 1, 1992, p. C4.

30. Michael Barnes, "What the House Didn't Know about TV," *Washington Post*, April 17, 1979, p. A21.
31. Michel, "Tenth Anniversary of C-SPAN," p. H-901.
32. *Congressional Record*, June 2, 1992, p. S-7402.
33. *Congressional Record*, July 14, 1994, p. H-5684.
34. *Congressional Record*, November 19, 1993, p. H-10275.
35. Colin Powell, remarks, Armed Forces Communications and Electronics Association, 1988, LEXIS-NEXIS database.
36. National Governors' Association Speech, February 1, 1994, LEXIS-NEXIS database.
37. Michel, "Tenth Anniversary of C-SPAN," p. H-901.
38. Mark West, "Your Calls Get Noticed," *C-SPAN Update*, November 16, 1987, p. 1.
39. Tim Gearan, quoted in Christopher Maloney, "A TV Guide to the Senate Workday," *C-SPAN Update*, September 19, 1988, p. 3.
40. Ibid.
41. Chapman quoted in Elizabeth Wehr, "Republicans Cry 'Foul' over Floor Tactics: Wright Finds a Vote to Pass Reconciliation Bill," *Congressional Quarterly Weekly Report*, October 31, 1987, p. 2653.
42. Mary Holley, "Impact of House TV Sometimes Subtle," *C-SPAN Update*, November 16, 1987, p. 4.
43. Cohen, "The Senate Show," p. 610.
44. Michel quoted in Donnie Radcliffe, "Many Hits, One Era," *Washington Post*, March 9, 1994, p. C4.
45. "The Health Care Controversy," *Congressional Record*, August 5, 1992, p. S-11587.
46. Gingrich quoted in American Enterprise Institute Symposium, "C-SPAN Tenth Anniversary Retrospective," telecast, April 3, 1989.
47. Adelman, "Real People," p. 32.
48. Mann quoted in American Enterprise Institute Symposium, "C-SPAN Tenth Anniversary Retrospective," telecast, April 3, 1989.
49. Hymel quoted in "How Do You Use C-SPAN?," *C-SPAN Update*, April 1, 1985, p. 3.
50. Sarah Pekkanen, "Technology Usurps Role of Capitol Hill Lobbyists, *The Hill*, March 15, 1994, p. 1.
51. Michael Hedges, "Were Marchers Just Too Far 'Out'?" *New York Times*, April 27, 1993, p. A1.
52. Howard Kurtz, "Television Has Trouble Bringing Congress's Revolution into Focus," *Washington Post*, January 24, 1995, p. A1.
53. Southwick, "Breathing New Life into Election Coverage."
54. *Congressional Record*, June 26, 1992, p. S-9043.

55. *Congressional Record*, September 14, 1993, p. S-11654.
56. James Reston, "Give TV Its Day," *New York Times*, June 4, 1986, p. A27.
57. Kurtz, "Television Has Trouble," p. A4.
58. Quoted in Burger, "GOP Rally Sept. 27."
59. Quoted in Tom Brazaitis, "GOP Promises Action; Its 'Contract with America' to Get Priority," *Plain Dealer*, November 10, 1994, p. A1.
60. Ibid.
61. Ibid.
62. Quoted in King and Stencel, *On the Line*, p. 60.
63. Tom Rosenstiel quoted in Dan Balz, "Soundbitten," *Washington Monthly* 25, nos. 7, 8 (July-August 1993): 45.
64. Jonathan Schell, "Real Washington Is Not in D.C.," *Newsday*, March 28, 1993, p. 39.
65. Jim Wooten, "Our Civilization Will Survive Perot's Candidacy," *Atlanta Journal and Constitution*, June 10, 1992, p. A10.
66. Don Kellerman quoted in Barbara Matusow, "Twinkle, Twinkle, Little Stars," *Washingtonian*, May 1993.
67. Don Oberdorfer, "Lies and Videotape: Watching Journalism Change in an Age of Suspicion," *Washington Post*, April 18, 1993, p. C1.
68. Russert quoted in Howard Kurtz, "The Public Eye," *Washington Post*, December 20, 1994, p. B1.
69. See Patricia Brennan, "C-SPAN: America's Town Hall, Marking Ten Years," *Washington Post*, April 2, 1989, p. Y5, and Howard Kurtz, "For Big Three Networks, Party Conventions Lose Their Luster," *Washington Post*, July 5, 1992, p. A6.
70. Jon Margolis, "Politicians Bypass Press, Go Straight to Voters," *Chicago Tribune*, February 28, 1994, p. 1.
71. King and Stencel, *On the Line*, p. 2.
72. Christopher Maloney, "Presidential Politics Interrupt House Floor Proceedings," *C-SPAN Update*, October 10, 1988, p. 11.
73. Janette Kenner Muir, "Video Verite: C-SPAN Covers the Candidates," in Robert E. Denton, *The 1992 Presidential Campaign* (Westport, Conn.: Praeger, 1994), p. 238.
74. Ibid., p. 242.
75. "Let The Naysayers Snipe," *Investor's Business Daily*, November 18, 1994. p. A2.
76. Robert Emmit Hoyt, "Voters Freed Clinton with GOP Landslide," *Arizona Republic*, November 13, 1994, p. E1.
77. Brian Lamb, interview on *The Derek McGinty Show*, December 7, 1994.
78. David Obey, "Lights, Camera, Congress," *Wall Street Journal*, December 21, 1994, p. A14.

79. Nader interview.

80. Matusow, "Twinkle, Twinkle, Little Stars."

81. Ely, "Strength of Belief," p. 2.

82. Wheeler quoted in Ely, "The Odd Couple," p. 3.

83. Phil Rosenthal, "PBS Needs a Lesson in Making Money," *Daily News*, January 4, 1995.

84. Gingrich quoted in John Carmody, "The TV Column," *Washington Post*, January 3, 1995, p. C4.

85. Moyers quoted in John Carmody, "The TV Column," *Washington Post*, January 6, 1995, p. C6.

86. Transcript, National Press Club Speech, January 17, 1995, LEXIS-NEXIS database.

87. Quotations in *C-SPAN Update*, March 19, 1984, pp. 10–11.

88. See Marshall McLuhan, *Understanding the Media: The Extensions of Man* (New York: McGraw Hill, 1964).

89. David Nyhan, "Lite News: Get Ready for a Deluge of Slime," *Boston Globe*, February 11, 1993, p. 23.

90. Ornstein, "Yes, Television," pp. 5–8.

91. *Congressional Record*, March 23, 1993, p. S-3439; *Congressional Record*, April 30, 1992, p. E-1199; *Congressional Record*, November 8, 1993, p. S-15264.

92. James P. Pinkerton, "The Message Has Become the Political Medium," *Nassau and Suffolk Edition*, February 4, 1993, p. 98.

93. Jill Lawrence, "Washington Dateline," *AP Newsfeatures*, June 16, 1985, LEXIS-NEXIS database.

94. O'Neill, *Man of the House*, p. 44.

95. Van Deerlin quoted in American Enterprise Institute, "Ten Years of Televising Congress."

96. See David Broder, "Din of Special Interests," *Washington Post*, January 25, 1995, p. A25; Robert Wright, "Hyperdemocracy," *Time*, January 23, 1995, p. 14; and Jonathan Rauch, *Demosclerosis* (New York: Times Books, 1994).

97. James Madison, "Federalist 10," paraphrased in Robert Wright, "Hyperdemocracy," *Time*, January 23, 1995, p. 14.

98. Ibid.

99. *Congressional Record*, October 1, 1992, p. S-16065.

100. Quoted in Ornstein, "Yes, Television," p. 6.

101. Greenfield, "C-SPAN Comes Close to National Treasure."

102. See Alvin Toffler and Heidi Toffler, *Creating a New Civilization: The Politics of the Third Wave* (Atlanta: Turner Publishing, 1994).

103. Madison, "Federalist 10," *The Federalist Papers* (New York: New American Library, 1961), p. 82.

104. Simon Tisdall, "Calling the Phone-In President," *Guardian*, February 1, 1993, p. 19.

105. Grover G. Norquist, "Looking Ahead," *American Spectator*, December 1994.

106. Jurek Martin, "Less a Honeymoon, More a New Affair: The Altered Relationship between Media and Present," *Financial Times*, February 3, 1993, p. 6.

107. Shales, "As the Hill Turns," p. E1.

Chapter 9

1. Ken Auletta, author of *Three Blind Mice*, CNN interview with Deborah Marchini and Stuart Varney, July 16, 1992, LEXIS-NEXIS database.

2. Gladstone, "Cable's Own Public Television Network," p. 33.

3. Max Frankel, "Mediology," *New York Times Magazine*, October 30, 1994, p. 34.

4. Rosenstiel, *Strange Bedfellows*, p. 346.

5. Nyhan, "Lite News," p. 23.

6. Times Mirror Center for the People and the Press, *The People, the Press, and Their Leaders* (Washington, D.C.: Times Mirror Center for the People and the Press, 1995), p. 15.

7. Wooten, "Our Civilization Will Survive Perot's Candidacy," p. A10.

8. America's Talking Channel was launched in 1994; by May 1995 it had fifteen million households subscribing, about one-fourth of the C-SPAN household total.

9. Diane Rehm, "Tower of Babble: How to Keep Talk Radio Democracy from Deepening America's Divisions," *Washington Post*, September 11, 1994, p. C3.

10. Timothy Egan, "Triumph Leaves Talk Radio Pondering Its Next Targets," *New York Times*, January 1, 1995, p. A1.

11. Cokie Roberts, "Theodore White Lecture with Cokie Roberts," Harvard University, 1994, p. 21.

12. Jonathan Alter, Howard Fineman, and Eleanor Clift, "The World of Congress," *Newsweek*, April 24, 1989, p. 28.

13. Quoted in Marc Gunther, "C-SPAN's Uphill Battle to Show Congress," *Philadelphia Inquirer*, May 21, 1995, p. A3.

14. Janet Hook, "House Cleaning Up Its Act for TV after Viewers Give It Thumbs Down," *Washington Times*, February 18, 1984.

15. See Karen Foerstel, "Special Orders Deal Is Struck," *Roll Call*, February 7, 1994, p. 1, and James V. Grimaldi, "Playing to Empty House, Speakers Get Tight Focus," *Washington Post*, April 6, 1994, p. A17.

16. Karen Foerstel, " 'Mud Wrestling' Opens House Oxford Debates," *Roll Call*, March 21, 1994.

17. Gingrich's view of the future has been deeply affected by the works of futurists Heidi Toffler and Alvin Toffler. See Toffler and Toffler, *Creating A New*

Civilization.

18. John McCasin, "Tear Down the Walls," *Washington Times*, November 22, 1994, p. 10.

19. Nolan Walters, "First Fifty Days Fill Gingrich with Optimism," *Miami Herald*, February 19, 1995 p. A12.

20. Ted Hearn, "C-SPAN Gets More Backing," *Multichannel News*, January 2, 1995.

21. See Mary Jacoby and Benjamin Sheffner, "House Cuts Candid TV Floor Shots," *Roll Call*, April 3, 1995, p. 1.

22. Quoted in Timothy Burger and Gabriel Kahn, "Aftershocks," *Roll Call*, May 18, 1995.

23. Peter Hoekstra on C-SPAN, December 5, 1994.

24. David Nielsen, "Many in House Protest 'Baseball-Style' C-SPAN," *Topeka Capital Journal*, April 25, 1995.

25. Charlotte Grimes, "With C-SPAN In Congress, Viewers Get Big Pictures," *St Louis Post Dispatch*, March 30, 1995.

26. Blankley quoted in Howard Kurtz, "Gingrich Plans to End Daily News Briefings," *Washington Post*, May 3, 1995, p. A7.

27. Jack Germond and Jules Witcover, "Overexposed Newt Rejects Daily TV," *San Diego Union and Tribune*, May 6, 1995, p. B7.

28. Swain quoted in Nancy Roman, "House Shies from Wandering Eyes," *Washington Times*, March 30, 1995, p. A11.

29. Kurtz, "Under Speaker's Order," p. 21.

30. From Brian Lamb's letter to the new Republican leadership, November 21, 1994, in C-SPAN files.

31. J. McIver Weatherford, *Tribes on the Hill* (Hadley, Mass.: Bergin and Garvey Publishers, 1985), p. 200.

32. Jack Anderson quoted in ibid., p. 201.

33. Stephen Barr, "House Moves Record Closer to Truth," *Washington Post*, January 9, 1995, p. A15.

34. Thomas Mann, of the American Enterprise Institute, has developed this idea more broadly, arguing that a regular transition of power in Congress would be good for the institution. Partisan posturing over all types of reform would be reduced, and the tendency to run for Congress by running against it would be tempered by candidates' realization that they would have a chance to run the institution. They would have a stake in inheriting a strong institution with public support as opposed to a weak institution with no respect or legitimacy.

35. Turton quoted in Marcia Gelbart, "Dems Learning to Use House Special Orders, but GOP Masters Genre," *The Hill*, March 8, 1995, p. 23.

36. Ibid.

37. Douglas Davis, "The American Voter Mounts the Stage," *Newsday*, January 3,

1993, p. 26. Davis's arguments are spelled out in more detail in Davis, *The Five Myths of Television Power.*

38. Brian Lamb, "No More Tears for Network TV," *Wall Street Journal*, October 15, 1991, p. A22.

39. Ray Archer, "Election Leftovers," *Arizona Republic*, November 14, 1994, p. B4.

40. Howard Fineman and Donna Foote, "The Power of TALK," *Newsweek*, February 8, 1993, p. 24.

41. Dan Balz, "Candidates Skirt News Media, Favor Direct Delivery of Message," *Washington Post*, May 19, 1992, p. A1.

42. Representative Bill Paxon (R-NY), press conference, November 9, 1994, LEXIS-NEXIS database.

43. Jonathan Alter, "Why the Old Media's Losing Control," *Newsweek*, June 8, 1992, p. 28.

44. Elving, "Brighter Lights, Wider Windows," p. 194.

45. Members of Congress join together in special-interest caucuses designed to exchange information and plan strategy. Until the 104th Congress, public funds were used to support a number of them. Groups such as the Black Caucus, the Hispanic Caucus, the Mushroom Caucus, and the Arts Caucus meet on a regular basis.

46. Peter Range, "Live from the White House, It's the All-Clinton Channel," *TV Guide*, April 10, 1993, p. 16.

47. Ibid.

48. Gonzales quoted in "A Proposal: Videos of Fed," *New York Times*, October 8, 1992, p. D10.

49. Kimberly Merline, "Media Groups Seek Televised Coverage of Supreme Court," *C-SPAN Update*, September 29, 1991, p. 8.

50. Senate Judiciary Committee hearing, July 22, 1993, LEXIS-NEXIS database.

51. Merline, "Media Groups Seek Televised Coverage," p. 8.

52. Joan Biskupic, "Vote on Cameras Reveals Judges' Deep Concern," *Washington Post*, September 23, 1994. p. A3.

53. Karen Fisher, "Legislatures in the Living Room," *State Legislatures*, August 1991.

54. Frazee interview.

55. Interview with the authors.

56. Jennifer Pendleton, "Tough Times of Cable Network Wannabe's," *Cable World*, February 27, 1995, p. 50.

57. Lolita Baldor, "Congress, Citizens Chat in Cyberspace," *Connecticut Post*, April 5, 1995, p. A1.

58. Frazee interview.

59. "Funeral Speech of Pericles," book 2, section 40, *Thucydides Translated into English*, 2d ed. trans. Benjamin Jowett (Oxford: Clarendon Press, 1900), vol. 2, pp. 129–30.

60. Thomas Jefferson, letter to Judge John Tyler, June 28, 1804, in *The Life and Selected Writings of Thomas Jefferson*, ed. Adrian Koch and William Peden (New York: Modern Library, 1944), p. 576.

61. Adapted from Thomas Jefferson, letter to Edward Carrington, January 16, 1787, in *The Papers of Thomas Jefferson*, ed. Julian P. Boyd (Princeton, N.J.: Princeton University Press, 1955), vol. 11, p. 59.

62. Alexis de Tocqueville, *Democracy in America* (1835; Henry Reeve text as revised by Francis Bowen and Phillips Bradley, New York: Alfred A. Knopf, 1963), p. 183.

63. Lamb interview, spring 1995.

Selected Bibliography

The most complete sources will be found in the notes for each chapter. This bibliography selectively lists a sampling of the key resources on which we relied. Relatively inaccessible sources, such as newsletters and internal documents, are not listed here. In some cases, source documents not quoted from directly are listed here rather than in the notes. These documents served as starting points for finding the sources actually quoted. The listing of articles is most selective. A LEXIS-NEXIS database search discovered over two thousand articles from the popular press mentioning C-SPAN. Citations from professional journals added hundreds more. The listing of articles in this bibliography is designed to give a flavor for the types of sources used and to provide a variety of resources for further inquiry.

Books

Bates, Stephen. *The Media and Congress*. Columbus, Ohio: Publishing Horizons, 1987.

Bennett, Lance. *News: The Politics of Illusion*. New York: Longman, 1988.

Byrd, Robert S. *The Senate, 1789–1989: Addresses on the History of the Senate*.

Washington, D.C.: Government Printing Office, 1989.

Chester, Edward W. *Radio, Television, and American Politics*. New York: Sheed and Ward, 1969.

Cirino, Robert. *Don't Blame the People*. New York: Vintage Books, 1971.

Congressional Research Service. *Congress and the Mass Media: An Institutional Perspective*. Prepared for the U.S. Congress, Joint Committee on Congressional Operations. Washington, D.C.: Government Printing Office, 1974.

Cook, Timothy. *Making Laws and Making News*. Washington, D.C.: Brookings Institution, 1989.

Crain, W. Mark, and Brian L. Goff. *Televised Legislatures: Political Information Technology and Public Choice*. Boston: Kluwer Academic Publishers, 1988.

Davis, Douglas. *The Five Myths of Television Power; or, Why the Medium Is Not the Message*. New York: Simon and Schuster, 1993.

Douglas, Susan. *Inventing American Broadcasting*. Baltimore: Johns Hopkins University Press. 1987.

Foote, Joe S. *Television Access and Political Power*. New York: Praeger, 1990.

Franklin, Bob. *Televising Democracies*. New York: Routledge, 1992.

Frantzich, Stephen E. *Computers in Congress: The Politics of Information*. Beverly Hills, Calif.: Sage Publications, 1982.

Garay, Ronald. *Congressional Television: A Legislative History*. Westport, Conn.: Greenwood Press, 1984.

Hall, Terry. *Congress on the Tube: Who's Watching?* Norman, Okla.: Bureau of Government Research, 1980.

King, Larry, and Mark Stencel. *On the Line: The New Road to the White House*. New York: Harcourt Brace and Company, 1993.

Lamb, Brian. *C-SPAN: America's Town Hall*. Washington, D.C.: Acropolis Books, 1988.

Mann, Thomas E., and Norman J. Ornstein. *Congress, the Press, and the Public*. Washington, D.C.: American Enterprise Institute and Brookings Institution, 1994.

O'Neill, Thomas P. ("Tip"). *Man of the House*. New York: St. Martin's, 1987.

Patterson, Thomas, and Robert McClure. *The Unseeing Eye: The Myth of Television Power in National Elections*. New York: Putman, 1976.

Reinsch, J. Leonard. *Getting Elected*. New York: Hippocrene Books, 1988.

Rosenstiel, Tom. *Strange Bedfellows: How Television and the Presidential Candidates Changed American Politics, 1992*. New York: Hyperion, 1993.

Toffler, Alvin, and Heidi Toffler. *Creating a New Civilization: The Politics of the Third Wave*. Atlanta: Turner Publishing, 1994.

U.S. Congress. Senate. Committee on Rules and Administration. *Television and Radio Coverage of Proceedings in the Senate Chamber: Hearings on S. Res. 20*. 97th Cong., 1st sess. Washington, D.C.: Government Printing Office, 1981.

Watt, Phyllis. *Television Cameras in Congress*. Columbia: University of Missouri,

School of Journalism, 1983.

Weatherford, J. McIver. *Tribes on the Hill*. Hadley, Mass.: Bergin and Garvey Publishers, 1985.

Whittemore, Hank. *CNN: The Inside Story*. New York: Little Brown, 1990.

Wolfson, Lewis. *The Untapped Power of the Press*. New York: Praeger, 1985.

Wright, Jim. *Worth It All*. Washington, D.C.: Brasseys, 1993.

Articles, Theses, and Dissertations

Adelman, Kenneth. "Real People: On C-SPAN Substance Can Be More Interesting Than Style." *Washingtonian* 28 (December 1992): 32–36.

Alter, Jonathan. "Why the Old Media's Losing Control." *Newsweek*, June 8, 1992, p. 28.

Arieff, Irwin B. "House TV Gets Mixed Reviews but Cancellation Isn't Likely." *Congressional Quarterly Weekly Report*, March 15 1980, pp. 735–37.

Bark, Ed. "How Does TV Fit into Your Future? Will the Tube Grow Up, or Just Grow?" *Dallas Morning News*, February 21, 1993, p. 1C.

Barnes, Michael. "What the House Didn't Know about TV." *Washington Post*, April 17, 1979, p. A21.

Biskupic, Joan. "Vote on Camera Reveals Judges' Deep Concern." *Washington Post*, September 23, 1994, p. A3.

Boakes, Janet Carol. "A Study of Television Viewership of Congressional Proceedings Carried on C-SPAN, the Cable Satellite Public Affairs Network." Master's thesis, University of Pennsylvania, Annenberg School of Communications, 1983.

Brown, Merrill. "C-SPAN, Almost Ten Years Old, Continues to Broaden Its Vision." *Channels of Communications* 8, no. 10 (November 1988): 16.

Calmes, Jacqueline. "Senate Agrees to Test of Radio, TV Coverage." *Congressional Quarterly Weekly Report*, March 1, 1986, pp. 520–21, 525.

Carney, Elizabeth Newlin. "Squeeze on for C-SPAN." *National Journal* 26, no. 24 (June 11, 1994): 1356.

Charen, Mona. "C-SPAN: Where TV Has Never Gone Before." *Reader's Digest*, May 1992, pp. 111–15.

Clancey, Maura. "The Political Knowledge, Participation, and Opinions of C-SPAN Viewers: An Exploration and Assessment of Mass Media Impact." Ph.D. diss., University of Maryland, 1990.

Cohen, Richard E. "The Senate Show." *National Journal* 18, no. 10 (March 8, 1986): 610.

Corry, John. "Why Convention Coverage Should be Gavel-to-Gavel." *New York Times*, July 8, 1984, sec. 2, p. 1.

Crain, W. Mark, and Brian L. Goff. "Televising Legislatures: An Economic Analysis." *Journal of Law and Economics* 29 (October 1986): 405–21.

Crouse, Chuck. "C-SPAN's Growing Role in a Tumultuous Presidential Year." *Communicator*, August 1992, p. 12.

Cummings, Jeanne. "Election '92 TV's Starring Role: Cable Allows Candidates to Target Their Messages." *Atlanta Constitution*, November 1, 1992, p. C4.

Duff, Marilyn. "C-SPAN: The Way TV News Should Be." *Human Events*, November 1994, pp. 1–5.

Ehrenhalt, Alan. "Media, Power Shifts Dominate O'Neill's House." *Congressional Quarterly Weekly Report*, September 13, 1986, p. 2135.

Elving, Ronald D. "Politics of Congress in the Age of TV." *Congressional Quarterly Weekly Report*, April 1, 1989, p. 722.

Fenno, Richard F., Jr. "The Senate through the Looking Glass: The Debate over Television." *Legislative Studies Quarterly* 14, no. 3 (1989): 313–48.

Fineman, Howard. "For the Son of C-SPAN, Exposure=Power." *Newsweek*, April 3, 1989, p. 23.

Fineman, Howard, and Donna Foote. "The Power of TALK." *Newsweek*, February 8, 1993, p. 24.

Frankel, Max. "Full-Text TV." *New York Times Magazine*, February 5, 1995, p. 32.

Franklin, Bob. "Televising Legislatures: The British and American Experience." *Parliamentary Affairs* 42 (October 1989): 485–502.

Geltner, Sharon. "Brian Lamb: The Man behind C-SPAN II." *Saturday Evening Post* 258 (January-February 1986): p. 46.

———. "Matching Set: Brian Lamb and Cable's C- SPAN." *Washington Journalism Review*, September 1984.

Gladstone, Brooke. "Cable's Own Public Television Network." *Channels*, September/October 1984, pp. 33–35.

Granat, Diane. "Televised Partisan Skirmishes Erupt in House." *Congressional Quarterly Weekly Report*, February 11, 1984, p. 246.

Greenfield, Jeff. "C-SPAN Comes Close to National Treasure." *Oregonian*, October 6, 1986.

Grimaldi, James V. "Playing to Empty House, Speakers Get Tight Focus." *Washington Post*, April 6, 1994, p. A17.

Harrington, Linda M. "C-SPAN: TV's Political Insider in Washington." *Chicago Tribune*, October 31, 1993, p. C5.

Harwood, Richard. "Sources Who Supply 'The News.'" *Washington Post*, January 7, 1995, p. A21.

Hearn, Ted. "C-SPAN Gets More Backing." *Multichannel News*, January 2, 1995, p. 1.

Hickey, Neil. "Eye on Democracy." *TV Guide*, March 30, 1985.

Hook, Janet. "An After-Hours Image Adjustment." *Congressional Quarterly Weekly Report*, February 12, 1994, p. 317.

Ivins, Molly. "Attention News Junkies: You Don't Have to Rely on Brokaw/Jennings/Rather Versions." *TV Guide*, December 3, 1988.

Jacoby, Mary, and Benjamin Sheffner. "House Cuts Candid TV Floor Shots," *Roll Call*, April 3, 1995, p. 1.

King, Susan. "Q & A: Brian Lamb, the Election's Real Winner." *Los Angeles Times*, January 17, 1993, p. 83.

Kinsley, Michael. "The Intellectual Free Lunch." *New Yorker*, February 6, 1995, pp. 4, 5.

Knoll, Steve. "C-SPAN Strives to Expand." *New York Times*, April 8, 1984, sec. II, p. 28.

Kolbert, Elizabeth. "The Longest Campaign Has Started." *New York Times*, September 19, 1993, sec. 4, p. 6.

———. "Some Cable Systems Are Cutting C-SPAN for Other Channels." *New York Times*, June 20, 1994, pp. A1, D5.

Kondracke, Morton M. "Why Are Voters 'Mad as Hell'? If You Ask, Duck!" *Roll Call*, October 20, 1994, p. 6.

Krolik, Richard. "Everything You Wanted to Know about C-SPAN." *Television Quarterly* 25, no. 4 (1992): 91–95.

Kurtz, Howard. "The Public Eye." *Washington Post*, December 20, 1994, p. B1.

———. "Television Has Trouble Bringing Congress's Revolution into Focus." *Washington Post*, January 24, 1995, p. A1.

Lamb, Brian. "Free(r) at Last." *Broadcasting*, August 11, 1986, pp.41–42, 44, 46.

Lardner, James. "The Anti-Network." *New Yorker*, March 14, 1994, pp. 49–55.

Long, Russell. "Keep the TV Cameras out of the Senate." *USA Today*, September 18, 1985, p.10A.

Maloney, Christopher. "Lights, Camera, Quorum Call: A Legislative History of Senate Television." Master's thesis, College of William and Mary, 1990.

Matusow, Barbara. "Made for TV." *Washingtonian* 23 (December 1987): 208–13, 215–17.

———. "Twinkle, Twinkle, Little Stars." *Washingtonian* 29 (May 1993).

Meyer, Thomas. "No Sound Bites Here." *New York Times Magazine*, March 15, 1992, p. 46.

Minzesheimer, Bob. "Enjoying Politics at the C-SPAN Pace." *USA Today*, February 19, 1992, sec. A, p. 9.

Morin, Richard, and David S. Broder. "Familiarity Is Breeding Contempt of Congress." *Washington Post*, July 3, 1994, pp. A1, A8.

Muir, Janette Kenner. "Video Verite: C-SPAN Covers the Candidates." In Robert E. Denton, *The 1992 Presidential Campaign*, pp. 226–45. Westport, Conn.: Praeger, 1994.

Norquist, Grover G. "Looking Ahead." *American Spectator*, December 1994.

Oberdorfer, Don. "Lies and Videotape: Watching Journalism Change in an Age of Suspicion." *Washington Post*, April 18, 1993, p. C1.

O'Daniel, Michael. "Cable Has a Secret." *Emmy Magazine*, November/December

1983, p. 67.

Ornstein, Norman J. "Yes, Television Has Made Congress Better." *TV Guide*, July 25, 1987, pp. 5–8.

Ornstein, Norman, and Michael Robinson. "The Case of Our Disappearing Congress: Where's All the Coverage?" *Congressional Record* (daily ed.), February 3, 1986, pp. S828–S830, originally in *TV Guide*, January 11, 1986, p. 4.

Pasdeloup, Vincente. "It's Back to Square One for Cable after Must-Carry Ruling." *Cable World* 6, no. 27 (July 4, 1994): 23.

Perry, James. "Beginning of House Television Marked with Skepticism." *Wall Street Journal*, March 26, 1979, p. 16.

Peters, Casey. "The Cable Satellite Public Affairs Network: C-SPAN in the First Decade of Congressional Television." Master's thesis, UCLA, 1992.

Polier, Rex. "See Congress Run." *Western World*, September 1983, pp. 72, 77.

Postman, Neil, and Jay Rosen. "When Television News Has a Breakdown: Now That's Entertainment." *Los Angeles Times*, January 31, 1988, sec. 5, p. 5.

Radcliffe, Donnie. "Many Hits, One Era." *Washington Post*, March 9, 1994, p. C4.

Range, Peter Ross. "C-SPAN: The Little Network That Could." *TV Guide*, July 18–24, 1992, pp. 12–15.

Rapping, Elayne. "Cable's Silver Lining." *Progressive* 58, no. 9 (September 1994): 36.

Rehm, Diane. "Tower of Babble: How to Keep Talk Radio Democracy from Deepening America's Divisions." *Washington Post*, September 11, 1994, p. C3.

Reston, James. "Give TV Its Day." *New York Times*, June 4, 1986, p. A27.

Roberts, Steven V. "House Democrats Seeking Better TV Image." *New York Times*, May 10, 1984, p. A29.

———. "Man in the News: George John Mitchell." *New York Times*, November 30, 1988, p. A1.

———. "Senators Ponder Value of Letting TV in the Door." *New York Times*, September 16, 1985, p. B6.

Robinson, Michael. "Three Faces of Congressional Media." In Thomas Mann and Norman Ornstein, eds., *The New Congress*. Washington, D.C.: American Enterprise Institute, 1981.

Rosenstiel, Tom. "Democrats Learn to Control Show." *Los Angeles Times*, July 21, 1988, sec. 1, p. 5.

Salant, Richard. "Network News: Prospects for Its Future—If Any." *Los Angeles Times*, May 20, 1988, pp. 1, 20.

Schell, Jonathan. "Real Washington Is Not in D.C." *Newsday*, March 28, 1993, p. 39.

Shales, Tom. "As the Hill Turns: C-SPAN's Riveting Mini-Series." *Washington Post*, May 17, 1984, p. E1.

———. "The Dems on TV: Going Full Bore." *Washington Post*, July 15, 1992, sec. F, p. 4.

————. "Network Uncoverage: Peeking in Periodically." *Washington Post*, July 18, 1988, sec. B, pp. 1, 3.

————. "On the Air: Live from New Orleans—It's RNC." *Washington Post*, August 15, 1988, sec. C, p. 8.

Sheridan, Patrick. "C-SPAN Holds Firm: No Tapes of Kerrey Joke." *Broadcasting*, December 2, 1991, pp. 38, 42.

————. "C-SPAN Plans 1000 Hours of Election Coverage." *Broadcasting*, November 11, 1991, p. 38.

Starr, Mark, and Gloria Borger. "Not Ready for Prime Time?" *Newsweek*, May 28, 1984, p. 24.

Stewart, Robert. "Lawmakers Perfect Art of One-Minute Zinger." *Los Angeles Times*, October 4, 1992, p. A13.

Stump, Matt, and Harry Jessell. "Cable: The First Forty Years." *Broadcasting*, November 21, 1988, p. 35.

Warren, James. "Taking the Long View." *Chicago Tribune*, March 15, 1992, p. 1.

Wehr, Elizabeth. "Republicans Cry 'Foul' over Floor Tactics: Wright Finds a Vote to Pass Reconciliation Bill." *Congressional Quarterly Weekly Report*, October 31, 1987, p. 2653.

Wertz, Diane. "C-SPAN: The Comfort Zone." *Newsday*, March 20, 1991, p. 73.

Winfrey, Lee. "TV and the Campaign Trail." *Philadelphia Inquirer*, July 15, 1992, sec. D, pp. 1, 5.

Wolfe, John. "C-SPAN's Persuasive Powers Face a Key Test." *Cablevision*, May 19, 1986, p. 30.

Wright, Robert. "Hyperdemocracy." *Time*, January 23, 1995, p. 14.

Zuckerman, Laurence. "The Raw Material News Network Hits Its Stride." *Columbia Journalism Review* 24 (March-April 1980): 42–44.

Index